The American Development of Biology

Ronald Rainger
Keith R. Benson
Jane Maienschein
Editors

The American Development of Biology

Rutgers University Press
New Brunswick and London

First published in paperback by Rutgers University Press, 1991
First published in cloth by the University of Pennsylvania Press, 1988

Printed in the United States of America

Library of Congress Cataloging-in-Publication Data

The American development of biology / Ronald Rainger, Keith R. Benson, Jane Maienschein, editors.
p. cm.
Based on papers presented at a conference held at the University of Washington's Friday Harbor Laboratories, Sept. 1986, commissioned by the American Society of Zoologists to celebrate its centenary.
"First published in cloth by the University of Pennsylvania Press, 1988"—T.p. verso.
Includes bibliographical references and index.
ISBN 0-8135-1702-8 (pbk.)
1. Biology—United States—History. I. Rainger, Ronald, 1949– II. Benson, Keith Ronald. III. Maienschein, Jane. IV. American Society of Zoologists.
QH305.2.U6A54 1991
574'.0973—dc20 *91–11533*
CIP

Contents

Contributors vii

Preface xi

Introduction 3

PART ONE Natural History to Biology

1 Museums on Campus: A Tradition of Inquiry and Teaching 15
SALLY GREGORY KOHLSTEDT

2 From Museum Research to Laboratory Research: The Transformation of Natural History into Academic Biology 49
KEITH R. BENSON

PART TWO Centers of Cooperation

3 Organizing Biology: The American Society of Naturalists and its "Affiliated Societies," 1883–1923 87
TOBY A. APPEL

4 Summer Resort and Scientific Discipline: Woods Hole and the Structure of American Biology, 1882–1925 121
PHILIP J. PAULY

5 Whitman at Chicago: Establishing a Chicago Style of Biology? 151
JANE MAIENSCHEIN

PART THREE Working at the Boundaries of Biology

6 Charles Otis Whitman, Wallace Craig, and the Biological Study of Animal Behavior in the United States, 1898–1925 185
RICHARD W. BURKHARDT, JR.

7 Vertebrate Paleontology as Biology: Henry Fairfield Osborn and the American Museum of Natural History 219
RONALD RAINGER

8 Organism and Environment: Frederic Clements's Vision of a Unified Physiological Ecology 257
JOEL B. HAGEN

9 Mendel in America: Theory and Practice, 1900–1919 281
DIANE B. PAUL AND BARBARA A. KIMMELMAN

10 Cellular Politics: Ernest Everett Just, Richard B. Goldschmidt, and the Attempt to Reconcile Embryology and Genetics 311
SCOTT F. GILBERT

BIBLIOGRAPHY 347

INDEX 365

Contributors

Toby A. Appel (Ph.D., Princeton University) teaches history of medicine at the University of Maryland, Baltimore County. She is the author of *The Cuvier-Geoffroy Debate: French Biology in the Decades before Darwin* (1987) and coeditor with John R. Brobeck and Orr E. Reynolds of *History of the American Physiological Society: The First Century, 1887–1987* (1987). She is currently working on contract to the National Science Foundation on a history of NSF and postwar federal patronage of biology.

Keith R. Benson (Ph.D., Oregon State University), Associate Professor of Medical History and Ethics, University of Washington. A teacher of the history of science, Benson has published papers on biology at Johns Hopkins University, the American natural history tradition, and the history of marine biology. He is currently completing a monograph on the role of American marine biology stations in the growth of biology in the United States.

Richard W. Burkhardt, Jr. (Ph.D., Harvard University), Professor of History, University of Illinois. He is the author of *The Spirit of System* (1977), a study of Lamarck's ideas on evolution, and has published other articles on the history of evolutionary biology. He is currently writing a book on the development of ethology as a scientific disipline, having already published several articles on the subject.

Scott F. Gilbert (Ph.D., Johns Hopkins University), Associate Professor of Biology, Swarthmore College. He is the author of a recent textbook, *Development Biology* (1988, and forthcoming edition), whose research involves the developmental regulation of stage-specific gene products in early embryogenesis. He has, in several articles, examined the history of American developmental biology and the relationship between embryology and genetics.

Joel B. Hagen (Ph.D., Oregon State University), Visiting Assistant Professor, Department of History and Committee on History and Philosophy of Science, University of Maryland. A biologist and historian of science, he has written articles on the history of ecology and plant biology, which will provide the basis for a monograph on the subject.

Barbara A. Kimmelman (Ph.D., University of Pennsylvania), Assistant Professor of History, Department of Humanities and Social Sciences, Philadelphia College of Textiles and Science. Kimmelman is currently revising her 1987 doctoral dissertation on the early disciplinary context of genetics at agricultural institutions during the Progressive period. Her articles (and current writings) continue to explore the cultural context of American genetics and of American agricultural science and education in the early decades of the twentieth century.

Sally Gregory Kohlstedt (Ph.D., University of Illinois), Professor in the History of Science and Technology Program, University of Minnesota. Her contribution to the volume was written while she was a fellow at the Woodrow Wilson Center, housed in the Smithsonian Institution. She has edited, with Margaret Rossiter, *Historical Writing on American Science* (1986) and, with Rod Home, *International Science and National Scientific Identity* (1991). She publishes on institutional history (including a book on the American Association for the Advancement of Science), on women's history, and on science in the culture of nineteenth-century America.

Jane Maienschein (Ph.D., Indiana University), Professor of Philosophy and Zoology, Arizona State University. She is the editor of *Defining Biology: Lectures from the 1890's* (1986), a study of biological lectures presented at the Marine Biological Laboratory, Woods Hole, Massachusetts. She also coorganized a conference and coedited a volume on the history of ecology and evolutionary biology, papers from which appeared as a special issue of the *Journal of the History of Biology* (June 1986). Her research focuses on modern experimental biology, and she has recently completed *Transforming Traditions in American Biology, 1880–1915*.

Diane B. Paul (Ph.D., Brandeis University), Associate Professor of Political Science, University of Massachusetts, Boston. Her early work was in political science and resulted in studies of such diverse topics as nineteenth-century Marxism and contemporary tax policy. In recent years she has focused on issues in the history of evolution and genetics. She has been a fellow in the Program for Science, Technology and Society, MIT, and an EVIST fellow in Richard Lewontin's laboratory at the Museum of Comparative Zoology, Harvard University, where she did research on the history of population genetics. She has published several articles on the history of genetics, eugenics, and the shaping of contemporary science textbooks, and is currently at work on a history of clinical genetics.

Philip J. Pauly (Ph.D., Johns Hopkins University), Associate Professor of History, Rutgers, the State University of New Jersey. He is the author of *Controlling Life: Jacques Loeb and the Engineering Ideal in Biology* (1987) and has written on the history of the biological and behavioral sciences and on popular science. This paper is part of an ongoing project on the cultural history of the life sciences in America.

Ronald Rainger (Ph.D., Indiana University), Assistant Professor of History, Texas Tech University. He has published a number of articles on the history of anthropology and paleontology. He is the author of *An Agenda for Antiquity: Henry Fairfield Osborn and Vertebrate Paleontology at the American Museum of Natural History, 1890–1935* (1991).

Preface

The American Society of Zoologists celebrates its centennial year in 1989. The same year also marks the ten-year anniversary of the symposium, sponsored by the History of Science Society, at which we first collaborated in our study of American biology. These two benchmarks are not unrelated. The American Society of Zoologists commissioned us to prepare a work commemorating one hundred years of American biology, and we decided, in organizing this volume, to utilize the rapidly expanding interest in turn-of-the-century biology. The papers in this volume represent original work to celebrate the centenary of the American Society of Zoologists, and they illustrate the impressive nature of historical scholarship that has subsequently focused on the development of biology in the United States.

We also decided to produce an edited collaborative volume on the development of biology in America because such work provides a broader perspective than any single monograph. But the danger of collaboration is that the whole may be no greater than the sum of the parts. To alleviate this potential pitfall, we instructed all of the contributors to prepare papers representing previously unpublished research that addressed our general theme of the early development of American biology. We circulated the papers among ourselves, read and prepared comments on the contributions, and, perhaps most critically and constructively, all met together at the University of Washington's Friday Harbor Laboratories in September 1986. Several rounds of revision followed the conference. As we worked together we became impressed by the cohesiveness behind our diverse offerings; indeed, we happily noted that the volume as a whole was more than the sum of its parts.

We would like to think it was due to our work and attention to detail that the Friday Harbor Conference succeeded and that this volume followed so smoothly. Anyone who is familiar with Friday Harbor realizes how conducive the location is for scholarly conferences. All the conferees resided in the comfortable apartment overlooking the island's harbor, and we gathered together to discuss our work in a room complete with a scenic vista and a roaring fire. Our caterer, Maryanne Rock, made sure we ate well and we reciprocated by eagerly consuming her culinary preparations; especially popular was her salmon dinner. As editors, therefore, we confess that the lively and active spirit of the conference was primarily due to the convivial nature of the group and the wonderful opportunity we all had to work together in the ambiance of Friday Harbor.

The cooperative and integrative nature of the Friday Harbor Conference and the resulting volume was made possible by crucial support of the American Society of Zoologists. Through its Division of the History and Philosophy of Biology, the society is actively encouraging the participation of historians and philosophers of science within the community of zoologists. We are extremely grateful for the financial support of the society, especially the efforts of Trish Moore, Bill Dawson, and Mary Adams-Wiley. Dean Joe Norman of the Graduate School at the University of Washington also made a generous contribution to support the conference. Of course we are thankful for the efforts of all the contributors who adhered strictly to our time demands, responded promptly to our requests for material, cooperated willingly to our calls for revision, and allowed themselves to be coerced to spend a busy weekend at lovely Friday Harbor in the San Juan Islands. Several of our colleagues provided their own financial support for travel to the conference, and we are indebted to them for their generosity. We would also like to thank Garland Allen for his valuable contributions to the conference.

The Editors

The American Development of Biology

Introduction

In his textbook *Life Science in the Twentieth Century,*[1] Garland Allen maintained that the years 1890 to 1915 witnessed a revolutionary break in the life sciences, as a generation of younger researchers rejected the older morphological work for experimental work in biology. Allen explicitly invited other historians of science to examine the phenomenon. Responding to that challenge, the editors of this volume organized a session of the History of Science Society annual meeting in 1979 to examine the particular claim that morphology had been rejected. We all saw a less revolutionary change than Allen did, as well as a good deal less uniformity in what went before and what came after the supposed break. Frederick Churchill proposed further study to determine just what had happened during this period.[2]

During the last decade, a number of other historians have also been examining the period and subject in question. Much of that work has, in fact, established the late nineteenth century as a time of important changes in biology. Historians have divided the period in different ways, focusing variously on individuals, institutions, particular research problems or programs, types of research, types of work done, work in particular countries, or on different subfields.

Our concern is the study of biology in the United States.[3] The implicit assumption behind the collected essays is that biology in the United States, due in part to its late nineteenth-century setting and its comparative isolation from Europe, developed a distinctive American character. Although the volume does not pretend to establish this claim explicitly, it does attempt to provide a careful study of biology during its formative years in America. In Americans' efforts to create a new social order generally, and to establish new means for organizing society after the Civil War, they also reorganized the scientific community.[4] At the turn of the century, for example, a host of new specialized societies developed, new university departments and graduate programs appeared, modern research laboratories were established, and the character of research changed as new instruments and new standards of reliable results emerged. In addition, a number of more specialized areas of biology developed out of what had been largely gathered under the rubric of a natural history tradition.

This volume focuses on the natural-history-based fields of morphology, embryology, evolution, ethology, genetics, and plant ecology. It does not pretend to provide a comprehensive survey of all biology. For example, we have not included areas such as biochemistry, physiology, and anatomy, which originated from other concerns by people trained for other, generally more practical goals; some of these topics have been examined elsewhere.[5] We are concentrating on a cluster of research areas within the life sciences that were considered around 1900 to form a central part of biology.

To establish what happened in these natural-history-based areas of biology around 1900, this volume examines the period 1880 to 1920 in greatest detail, but looks also at the background in the mid-nineteenth century and the subsequent developments into the mid-twentieth century. Sally Gregory Kohlstedt and Scott Gilbert provide these anchor points in their papers in which they consider, respectively, natural history museums as an established place of research from the 1860s, and the attempts by researchers imbued with the ideals left over from the 1890s to make their ways in a twentieth-century environment increasingly skeptical of that earlier program. Kohlstedt's and Gilbert's papers show that the years 1880 to 1920 were not insulated; on the contrary, these earlier and later traditions helped shape and extend the changes of that pivotal time.

Yet the primary emphasis of this volume remains the period around 1900, for during that time Americans sought to identify a "core" for biology, if only as an ideal. Leading practitioners made self-conscious reference to biology as a distinct discipline, especially at the Marine Biological Laboratory (MBL), the Johns Hopkins University, and the University of Chicago. To each of these places, eager students were attracted by the rhetorical proclamations for a biological program, offered in the environment of a progressive and exciting community committed to the ideals of doing research. Henry Newell Martin and William Keith Brooks at Hopkins and Charles Otis Whitman at the MBL and Chicago lobbied for biological work with the conviction that this represented something new, productive, and exciting.

At first, the biology department at Johns Hopkins, which embraced both morphology and physiology, provided the most convincing program along the new lines. Then, for various reasons, Whitman and the MBL and Chicago gained preeminence. Keith Benson, Philip Pauly, Jane Maienschein, and others have examined in detail elsewhere the way in which these programs developed and how the identity of biology, even as it changed around 1900, depended on these important programs.[6] Pauly's comparative study of the relation of biology to medicine is particularly instructive for illuminating the process of definition of biology in the late nineteenth century. In this volume, Pauly adds another dimension to that definition by examining the MBL as a "biologist's club" that provided, over the course of years, the social and research experiences that were central to biology. Maienschein's examination here of the research efforts at Chicago, which Whitman established, further explores the character and identity of biology, as does her edition of MBL lectures.[7]

In the process of self-definition, an ideal for biology emerged. This ideal, which served as a "core," centered on marine study, particularly of invertebrates, and was primarily concerned with questions of embryology, heredity, and evolution (all of which the researchers thought were intricately interconnected). The core embraced experimental methods and approaches, although experimentation was never claimed as the necessary or only approach. Neither did the core reject the traditional concerns of natural history or morphology. In their papers here and elsewhere, Kohlstedt and Benson have both demonstrated the nature of work in natural history and in morphology—work that clearly contributed to the foundation for the programs at Hopkins, the MBL, and Chicago. Thus, the self-conscious efforts by Americans around 1900 to define biology and to effect new programs did not involve a rejection of those older problems and methods. Instead, those efforts centered on the identification of a cooperative, comparative program of research into questions of embryology, heredity, and evolution.

A core did exist, then, as identified by those in leading positions. Yet although it provided a coherent focus, this ideal core did not define or constrain all of biology—and this is perhaps the most important conclusion of this collaborative volume. Biology was not, either before or after 1900, so monolithic or unified as the core advocates or popular impressions might suggest. Even the primary proponents of the core did not claim that biology must necessarily include only the core work. Rather the core served as a body of issues and approaches that the participants agreed represented legitimate biology. Concern with definition, legitimization, and identification permeated American society at this time. Biologists shared the concern to identify their enterprise and themselves, as Toby Appel's paper so clearly indicates. Yet rhetorical excesses aside, they also recognized that they had not provided a perfect definition that stood for all work in biology. They sought to identify a core but not to circumscribe biology thereby: biology must consist of more than a core, just as a cell requires more than a nucleus.

A wide variety of researchers were active around 1900; many would have been explicitly called biologists, while some would have had other labels. Many of these biologists also self-consciously considered how and where their research fit into the scheme of things. These researchers worked within the bounds of biology, but many operated away from the center or idealized core, exploring a wider range of problems or employing alternative approaches. This larger community may be metaphorically seen as residing in the cytoplasm rather than in the nuclear core, with a few at the cell walls or even passing through those walls as conditions changed. Because most biologists around 1900 considered the cytoplasm at least as important as the nucleus, since the entire organism must necessarily work together as a whole, this existence away from the nuclear center implied no such inferior status as might be suggested to some readers today.

Regardless of their positions within biology, all were concerned with understanding living organisms by doing research in a recognizably professional way, using professional research tools. The biologists considered in this volume were trained professionals with college degrees and Ph.D.'s. Yet professionalism meant different things to different groups of people, so that the research work of the whole community included a variety of interests. As Joel Hagen has suggested, perhaps the work defies any all-encompassing catholic definition; perhaps what falls within biology was connected in only a loose way, as is the material in the cytoplasm. But loose though it may have been, there was a connection, and there was a sense that certain work fell within biology, whereas other sorts of work did not. This volume hopes to offer a sense of what the advocates believed lay inside the bounds of biology, by showing the complexity and the diversity of turn-of-the-century American biology in a way that no one paper or monograph alone can achieve.

The essays in this volume go far toward identifying what counted as biology. The papers consider important individuals, research problems, and institutional settings. As a group, they offer important insights not only into what characterized research and methods but also into factors that influenced the development of biology. Taken as a whole, therefore, the volume demonstrates the richness and variety of research being done, how and where such work was organized and pursued, what techniques and methods were employed, and what factors promoted or retarded different programs of biological research. The conclusion of the volume is that something important did happen in biology around 1900—namely, a conscious concern with self-definition emerged as the biologists explored a wide range of related researches. This concern found support from the concomitant enthusiasm for institution building and professional definition that was characteristic of society in general, a fortunate juxtaposition that produced the important development of biology at a pivotal time in American history.

The papers in this volume fall into three general groups, each group clustered around a theme. The first demonstrates the character of work in natural history during the last part of the nineteenth century and illustrates the ongoing and dynamic changes within that natural history tradition. The second group concentrates on several centers of cooperation: societies, laboratories, and universities. The third explores the boundaries of biology and the process of definition.

Some of the papers examine what it meant to do research for the biologists involved. Kohlstedt's and Benson's contributions are particularly important in this regard. They demonstrate the way in which a sense of research and teaching, which served to spread the ideal, developed out of roots in natural history collecting. Museums, for example, provided a place for naturalists to pursue the use of extensive collections as well as opportunities for fieldwork. Ronald Rainger's study of the American Museum of Natural History shows the parallel importance of doing research for the participants in the Department of Vertebrate Paleontology and the way in which those researchers explicitly identified their work as biological. The papers also demonstrate what sorts of research programs and traditions succeeded, as well as those that failed.

Several of the papers examine the cooperative or collaborative nature of the research done by biologists who thought that comparative studies of their respective research projects could produce larger and more important results than consideration of their individual studies alone. In particular, the American Museum researchers under Henry Fairfield Osborn adopted that attitude, as Rainger explains. So did the workers at the MBL and at Chicago, as Pauly and Maienschein demonstrate. In contrast to the usual picture, the work of breeders both inside and outside the universities took on a cooperative character that influenced the kinds of work they produced, as Diane Paul and Barbara Kimmelman so effectively illustrate.

Different ways of doing the work also influenced what fell within the definition of biology and what did not. Just because a researcher worked at a museum of natural history, or a hospital, or an agricultural station, for example, did not mean in itself that the work was not biology. The boundaries were not so firmly set, and some sorts of research apparently at the periphery could at times be considered biological. Biologists did not all have to work in laboratories, although that became increasingly the norm. Nor did the research have to take place in universities, as Paul and Kimmelman show in their examination of breeding studies. Work in the swamp marsh, the fossil bed, or the cornfield could be biological as well, if the purposes involved doing research and understanding life in terms of what the community at the time considered a properly professional and scientific way.

Sometimes new techniques served to reinforce the professional nature of biological research. For example, the improved cytological techniques of staining, sectioning, and preserving specimens, acquired at the Naples Zoological Station, then at Johns Hopkins and the MBL, provided a sense of what was current and professional.[8] Yet outside the core areas of research in embryology, heredity, and evolution, the existence of such standardized techniques was not so obvious. Although Clements prescribed proper techniques and research approaches for understanding communities of organisms, as Hagen shows, plant ecologists generally did not agree that those were the only or even the best ways to do biology. Thus, many of the most able biologists realized the exploratory nature of their methods and maintained acceptance of relatively open boundaries for biology.

Institutional settings provided another way to establish constraints and conditions for defining biological work. Toby Appel discusses the setting provided by various societies, for example. Pauly considers social conditions shaping the MBL's emergence as one of the leading institutions for biological research, while Maienschein discusses individual and institutional factors that promoted the emergence of biology at the University of Chicago. Studies by Paul and Kimmelman and Gilbert consider the ways in which economic and political opportunities and constraints influenced the character of work done. Richard Burkhardt illustrates how economic as well as conceptual factors limited behavioral study at Chicago and elsewhere, while Rainger's analysis of Osborn's powerful social, economic, and political connections shows how those factors allowed for the organization and elaboration of a successful program of research in an apparently peripheral field of research. The detailed discussions of institutions, intertwined with consideration of the individuals who helped to shape those institutions, goes far toward establishing what was included as biology in America and toward suggesting how, where, and why American biological research was organized and pursued.

Several papers illustrate work at the edges of what was considered mainstream biology. Burkhardt shows that the study of animal behavior, pursued through both biology and psychology, failed to become established within either area, despite its influence on such well-known ethologists as Konrad Lorenz. Burkhardt and Maienschein show that the behavioral studies by Whitman and Wallace Craig remained on the periphery, despite Whitman's own efforts to define biology. Similarly, Hagen shows that the particular organismic metaphor and research program in plant ecology that Clements pursued failed to achieve the results envisioned for them. The research remained within biology but, again, at the periphery. These papers help significantly to illuminate what the boundaries of biology were.

Some papers examine the difficulties encountered by individual scientists whose research programs were not considered central. Burkhardt's discussion of Whitman's and Craig's financial sacrifices to maintain their pigeon work illustrates that research programs of that sort required institutional support. Gilbert's study of the American expatriated to Germany, Ernest Everett Just, and the German expatriated to America, Richard Goldschmidt, provides further insight into the changing configuration of biology. These two men did work that had represented the very core of biology around 1900, but that became less central as the twentieth century advanced. Looking at their views and their respective tenuous places at the edges of the biological community,

Gilbert considerably strengthens our collective conclusion that, despite many attempts at definition, the discipline remained indistinct.

The combined result of the various studies in this volume, each with its distinctive emphasis and approach, is the demonstration that biology was rich and diverse at the turn of the century. Although there was disagreement about precisely what that meant and about what was inside or outside the boundaries, the general sense remained that something important was happening in the study of living organisms. There was a sense of "biology," even if it was better defined in the ideal rather than in actual practice. To thrive, biology needed expanded institutions, more obviously productive ways of working, more modern techniques, less speculative theories, more productive approaches, and more accessible research subjects; it needed to be, perhaps above all, professional and scientific in its research. Despite differences of opinion, the sense of something positive, new, and important persisted.

Beyond establishing the diverse character of American biology around 1900, this volume has further significance for the history of science. It offers a revisionist view of what happened, of course, and calls into question any historical accounts that suggest a unified or monolithic character for biological work. It emphasizes that in order to understand the roots of what is considered biology today, as biologists are increasingly seeking to do in the context of their various centennial celebrations, appreciation of the complexities and the varieties in the past is essential. Any effort to understand biology must respect the indistinctness of the boundaries and the way in which some areas of interest have passed in and out of what is considered proper, as well as the multitude of other changing factors suggested in this volume.

The collaborative nature of this volume allows it to go beyond many of the limitations of individual studies; however, its focus on particular, well-defined questions about the changing nature of biology in America around 1900 means that many promising avenues of research remain unexplored. In particular, further work in other fields of research such as anatomy, biochemistry, botany, and physiology could be instructive in helping to outline the boundaries. The impetus and support for specialization, professionalization, and organization all warrant further careful analysis, as does the impact of changing techniques and equipment. Applying the results of existing work on such influences to the case of biology in America should help to determine the extent to which it was like, or different from, other scientific fields, as well as other academic or intellectual pursuits around the same time. Comparison with other work in the history of science more generally may suggest whether this case of change follows the patterns of other scientific disciplines.

Another avenue for study could involve extending our focus to the present day, to determine further changes in the boundaries of biology. Still another consideration is nationalism in science. Is it just an illusion or does biology have a national character that significantly affects the content as well as the context of work done? This volume does not pretend to answer such questions definitively, for such answers will require careful comparison of this American case with study of biological science in other countries. Additional collaborative and comparative study will probably prove most effective for addressing questions about national styles in science.

Further examination of the centers of scientific work offers promise, as historians consider the role of government, universities, or separate research laboratories in directing science. Who paid for the biological work and what difference did the support make? How was the work organized? Was research undertaken individually or collaboratively, and how has that changed? What role have research schools or styles or programs or traditions played? These units of study afford a perspective on science that has been typically underexplored by historians of science despite some solid beginnings in recent years.

All these areas for further exploration arise from the studies in this volume. As the papers here concentrate on establishing what happened at the relevant time and place, they go beyond any attempt simply to define biology or to establish its emergence around 1900. Although the papers purposefully remain largely descriptive in orientation, they also begin to explore the causes of the particular cleavages and divisions. They begin to ask not only how, but also why things changed. This volume is just a beginning, but it is an unusual one. Its collaborative nature takes a major step towards addressing Churchill's charge of ten years ago that historians should seek to understand the nature of biology—in this case biology at a critically important time in the changing American setting.

Notes

1. Garland Allen, *Life Science in the Twentieth Century* (Cambridge: Cambridge University Press, revised ed., 1978).

2. Frederick Churchill, "In Search of the New Biology: An Epilogue," *Journal of the History of Biology,* 1981, *14*: 177–191; the volume includes "Special Section on American Morphology at the Turn of the Century," pp. 83–191.

3. For study of American science, see Sally Gregory Kohlstedt and Margaret Rossiter, eds., *Historical Writing on American Science: Perspectives and Prospects* (Baltimore: John Hopkins University Press, 1986).

4. Studies of changes in America during the period include, for example, Laurence Veysey, *The Emergence of the American University* (Chicago: University of Chicago Press, 1965); Robert H. Wiebe, *The Search for Order 1877–1920* (New York: Hill and Wang, 1967); James Weinstein, *The Corporate Ideal in the Liberal State 1890–1918* (Boston: Beacon Press, 1968); Burton J. Bledstein, *The Culture of Professionalism: The Middle Class and the Development of Higher Education in America* (New York: Norton, 1976); Peter Hall, *The Organization of American Culture, 1700–*

1900: Private Institutions, Elites and the Origins of American Nationality (New York: New York University Press, 1982).

5. For example, Robert Kohler, *From Medical Chemistry to Biochemistry. The Making of a Biomedical Discipline* (Cambridge: Cambridge University Press, 1982); Gerald L. Geison, ed., *Physiology in the American Context 1850–1940* (Bethesda, Md.: American Physiological Society, 1987).

6. Keith Benson, "American Morphology in the Late Nineteenth Century: The Biology Department at Johns Hopkins University," *J. Hist. Biol.*, 1985, *18:* 163–205; Philip Pauly, "The Appearance of Academic Biology in Late Nineteenth Century America," *J. Hist. Biol.*, 1984, *17*: 369–397; idem, *Controlling Life: Jacques Loeb and the Engineering Ideal in Biology* (New York: Oxford University Press, 1987); Jane Maienschein, *Defining Biology. Lectures from the 1890s* (Cambridge, Mass.: Harvard University Press, 1986); Jeffrey Werdinger, "Embryology at Woods Hole: The Emergence of a New American Biology" (Ph.D. dissertation, Indiana University, 1980).

7. Maienschein, *Defining Biology.*

8. Keith Benson, "The Naples Stazione Zoologica and Its Impact on the Emergence of American Marine Biology: Entwicklungsmechanik and Cell Lineage Studies," *J. Hist. Biol.* (forthcoming).

Part One

Natural History to Biology

Sally Gregory Kohlstedt

1 Museums on Campus: A Tradition of Inquiry and Teaching

A report on the proposed state university for California in 1864 called for a state museum at the center of an institution for "agriculture and the mechanic arts," to use the vocabulary of the Morrill Act passed just two years earlier. The so-called museum plan reflected the hopes of Josiah Dwight Whitney for establishing some secure repository for the collections of the California State Geological Survey under his supervision. His proposal, however, was less self-serving than it might appear. Whitney himself had had ample access to natural history specimens while a student at Yale in the late 1830s.[1] Moreover, he knew that collections of minerals, animals, and plants had become, during the previous two decades, well-established features of many college campuses. Whitney's plan for a polytechnical school in the 1864 report contained a narrower curriculum than that eventually adopted for the University of California at Berkeley (in 1868), but the final plans included a museum building that would house the findings of the State Geological Survey along with materials acquired and arranged by the "resident professors of the University."[2] These collections and others like them on college campuses across the country extended and helped restructure the place of science and technology in higher education.

The regular development of museum resources by colleges during the nineteenth century has only infrequently been noted by historians of science or education, whose attention has been drawn more to the efforts of graduate programs building experimental laboratories at the end of the century.[3] In fact, when noted at all, nineteenth-century college museums are not uncommonly portrayed as obstacles to be overcome by faculty members seeking to find space for such research facilities. Yet museums existed, expanded, and were newly created at the same time that courses named "biology" appeared on campuses. Contemporaries were well aware that the fundamental breakthroughs in geology, zoology, and botany in their lifetime depended in large measure on fieldwork and taxonomic methods of inquiry.[4] The ordering of specimens, often according to external characteristics, provided the essential configurations around which debate and early work on evolution was elaborated. Collections therefore became a fundamental tool for teaching natural history in undergraduate curricula by the 1860s. They predate the "revolution" in higher education in the 1870s, which is presumed to have largely introduced science into the curricula.[5] College museums—visible, physical entities—by that decade were already a symbol and a means of instruction in the natural sciences.

Student natural history societies and faculty-led field trips, along with various donations from alumni and government surveys, equipped numerous campuses in the first half of the century with specimens and cabinets. I have argued elsewhere that the displayed materials aroused local interest and in many cases provided an incentive for the inclusion of mineralogy, geology, zoology, and botany in the college curricula.[6] The efforts were informal and the results collaborative, if sometimes uneven, because students and faculty members at academies and colleges together did much of the collecting, mounting, and maintenance. The results might be closely related to the research interests of particular professors. Sometimes collections documented a state survey and at other times they became the basis for a handbook on local flora or fauna. On campuses, as elsewhere, they were at the heart of systematic investigation in natural history throughout the century, even as attention shifted from identification to geographic distribution and questions of development.[7]

College collections were often modest in scale and organization, and only gradually were they linked to formal instruction. In the years following the Civil War, however, colleges and universities gave administrative sanction, allowing such collections to grow at sometimes startling rates and to acquire new, larger, and more prominent space on campus. How this occurred and what it meant for students and faculty provides a backdrop essential for understanding those men and women who self-consciously worked to define a new biology.

Mid-century College Collections

By mid-century, courses in natural history were evident in the general curriculum of many colleges, academies, and normal schools and a fundamental part of special programs leading to a bachelor of science degree. In 1846 Louis Agassiz, a naturalist from Switzerland who had studied in various German universities and in Paris and used museums' collections for the basis of his major publications, arrived to lecture in the United States. Agassiz became professor at the Lawrence Scientific School of Harvard and affiliated with the

Saturday Club of Henry Wadsworth Longfellow and Oliver Wendell Holmes in Cambridge. He drew large crowds to his public lectures and attracted patrons for his projected museum of comparative zoology at Cambridge. His students, recalling their study with Agassiz, inevitably emphasized their intensive research with fish and other specimens that Agassiz initially housed in an old boathouse along the Charles River.[8] His museum, affiliated with a college (in this case quite loosely, because the Massachusetts legislature provided an unprecedented $100,000 in 1859 and worried about granting so much to the private corporation Harvard had become), provided a model admired but not easily emulated. His colleague, Jeffries Wyman, had a smaller although better organized collection for anatomy and physiology on the third floor of Boylston Hall.[9]

Harvard established (in 1849) a separate school for science, as did Yale and Amherst (both in 1854), separate because the sciences were presumed appropriate (sometimes vocational) for students who could or did not aspire to more prestigious careers in medicine or law. These prominent New England schools also attracted postgraduates, including trained physicians like Joseph LeConte, who sought advanced scientific education. Other colleges offered certificate programs, while still others began to offer bachelor of science degrees. LeConte, for example, probably helped design a two-year course at Oglesthorpe College in Georgia in 1853, "for young men whose time, means, or other circumstances do not admit of their pursuing a regular college course." As another historian put it, "those who were allowed to take the Scientific Course were boys who had no career aspirations other than farming or local trades and business."[10] Despite the implicit class overtones, such programs seem to have proliferated across the country. Indeed, higher education was becoming essential for those pursuing careers in science, and colleges found ways to provide, if not systematic training, at least an introduction to the relevant subjects.[11]

By the 1850s, women sometimes had rather significant opportunities to study various natural sciences. Enterprising headmistresses and headmasters enjoyed considerable flexibility in creating curricula and in choosing texts as they organized academies and later colleges for young women. Their interest in botany and chemistry could be sustained through a more innovative curriculum, because, in distinction perhaps from some all-male academies, the young women's studies were not bound to prepare them for college work, or for particular professions. Moreover, the students of Emma Willard at the Troy Female Academy and similar schools did eventually turn their knowledge into employment considered appropriate for women, including teaching, textbook writing, and scientific illustration. As full-fledged women's colleges were established, most included science from the outset, although the women faculty members, like the men a generation before them, found it necessary to teach themselves informally through self-study, private tutoring, and summer institutes. They, too, acquired collections more often by donation than purchase.[12]

Most of the antebellum college collections were small, focused on one or two areas from among mineralogy, botany, aspects of zoology (birds, insects, and mollusks were most common), or paleontology. They might be housed with the college library, along the walls of a classroom, in the office of a faculty member, or in such other spare space as a refectory or (in the secular spirit of the later century) in a former chapel.[13] Only a few colleges, including Brown, Gettysburg, Amherst, and Williams, had buildings with permanently designed or designated museum rooms or wings. Others, like the College of Charleston, collaborated with a local society and also gained direct public financing for collections housed on campus.[14] By mid-century, college catalogs frequently mention the existence of such collections, along with the library and philosophical (i.e., physical and chemical) apparatus, as evidence of their resources and their commitment to current educational trends.

Collection Development

Strategies to build collections, usually justified and described as essential to teaching, were several. Although the early source of collections had been internal at individual colleges, based on the efforts of students, faculty, and alumni, they expanded and consolidated in the last half of the century, thanks to government and individual patronage and to the natural history dealers who sold well-designed and labeled series. The older accession techniques of hiring professors with personal collections and soliciting donations from students and alumni continued, but college faculty members and administrators turned to trustees and other donors with increasing frequency to supplement the fieldwork of teachers and students. Miscellaneous gifts were increasingly replaced by planned purchases and by sponsored expeditions. Once established, collections often became a magnet that attracted additional contributions, including materials on loan or deposit from prominent naturalists and state agencies. The early pattern of voluntarism remained, especially within smaller colleges, but commercial and governmental distribution practices encouraged the use of systematic designs and formality.

Many of the earliest collections arrived with faculty members. Subsequent acquisitions during summer expeditions or on student field trips sometimes blurred the distinction between their private holdings and those of the college. In a sense, some were hired as a package, with the college acquiring a significant collection as well as a teacher; yet when faculty left with their private cabinets, the college cabinets might be "destitute."[15] Taxidermist J. W. P. Jenks brought some 4,500 birds and "such specimens in Mammalogy, Herpetology, Ichthyology, Conchology, and Comparative Anatomy, as will meet the want of instruction" when appointed at Brown in 1871–72.[16] Sometimes the collections were purchased after the retirement or death of a faculty member.[17]

College administrators after mid-century often presumed that professors would build collections relevant to their teaching and their independent research, especially at ambitious, endowed institutions like Cornell and Vanderbilt. Faculty members assigned to the "constant upgrading of the museum" found the work demanding, and sometimes turned to friends for help in working with unfamiliar specimens.[18] A growing number of scientists went abroad for study, and their experiences in Paris and in German universities and museums aroused their interest in developing collections as well.[19] Faculty aspirations at larger institutions moved beyond the often miscellaneous quality of donated and locally acquired material, and they looked toward more comprehensive, if selective holdings. Pragmatic faculty members like Burt G. Wilder at Cornell concluded, "true economy consists in paying liberally for what is wanted rather than in taking what is not wanted as a gift."[20]

To gain well-designed collections, instructors turned to college administrators who, in turn, asked trustees for funds. With financial support, they could purchase supplementary materials from one of several dealers whose business operations helped establish, to use an economic term, a market economy in natural history.[21] The old, informal methods for gathering and exchanging materials used effectively by student societies and individual faculty members were supplanted by sales coordinated through merchant naturalists like Henry A. Ward. He provided individual specimens of minerals, shells, and fossils; prepared both skins and skeletons for display; and organized synoptic teaching collections. Specialized catalogs allowed those with particular needs to select specific materials. Ward's business depended heavily on the sale of sets and individual items to schools and colleges, although his most remembered activities were dramatic displays at international expositions and large sales to major museums.

Using collections Ward had gathered in Europe or exchanged from around the world, his staff could outfit entire museums with casts of famous fossils, large mineral displays, mounted skeletons and skins, and even the appropriate glass and wooden cases for their display.[22] Although his museum at the University of Rochester, which was developed with $20,000 from local businessmen, never acquired the separate building for which he had hoped, it did demonstrate the possibilities of planned collegiate collections. About the same time, philanthropist Matthew Vassar hired the young but well-supplied naturalist dealer from Rochester to outfit (for $8,000) an entire natural history museum before his college in Poughkeepsie, New York, even opened. New colleges, like Vassar, had an advantage in that they were just developing their facilities and hiring a faculty. Buildings could be, and typically were, designed to accommodate cabinets for specimens and scientific apparatus. Thus, the rooms above the principal floor became the cabinet of natural history and above that the cabinet of geology and mineralogy in the large building that constituted Vassar College in 1869.[23]

Even more impressive was the independent museum at the University of Virginia sponsored by a Rochester philanthropist sympathetic to the South after its defeat in the Civil War.[24] Persuaded by Ward, Lewis Brooks sought a display that would "cover the ground *very* thoroughly, without redundancy. While other Cabinets in the United States may be larger or more exhaustive in the illustration of species, or special section of Nat. Sciences (e.g., the Meteorites at Amherst, *Birds* at Philadelphia, or *Shells* at Cornell)" [italics in original], he intended that none would exceed the University of Virginia "in capacity for scientific illustration in the highest University instruction."[25] Teaching collections were now being drawn directly into plans for development and expansion.[26] Their scale was larger, and growth coincided with the intermittent periods of prosperity in the last half of the century.

Ezra Cornell (the subsequent founder of the Ithaca-based university that bears his name) was importuned by the administrators at the University of Rochester, Hamilton College, and other upstate New York colleges for financial support. Most successful and persuasive was Cornell's first president, Andrew Dickson White, who was himself an "impassioned collector and had great faith in museums." Earlier, White had tried to interest Gerrit Smith in founding a college and wrote: "To support pure science by a struggling College with small endowment against the pressure of existing colleges would be vain" because any projected comprehensive college should have the best of libraries, collections in different departments, laboratories, a botanical garden, and an observatory.[27] Through the more generous patronage of Cornell, who agreed with White that the scale of facilities was related to the reputation of a university, he eventually acquired major fossil and shell collections for work on paleontology, for example. A view of the Cornell campus in the early 1870s shows the principal buildings, Sibley, McGraw, and White halls, lined upon the crest of a hill overlooking Cayuga Lake. The second building on campus, McGraw Hall, had a three-storied center section planned as a museum devoted to natural history.[28]

Burt G. Wilder studied with Agassiz, taught natural history and physiology, continued to participate in summer school projects in Massachusetts, and followed his mentor's example of building collections when he went to Cornell. The solicitation of private money would continue throughout the so-called Gilded Age, successful when a collection or even a building like McGraw might honor its donor by name. Such scientific support would eventually be systematized although never fully supplanted when turn-of-the-century philanthropists established foundations on Carnegie's model to oversee specific requests for scientific resources.[29] Not all institutions could aspire to the large museum established at Cornell, but a surprising range of secondary and postsecondary educators established collections for teaching students.

An Apparatus for Study

In the years following the Civil War, increasing if intermittent administrative attention was given to museums as colleges continued the process of redefinition begun decades earlier. Economic instability, punctuated by periods of labor unrest and financial panic, coincided with industrial expansion. In this context it is not surprising that there was also ongoing confusion about curricula, faculty preparation, and indeed the entire mission of colleges. None could be insulated from either the acquisitiveness or the instability of the period, and indeed some grew dramatically into universities while others failed completely.[30] Unfortunately, there has been relatively limited research on undergraduate teaching in this period—in contrast to histories of undergraduate study in the antebellum period and of graduate study at the end of the century—even in the accounts of individual colleges.[31] Yet it was the availability of science education in four-year college programs over the course of the century that made postgraduate education here and abroad possible. Each institution went through its own conversion, experimenting with different degree titles and trying special programs designed to take two, three, or four years. More than twenty-five colleges created such departments in the 1860s and 1870s and from 1870 to 1873 the number of "scientific schools" affiliated with but not part of the regular colleges jumped from seventeen to seventy.[32] Nearly all colleges made it clear, however, that science was part of their program, and they used scientific collections and equipment to emphasize the point.

Rationalization of curricula became a principle goal from the elementary to postgraduate levels, even as the range of educational opportunities expanded remarkably in the last half of the century. Tendencies toward professionalization among college faculty members and in scientific institutions like museums and herbaria also helped shift a casual, inclusive, and eclectic style toward one more systematic and standard.[33] Curricular uniformity was made possible by the availability of textbooks such as Louis Agassiz's *Essay on Classification* (1857), James Dwight Dana's regularly updated *System of Mineralogy* (originally published in 1837), and John W. Draper's *A Textbook on Chemistry* (1866).

Collections themselves gained new visibility and prestige as an apparatus for study. Professors used specimens of minerals, plants, and animals in coursework, and students could collect and handle them in conjunction with courses in both undergraduate and postgraduate programs. Demonstrations were important as the means of instruction shifted from recitation to lectures, or, viewed another way, as successful informal techniques found their way into the classroom.[34] A few institutions aspired to imitate the comprehensive museums in Rochester, Poughkeepsie, and Charlottesville, and especially Agassiz's specialized museum in Cambridge. James Dwight Dana, for example, disclaimed the possibility of rivaling the Museum of Comparative Zoology in paleontology, but nonetheless was hopeful that Yale's appointment of Othniel C. Marsh would put their museum "among the No. 1s both in Geology & Zoology."[35] Trustees and administrators responded to internal demands and outside opportunities—in this case, a gift from Marsh's uncle George Peabody—in order to develop college museums.

All of this was encouraged by the more evident emphasis on science. Some colleges reoriented a by-then traditional collegiate outlook based on arts and sciences, which had (and continues to have) considerable flexibility.[36] Agricultural editor Orange Judd established a museum at Wesleyan, in Connecticut, under George Brown Goode who was later director of the National Museum. Others joined new larger technical schools and state universities whose curriculum was, among other things, attentive to industrial technology, geological resources, and scientific agriculture, in part because of the incentives provided by the land grant act to states for colleges in 1862 and a subsequent federal grant in 1890.[37] Permanent research stations for agriculture might be established on or near the campus, while legislators might assign funded colleges the responsibility to maintain collections acquired by state surveys or bureau investigations. The prospects for collegiate museums were enhanced, as well, by the growing authority of scientists in public and political circles and by their appointment to administrative posts at colleges and universities.

The variety of programs (and their tentative nature) is evident in college and university catalogs that reflect frequent changes in both courses and general programs. Schools of mines inevitably had geological, mineralogical, and paleontological collections. Schools of agriculture emphasized zoological and botanical collections, which might be used for a diverse set of programs from summer school short-term classes to postgraduate programs in veterinary medicine and science.[38] There was often a duality to the holdings, making them at once local and at the same time universal, intended to represent "typical specimens" and to illustrate concepts of development in science.[39] The local specimens could serve research and teaching interests that had been generated by and were attentive to particular state and regional concerns in agriculture, mining, industry, and transportation.

An enterprising college president might use the collections for rather dramatic advertising. President Hartshorn at Mount Union College acquired thousands of minerals and small specimens, casts of fossils from Ward's, a thousand models of invention from the Patent Office, and more than a hundred mounted zoological specimens from "aardvark to zebra" in the 1860s. Reacting to the promotionalism, the *Nation* commented sarcastically in 1877, "Mount Union College has a stuffed gorilla which travels to public gatherings and agricultural fairs, and which was sent to the Centennial as representative of the institution. But whether the gorilla represented the faculty, or the students, or the ideal graduates, or the stage of development at which the process of evolution the college has arrived, nobody seems able to say."[40] Most administrators were more subtle.

Faculty members only rarely articulated their goals regarding collections, but in the expansionist mood of the 1870s and 1880s, those working in the natural sciences relished the possibility of organized and well-housed collections on which they could both base research projects and teach courses. At the same time, they increasingly appreciated that maintaining the collections was, in fact, more difficult than acquiring them—a discovery made earlier by volunteer curators in urban or regional natural history societies.[41] Responsibility was especially onerous if the collections left by a predecessor were quite narrow or far from current faculty interests. Some were impatient, as well, with "marvelous collections of birds and beasts and insects which some trustee with more zeal than knowledge had squandered hundreds of dollars upon."[42]

Faculty in the last half of the century emphasized thoroughness of coverage and direct applicability to coursework in describing their collections.[43] As courses on natural history became standard, teachers at colleges and universities often requested materials from Ward to provide specimens missing from a series or establish entire series that they and their predecessors had not yet acquired through the less expensive means of exchange or donation. Those faced with old, unidentified, miscellaneous holdings also tried to upgrade them, although few were in such desperate straits as the director of an academy in Tennessee who asked for help in identifying and organizing "a lot of fossils (a half bushel, perhaps, if your scientific sensibilities are not shocked by such a term)."[44] In some cases the details in Ward's catalog were sufficiently thorough that they were used as the basis for labels and even for textbooks. Ward's records over the course of fifty years detail the array of colleges and universities, small and large, secular and religious, comprehensive and specialized, that created or expanded collections in the last half of the century.[45] In the 1860s and 1870s, the Smithsonian Institution also distributed duplicate sets of shells, minerals, and other specimens to colleges as well as local societies. Its donations, often well publicized, helped attract further contributions.[46]

Duplicate sets indicate clearly the tendency toward standardization, and widespread interest in basic or synoptic collections prepared for didactic exhibition or for student reference. Properly used, they might present "fundamental ideas by concrete examples," demonstrating the relationship between species based on their "structural plan."[47] Small colleges with limited resources were attracted to Ward's organized series of minerals, shells, casts of famous fossils, and similar materials. Compact and portable, small wooden cases containing virtually identical specimens went to academy, college, and medical school classrooms. Wealthier institutions ordered models of invertebrates by the Auzoux brothers in Paris, the glass models of invertebrates by Leopold and Rudolf Blaschka of Dresden, and papier-mâché models of animals (including humans) from Germany and France, which could be disassembled in such a way as to reveal anatomical characteristics. The Smithsonian Institution, too, culled duplicates from its holdings for colleges.[48] Such standardization had a number of attractions, but it also meant that teachers and students were less mutually involved in building collections that related to their concerns and their experience with the local environment.

The shift during the 1870s and 1880s, however, was not immediate and some students worked with faculty members in ways that added to their own credentials. Where there were collections to be arranged or rearranged according to more abstract principles than the natural system or when new materials were added, students might be involved, typically assisting rather than initiating projects. At Princeton, there were summer "scientific expeditions" intended to discover specimens new to science, increase the college museum collection, and provide "invaluable practical training to the students who took part in them."[49] The trend was national, affecting not only major state-supported colleges but also private religious ones, such as Elon College, chartered by the Christian Church in North Carolina in the 1880s, whose original design included a museum as well as a library.[50] A student was put in charge of the collections and a local congressman arranged for five hundred specimens to be donated by the Smithsonian Institution. Some students were individually enterprising in ways reminiscent of the antebellum period. T. Gilbert Pearson, who was later president of the National Association of Audubon Societies, earned his tuition and board at Guilford College by mounting and arranging birds there in the 1890s.[51] At Denison College, an experienced faculty member like Clarence Herrick could lead students, including Frederic Clements (discussed by Hagen in this volume), on field trips to a local stone quarry and flint ridge as well as show them how to use microtomes, a freezing section cutter, and other specialized equipment to prepare and observe microscopic sections in a histology class.[52]

By the 1880s a substantial number of new and revived student scientific societies were locally important, but their activities were dwarfed by the larger departmental resources and more direct faculty involvement than societies created a half century earlier (Appendix 1). In fact, some (e.g., the Natural History Society at the University of Kansas) were founded by faculty members who recalled the importance of student-initiated societies in their own formative years.[53] Part of the change was also in the student profile. Most student classes at the end of the century were concentrated in the eighteen- to twenty-year-old category, rather than representing the wider age range (fifteen to thirty years old) evident in the first half of the century.[54] Perceived as requiring supervision, students were sponsored in systematic and planned activities. Professor Burt Wilder at Cornell, for example, made it clear that each student in his physiology course would be "advised to arrange his [or her, for Cornell was coeducational] time so as to enable him to spend a part of the whole of Saturday in the laboratory" working with specimens.[55] In addition, he urged the students to attend bimonthly meetings of the Natural History Society, by implication another part of their assignment. Such groups might plan excursions, but they were now almost exclusively consumers rather than builders of the university collections, which had been acquired primarily by purchase or large scale donation. Local expeditions continued, however, because they "aroused the faculty of observation and widened our knowledge of nature."[56]

The tone changed as well, as when the Franklin Scientific Society at the University of Pennsylvania advertised for new members by emphasizing the sociability of the group, open to those in the arts as well as the sciences, and only later mentioning their rooms with a library and reference collection of natural history items.[57] Frederick Rudolph observed the shift at Williams College in the 1870s: "The professionalization of scientific study destroyed the [student] Lyceum. . . . For as the college came to do more and more for the student[s], there was less and less that they needed to do for themselves." By the late nineteenth century, moreover, local college societies seemed perhaps quite unimportant in the context of stronger departments, local and state academies of science, and national specialized associations. Thus, it was a national honors society, Sigma Xi, founded at Cornell in 1886 and aspiring to be a "scientific Phi Beta Kappa," that became prominent on campuses in the twentieth century.[58]

Students took the museum effort more, but sometimes less, seriously. Perhaps the unusual items were too tempting to be overlooked as the basis for local pranks. Histories of individual colleges, although they contain little account of museums, do record stories of peripatetic stuffed alligators and of student pranksters who took a gorilla to the college library and left it there with *Origin of Species* in its hands.[59] More serious students, as some at Northwestern, presented specimens; the class of 1872 donated an elephant skeleton reputed to be larger than Jumbo, and the class of 1878 contributed a whale skeleton. Similarly, the college faculty at Syracuse voted in 1871 to purchase "a fine specimen of *Ichthyosaurus*" from their limited budget and solicited gifts from alumni to furnish the west wing of the basement as a museum, which opened in 1873.[60]

Increasing formality was in part related to the administrative interest in what the collections (like the students) required and how they might be used. College presidents, heads of normal schools, and directors of mining and scientific schools were all persuaded that their graduates with scientific and technical training enjoyed expanded employment opportunities in government, industry, and education. The growing authority of scientists among the public and politicians, and their representation in administrative posts at colleges and universities in the last third of the century, also increased the prospects for collegiate museums among other competing resources.

Museums could serve not only "for the illustration of lectures but also for the general education of the public."[61] David Starr Jordan, President of Indiana University and later of Stanford, wrote to Ward from Bloomington, "please send me some data as to your collection of mounted mammals, especially apes and monkeys. These please our constituents."[62] The few pictures that remain of these early college museums often show exotic specimens in an entry hall or in the middle of a room lined with upright wooden and glass cases for systematic display and storage. Dramatically mounted gorilla or elephant skins or large dinosaur skeletons might be used as centerpieces. Colleges with the generous budgets purchased mounted casts of the large and dramatically posed *Megatherium* from Henry Ward. Thus, the college museums in upstate New York at Syracuse, Rochester, and Cornell universities resembled in some features a scaled-down model of larger public museums in New York City and Chicago by the end of the century. A faculty member at Brown sought a mounted buffalo for public display, specifically demanding a male, because, he asserted, the male was most "typical" (did he mean "ideal"?).[63] There is no way to measure public response to these displays, but college museums continued to be a cultural attraction in small towns and part of a scientific network in larger ones. Given the educational and promotional possibilities, new science buildings with significant museum components (as indicated in Appendix 2) made good sense as an investment.

However much the museum as an apparatus for study and a public attraction could be assumed, the administrative structure was less clear. Cornell President Andrew Dickson White envisioned that the various collections that he and his faculty acquired would be "united in a spacious and convenient building, forming the central edifice of the university group."[64] More often, new colleges repeated a familiar pattern of eclectic early growth, appropriation of marginal or underutilized space (such as the college chapel), and neglect by faculty members with limited resources for maintenance; but these circumstances could also provide an argument for later action. The University of Kansas, for example, when it opened in 1866 could mention only "a museum room on the third floor [of Old North College with] a few geological specimens collected by the faculty in their travels." This base grew rapidly. During the next two decades, thanks to donations in botany, zoology, and geology from faculty member and state entomologist Francis H. Snow, the University had ample reason to petition the state legislature for a hall of natural history. By then, a visitor could record enthusiastically, "the cabinets of birds and insects, of every color, size, and shape are a fascinating study, and we speak with a feeling of pride, mixed with not a touch of wonderment at the untiring energy displayed when we say that these cabinets have been collected almost entirely by Professor Snow, and his corps of students."[65] When the Snow Hall of Natural History opened in 1886, the two-story building had four museums (geology on first, zoology and entomology on second, and anatomy in the attic) with various workrooms for specimen preparation in the basement; in addition, it housed a lecture hall and laboratories.[66] The tribute was appropriate for a teacher and scientist who pioneered in the economic and systematic studies of natural history in Kansas.

Even as the museums became an assumed part of higher education, they were undergoing a change in orientation and role. The awkwardness of affiliating a major museum with a college had been evident years before when Louis Agassiz had established the Museum of Comparative Zoology, physically adjacent to but with trustees independent of Harvard. The Massachusetts legislature had granted $100,000 toward the enterprise with the explicit understanding that there would be regular public access to the collections. Equally important, Agassiz himself sought a certain degree of autonomy for his enterprise.[67] The museum was unusual in its initial contractual arrangements, but elsewhere the relative independence of the museum and the relationship between the faculty curator and the collections remained ambiguous in a period when few ethical and legal standards were articulated. Thus, for example, there was the problem of ownership of specimens gathered by the professors as part of their college responsibility, which collided with the tradition of individuals having the right to exchange or even to sell at least duplicates of what they collected on field trips. Increasingly, the tendency was to make collections institutional, and the sale of personal collections at the retirement of a faculty member diminished toward the end of the century, although a generous donation (even before there were tax incentives) often eased the potential tension.[68]

The ad hoc acquisition and maintenance probably served individual faculty members and smaller colleges reasonably well when lecture halls, faculty offices, and collections were in close proximity. As facilities expanded in the 1870s and 1880s, resources reached unprecedented scale.[69] Existing collections were regrouped and systematized during this period of expansion, often with the help of students or paid staff. Following the Agassiz model, instructors prevailed on serious students to assist them in identifying, describing, arranging, and analyzing college holdings. Sometimes a "practical naturalist" would be hired as curator, assigned to mount, arrange, and preserve collections.[70] In 1870, Northwestern University in Chicago hired a taxidermist rather than continue to depend on student and faculty volunteers to mount specimens and add to the collection begun by Robert Kennicott in 1857. Kennicott had been given room, board, and tuition for his early work at the Northwestern museum, a practice that was exploitative by some measures but more generous than that at many colleges where curatorial work was strictly voluntary.[71]

Other practical concerns about fire and ventilation helped substantiate arguments for new buildings.[72] There were competitive opportunities on growing campuses. The commitment to maintain all museum collections was debated, especially when, as at Princeton, the original endowment for the museum of geology and archeology proved inadequate to pay the curator's salary.[73] Should they be kept as departmental holdings or be combined into a single comprehensive museum? The latter seemed attractive to many administrators who were already consolidating libraries in an effort to economize on staff and maintain oversight of collections that, in some instances, had become faculty fiefdoms.

Uneasy Alliances

The proliferation of buildings for collections across the country after 1880 or so responded to but did not end the problems of organization and, in fact, revealed the tension between specialization and centralization, between education and research, and between emerging theoretical and methodological differences, particularly in the biological sciences. The issues were not unique to academia, of course, but had evident impact on the activities and status of the now well-established tradition of museums on campus.

Early museums and libraries had arranged collections on the basis of donor, student, or faculty interests, or even date of accession, while individual faculty members had often maintained private and even working collections of the college or university near their offices and classrooms.[74] There was certainly a logic to keeping functioning units together; moreover, proprietary sensibilities of individuals and departments with regard to collections remained strong. Thus, although the university museum at the University of Michigan was built in 1881, the collections housed there remained under departmental jurisdiction until the 1920s.[75] At the University of Chicago, George C. Walker donated $100,000 to establish a comprehensive museum building related to the departmental museums, as the general library was related to the departmental libraries. The forces, even at this new university, were too centrifugal. The paleontologists moved in, but within a decade there was a faculty committee charged to answer the question: "What is to be the future policy of the University regarding museums?" Other departments at Chicago had found it practical to maintain collections close to offices and their classrooms and ignored the master plan. Efforts to consolidate, with mixed success, would persist well into the twentieth century.[76]

Adding to the complexity of maintaining individual collections was the rapid growth of urban museums for art and for science in what has been billed a golden age of museum development. Major new natural history museums appeared in Boston, New York, Philadelphia, Washington, D.C., Pittsburgh, Chicago, and San Francisco, not to mention scores of others whose tally reached over fifty in a survey completed in 1900.[77] That these public museums could be an asset was shown in a college promotional pamphlet that pointed out: "New York is Barnard's laboratory."[78] The American Museum of Natural History, built alongside Central Park West in New York City, had the largest museum building in the United States when it opened in 1873. The Smithsonian subsequently acquired a separate natural history museum (opened in 1881); proponents argued for its establishment on the basis of substantial donations from foreign visitors at the Centennial Exposition. Only a few specialized college museums could compete in scale or facilities, such as the Peabody museums at Harvard and Yale or, somewhat later, the Museum of Anthropology founded by Phoebe Appleton Hearst and Museum of Vertebrate Zoology sponsored by Annie Montague Alexander, both at the University of California at Berkeley.

Not surprisingly, explicit connections between neighboring facilities, such as the Massachusetts Institute of Technology and the Boston Museum of Natural History, were developed; museum curators taught university courses, and the museum became, in effect, a student workshop.[79] In other circumstances, there were more informal relationships, with professors serving as curators and advanced students studying museum collections, although these arrangements always depended on mutual good will. Jealous of its tradition, the Academy of Natural Sciences of Philadelphia refused an invitation to relocate on the grounds of the University of Pennsylvania in 1889, but it did offer students opportunities to use Academy collections, as well as take special courses in paleontology, geology, and mineralogy. A decade later, Philadelphia philanthropists built the University Museum on the campus, complementary to the Academy in its emphasis on archeology and ethnology.[80] By contrast, the Maryland Academy of Sciences donated its entire collection to Johns Hopkins in 1888. President Gilman and his professors had grown cautious about accessions, however, and reminded both students and prospective donors that "the proximity of Baltimore to Washington makes it easy for the specialist to visit the great collections of the capital."[81] In Chicago, the situation varied considerably from discipline to discipline, with some faculty members from the University of Chicago competing and others cooperating with departments at the new Columbian Field Museum and with those at the Chicago Academy of Sciences.[82] Urban museums, as one way of justifying their public funding, established educational programs on every level, serving schoolchildren and their teachers, as well as advanced students by the turn of the century. This function, like that of specimen maintenance, did not always attract the commitment of the publishing academicians.[83]

Somewhat different circumstances led to similar results in the case of agriculturally oriented states where scientific surveys had led to large-scale collections. Already in the 1840s New York State provided a precedent by establishing a state museum in Albany, based on James Hall's collections in geology, mineralogy, and paleontology.[84] Where the state university and state capital were in the same city, the legislature might actually mandate collaboration at one level or another. Thus, the bill for the Minnesota Geological and Natural History Survey in 1872 explicitly placed the results of that survey at the University of Minnesota.[85] Accounts of museum building vary for each city and state, one of the cultural effects of our federated political system. Reciprocal relations between public and learned society museums and local colleges or universities, however, could potentially be to their mutual advantage. Where competition or even disinterest existed, college museums had particular incentives to expand independently, but the result could be a debilitating financial effort to build separate resources.[86] As universities developed research programs in new areas, however, museums came to be viewed less as centers for research than as adjacent places for student and faculty reference and public display.[87] Over time, cooperation would depend more on personal and intellectual compatibility than on formal institutional arrangements.

By the late 1880s, college and university collections in geology, mineralogy, paleontology, biology, zoology, entomology, botany, or anthropology might be under any one of a number of administrators or faculty members. As responsibility was handed on from the original collectors, whether teachers or students, the successors might resent the time required to maintain and explain materials of limited value to their research.[88] Reference to neglect, indifference, and even disdain for some of the materials displayed but not used became more common. George Brown Goode at Wesleyan, having worked on an international exposition at the Smithsonian, expressed his discontent, "I don't feel quite satisfied to spend as much time on so unimportant a museum and to carry so much good material there even if I am 'paid for it.'"[89] A revealing student view was offered in an editorial in the *Syracusan* in 1883: "Over one-half the students regard the Museum as a mysterious, gloomy, dusty affair, that is a necessity to a University but of no meaning or use to them. . . . the back of the Armadillo is to them only a favorable place on which to sit and yawn over Greek roots and the huge legs of the *Megatherium* afford a support for a half-hour's undisturbed study of metaphysics—or the cob-webs on the ceiling (generally the latter)."[90] Although the editorial concluded that more faculty use of the specimens and attention to the musty air and dust problems could counteract student attitudes, the reality was that the static and sometimes outdated displays in the 1880s were peripheral to formal and informal education at Syracuse University. Within a decade plans were underway for a new science facility with space to reorganize the museum; when Lyman Hall was completed in 1906 in Syracuse, ten rooms were devoted to college collections. Natural history museums continued to be built, expanded, and publicized as part of the claim to be a comprehensive college or university.

The Legacy

In retrospect, it is evident that during the first half of the century and rather later in the Midwest and Far West, the very process of collecting and arranging specimens was intellectually productive for students and for their instructors. Then, those who labored, to use nineteenth-century terms, "in the field," learned to accommodate those who worked only "in the closet." By the middle of the century, faculty members with broad assignments used collections as an auxiliary to both private research and the public visibility of their programs and, subsequently, as a way to enhance courses. The flexible antebellum colleges thus nurtured the personnel and resources on which colleges and universities would rely in subsequent decades. The positive effect is observable in the strength of the natural sciences, and reflected in part by the comparative lack of interest (by contrast with physical and social sciences) shown in studying abroad, as well as by the relatively large numbers of natural scientists at the end of the century.[91]

Yet the establishments on any given campus of a comprehensive museum of natural history did not assure its value. The collection could become a static display, used in routine fashion or not at all. These were often criticized by professors advocating other approaches in scientific teaching and research. In different circumstances, however, the tradition of fieldwork and study series in the museum became a complement to ongoing discussions of evolution and used in conjunction with laboratory work.[92] Other essays in this volume suggest ways in which the natural sciences changed, specialized, and developed methods that emphasized quantification and duplication. Laboratories built when German-trained scientists returned to faculty positions in the United States were used for experimental work and work with living forms.

Vocabulary shifted, too, with workrooms becoming laboratories and older designations in zoology, for example, taking on new names in smaller, more narrowly defined specialties. Fieldwork and systematics, except in museums, seemed increasingly outdated as compared to cytology, physiology, and embryology. In this process, there was a denigration of natural history and naturalists, terms that by the twentieth century became pejorative as too inclusive, too informal, and too descriptive. Yet elements of tradition continued to be valued, as in the American Society of Naturalists (discussed by Appel in this volume).[93]

In the late nineteenth century, museums and laboratories coexisted in most colleges, not cooperating extensively but competing only in a general way for space and resources.[94] Together they provided the inspiration and, to some extent, the training for self-conscious biologists, as discussed later in this volume. Thus, we must not let efforts by a new generation to establish an independent identity at the turn of the century distort our historical awareness of the ongoing activity that updated older methods in systematics and provided essential environmental (later to be called ecological) information. Museums, in fact, went through their own period of dynamic growth, punctuated periodically by new faculty positions, important publications, and expanded collections and physical facilities. College museums reached a kind of plateau at the end of the century, although in some parts of the country the pattern of development would continue well into the twentieth century.[95]

Situated as a very physical presence, museums remain on college campuses, of necessity changing with regard to content and presentation.[96] The initial and fundamental functions of collegiate museums regarding education and public display, identified by the middle of the nineteenth century, are still visible. Undergraduate majors typically work with geological, zoological, and botanical specimens and in advanced classes do fieldwork in their region. In museums, and in laboratories, they work with each other and their professors in a kind of intellectual intimacy perhaps less available elsewhere on the campus, thus providing the basis for subsequent collaborative work.[97] Much has changed, however, as the museums hire professional staff, follow specific models of exhibition and storage, and have explicitly defined relationships to the colleges that house them.

Some conclusions about the changing yet continuing role of museums might be drawn by returning again to the University of California at Berkeley nearly a half century after its founding. The university itself had grown remarkably and had by the early twentieth century an institutional configuration and faculty that reflected higher education generally, as indicated in discussions of the University of Chicago and of the research efforts at Woods Hole (see Maienschein and Pauly in this volume). Skillfully guided by President Benjamin I. Wheeler, western philanthropists supported the state-affiliated school and showed particular interest in specialized museums as well as other substantial building projects.[98]

The Museum of Vertebrate Zoology was established by the initial donation and then the long-term sponsorship of a remarkable woman, Annie Montague Alexander. Interested since childhood in the outdoors, she became an expert sharpshooter, an avid hiker and collector, and a skilled amateur in paleontology. Apparently first inspired by lectures of John C. Merriam at the University of California, she eventually established a museum at Berkeley and personally chose Joseph Grinnell, an ornithologist, editor, and teacher, as director in 1907. Grinnell decided that a tour of eastern public and collegiate museums would help him learn current management, exhibition, cataloging, and preservation techniques. He returned with professional contacts and a clear set of priorities.[99] The museum had a focus evident in its title, Museum of Vertebrate Zoology, and concentrated particularly on state (actually regional) mammals and birds. The rationale (reprinted as Appendix 3) embodied arguments for college museums that were by then well accepted: the museum would complement collections elsewhere and could be used for teaching in the university as well as in the secondary schools, for research, and for practical applications. Under Grinnell's supervision, the museum staff established a seasonal rhythm of work that included field trips, laboratory work, lectures, student supervision, museum maintenance, and publication. Alexander provided ongoing financial support, regular oversight, and external negotiation with administrators.

Museum of Vertebrate Zoology records demonstrate certain principles regarding museums on campus now in place. Directors and scientists are members of the faculty, but support staff are not. All collections are owned by the university, for Grinnell donated his personal collection in 1909 and Alexander continuously deposited the products of her various expeditions from California to Alaska. The museum's contract is with the trustees and, after years of being housed in a corrugated iron structure west of Faculty Glade, the collections were transferred in the 1930s to the new life sciences building. There the collections could be an integral part of ongoing taxonomic and theoretical work, both important to the old and the new biology.[100] Museum activity today—at times informal, collaborative, and field oriented—has elements of effective past educational methods now supplemented by departmental laboratories well established in the bureaucratic structure of a modern university. At Berkeley, as elsewhere, the specialized museum has become both a reflection of past knowledge and a resource for ongoing inquiry.

Appendix 1: Natural History Societies Established on Selected Campuses, 1868–1902

The following list is derived from reading college catalogs and histories found in the Library of Congress and is certainly not comprehensive. It does, however, suggest something of the diversity of institutions of higher education that established natural history societies (among a great variety of other clubs) in the late nineteenth century. An asterisk indicates a date in operation with the date of establishment not identified.

1868	Natural History Society	Mount Holyoke Seminary
1869	Natural History Association	Mount Union College
1871*	Franklin Scientific Society	University of Pennsylvania
1873	Natural History Society	University of Kansas
1874*	Natural History Society	Cornell University

1876	Toner Circle [sometimes Society]	Georgetown College [University]
1880	Naturalists Field Club	Johns Hopkins University
1881	Natural History Society	Mississippi State University
1883	Naturalists Club	Carleton College
1883	Elisha Mitchell Scientific Study	University of North Carolina
1887	Scientific Association	Denison College
1887	Natural Science Association	North Central College
1888	Natural History Club	Knox College
1893	Science Club	Pomona College
1894	Natural History Society	Bucknell University
1895	Science Association	University of Southern California
1895*	Zoology Club	Stanford University
1902	Natural History Society	University of New Hampshire

Appendix 2: Buildings Established for Scientific Study on Selected Campuses, 1871–1910

The following list is compiled from various college catalogs and histories and is intended primarily to demonstrate the construction (space was allocated for collections in those listed) that took place around the turn of the century. Note a significant break around the time of the panic and depression of 1893. The next wave of building for science on campuses came in the 1920s, especially at former seminaries and at black colleges; another unprecedented period of campus construction (with an emphasis on the sciences) came, of course, in the 1950s and 1960s.

1871	Culver Hall	University of New Hampshire
1872	Coburn Hall	Colby College, Maine
1872	Geological Hall	Rutgers University
1872	Scientific Building	University of Missouri
1873	Science Hall	Ohio Wesleyan
1875	Knowlton Hall	Hillsdale College, Michigan
1876	Lyman Williston Hall	Mount Holyoke College
1876	Peabody Museum	Harvard University
1879	Brooks Museum	University of Virginia
1881	University Museum	University of Michigan
1885	Tomes Scientific Hall	Dickinson College
1885	Wylie Hall	Indiana University
1886	Stephens Scientific Hall	Central College, Missouri
1886	Snow Hall of Natural History	University of Kansas
1886	Lilly Hall of Science	Smith College
1886	Chemistry Building	University of Nebraska
1888	Science Hall	Fordham University
1888	Science Hall	Hamline College, Minnesota
1890s	Wilkins Science Hall	University of Vermont
1892	Science Hall	Drake University, Iowa
1892	Science Hall	University of Tennessee, Knoxville

1894	Barney Science Hall	Denison College, Ohio
1897	Science Hall	Hanover College, Indiana
1898	Science Hall	West Virginia University
1898	Natural Science Building	University of Kentucky
1898	Science Hall [Morrill]	University of Maryland
1898	Fayerweather Science Hall	Maryville College, Indiana
1899	Science Hall	University of Richmond, Virginia
1901	Science Building	Stetson College, Florida
1901	Warner Science Hall	Middlebury College
1901	Science Building	University of North Dakota
1902	Science Hall [Montgomery]	Mississippi State University
1902	Museum Building	University of Utah
1904	Science Hall	Wofford College, North Carolina
1906	Science Hall (Carnegie)	Elmira College, New York
1906	Science Hall (Carnegie)	St. Lawrence University
1908	Science Hall	Kentucky University (Transylvania)
1910	Science Hall	University of Denver

Appendix 3: Value to the University of California of a Collection Representing the Mammals and Birds of the State

This memorandum (copied verbatim from a typescript manuscript, undated, c. 1910, from the Records of the President of the University of California [CU-5], Box 25: Annie Montague Alexander, 1905–1909, University Archives, Bancroft Library, University of California at Berkeley) provides an early twentieth-century rationale for establishing a state-based museum collection on the University campus.

1. *Instruction in the University.* The collection will be of great assistance in illustrating the following courses of undergraduate instruction.
   a. General introductory course in Zoology
   b. Comparative anatomy, particularly comparative study of skeletons; paleontology, in making available for comparison a representation of the recent life, through which all fossil forms must be interpreted.
   c. Courses in physical geography, in so far as the geographic distribution of animals is concerned.
   d. Special courses in animal life of California
2. *Research.* The collection will be of great value in research conducted along the following lines.
   a. Variation and evolution of species.
   b. Geographical distribution.
   c. Systematic work on West-American faunas.
   d. Basis for all palaeontological, geological and anthropological studies

in which reference is made to the animals of the later geological periods.

While the work of bringing the collection together is in progress many scientific results of value will certainly be obtained. This material will be published as the result of scientific research at the University, and will appeal to a class of scientists not previously interested in our University publications. The collection will be available for use by any qualified student of mammals or birds, and should make the University of California a recognized center for the study of mammals and birds of America. At the present time there is no good collection for comparison available in California.

3. *Nature Study in Secondary Schools.* The attempt is being made to interest students in secondary schools in the animals in the regions immediately about them. Work of this nature is possible as there are good herbariums here in which comparisons and determinations of specimens collected can be made. This collection would make it possible to do satisfactory work with the mammals and birds, which are not less important than the plants.
4. *Economic Value.* The relation of the mammal and bird faunas to agriculture and to other industries is considered of sufficient importance to justify the Federal Government in organizing a Biological Survey, in which a large portion of the work is based on investigations such as would be carried on for this state in connection with the museum. The work of the museum will also be of assistance to the state fish and game commission by giving us much information concerning the distribution, habits, and best means of protection of game.

Acknowledgments

Some data in this essay were presented at Dickinson College and subsequently published as "Natural History at Dickinson and Other Colleges in the Nineteenth Century," *John and Mary's Journal,* 1986, *10*: 27–48. I wish to thank the participants in the Friday Harbor conference, especially the three organizers and editors, for their friendly advice. Margaret Rossiter, Clark Elliott, and Ralph Dexter offered comments and new information after reading a preliminary draft. Carol Gregory Wright helped me cull college histories and catalogs. A fellowship at the Woodrow Wilson Center, Washington, D.C., provided time for me to rethink the entire project and for Alexandra Gerson to check footnotes at the Library of Congress.

Notes

1. "Report Relative to Establishing a State University," in *Journals of Senate and Assembly of the Fifteenth Session of the State of California, Appendix* (Sacra-

mento: State of California, 1864), pp. 3–29, written by Josiah Dwight Whitney, the State Geologist, J. F. Houghton, the Surveyor General, and John Sewell, the State Superintendent for Instruction.

Verne A. Stadtman, *The University of California, 1868–1968* (New York: McGraw-Hill, 1970), p. 26.

On Whitney's education at Yale, under Robert Hare, and in Europe, see Edwin Tenny Brewster, *Life and Letters of Josiah Dwight Whitney* (Boston: Houghton Mifflin, 1909), pp. 14–90. On his activities on the California Geological Survey, see Gerald T. White, *Scientists in Conflict: The Beginnings of the Oil Industry in California* (San Marino, Calif.: Huntington Library, 1968), esp. chap. 2.

2. Superintendent of Public Instruction of the State of California, *Third Biennial Report (1868–1869)* (Sacramento: State of California, 1870), p. 199. Whitney returned to teach at Harvard after 1865, but he remained head of the intermittently funded survey of California until 1874.

3. Thus, William Coleman, ed., *The Interpretation of Animal Forms* (New York: Johnson Reprint Co., 1967), p. xxvii, argues that "museums, once the symbol of the study of biology, were never created" at Johns Hopkins. Annual reports (for 1876, 1884, 1886, and 1888) make clear the growing collections at the university, including the relatively large donation of the Maryland Academy of Sciences. I thank Julia Morgan of the Ferdinand Hamburger, Jr. Archives at Johns Hopkins for her assistance; also see John C. French, *A History of the University Founded by Johns Hopkins* (Baltimore: Johns Hopkins University Press, 1946), p. 227. No provision, however, was made for a museum when a new campus was built in the 1910s.

4. Darwin's theory grew out of taxonomic study, and the two were not inconsistent according to Stephen Jay Gould, *Ontogeny and Phylogeny* (Cambridge, Mass.: Belknap Press of Harvard University Press, 1977), and Michael T. Ghiselin, *The Triumph of the Darwinian Method* (Berkeley: University of California Press, 1969).

5. See Robert A. McCaughey, "The Transformation of American Academic Life: Harvard University 1821–1892," *Perspectives in American History,* 1974, *8:* 239–334; and Colin B. Burke, *American Collegiate Populations: A Test of the Traditional View* (New York: New York University Press, 1982), pp. 1–9. For a comparative international perspective, see Konrad H. Jarusch, ed., *The Transformation of Higher Learning, 1860–1930: Expansion, Diversification, Social Opening, and Professionalism in England, Germany, Russia, and the United States* (Chicago: University of Chicago, 1983).

6. For a discussion of these earlier collections, see Sally Gregory Kohlstedt, "Natural History on Campus: From Informal Collecting to College Museums" (unpublished paper).

7. Jane Maienschein, ed., *Defining Biology: Lectures from the 1890s* (Cambridge, Mass.: Harvard University Press, 1986), pp. 3–50.

8. Edward Lurie, *The Founding of the Museum of Comparative Zoology* (Cambridge, Mass.: Museum of Comparative Zoology, 1960).

The students also gained camaraderie in a weekly Agassiz Club in the late 1850s, as documented in Dorothy G. Wayman, *Edward Sylvester Morse: A Biography* (Cambridge, Mass.: Harvard University Press, 1941), pp. 117–121. For subsequent difficulties, see Ralph W. Dexter, "The Salem Secession of Agassiz Zoologists," *Essex Institute Historical Collection,* 1965, *101:* 27–39.

9. James Dwight Dana, for example, wrote on 12 August 1863, to Spencer F. Baird that he did not hope to rival the Cambridge museum in paleontology; RU 7002, Smithsonian Institution Archives, Washington, D.C. (hereafter SIA). On a similar effort at McGill University in Montreal, see Susan Sheets-Pyenson, "Stones and Bones and Skeletons: The Origins and Early Development of the Peter Redpath Museum (1882–1912)," *McGill Journal of Education,* 1982, *17:* 49–62.

Toby Appel, "Jeffries Wyman, Agassiz's Colleague and Counterpart: A Study in Character and Reputation" (unpublished paper).

10. At Amherst the courses were for postgraduate students and a kind of extension program for "those young men who desire to study some subjects without joining the regular classes." W. S. Tyler, *History of Amherst College during the First Half Century, 1821–1871* (Springfield, Mass.: Clark W. Bryan, 1873), p. 327. Also see Russell H. Chittenden, *History of the Sheffield Scientific School of Yale University, 1848–1922,* 2 vols. (New Haven: Yale University Press, 1929).

Allan P. Tankersley, *College Life at Old Oglesthorpe* (Athens: University of Georgia, 1951), p. 113.

Ellen Langill, *Carroll College: The First Century, 1846–1946* (Waukesha, Wis.: Carroll College Press, 1980), p. 36. At Oberlin College the scientific course was originally intended for those who could not meet the requirements of the classics course; see John Barnard, *From Evangelicalism to Progressivism at Oberlin College, 1866–1917* (Columbus: Ohio State University Press, 1969), p. 83.

11. Robert V. Bruce, *The Launching of American Science, 1846–1876* (New York: Knopf, 1987), pp. 80–93. On p. 83, Bruce states that of all the scientists in the *Dictionary of American Biography,* eleven out of twelve scientists attended colleges, and four of five had a bachelors or other (often M.D.) degree.

12. Deborah Warner, "Science Education for Women in Ante-bellum America," *Isis,* 1978, *69:* 58–67; Mable Newcomer, *A Century of Higher Education for American Women* (New York: Harper, 1959), pp. 21–25, 81. Anne Firor Scott, "The Ever Widening Circle: The Diffusion of Feminist Values from the Troy Female Seminary," *History of Education Quarterly,* 1979 *19:* 3–25; Sally Gregory Kohlstedt, "In from the Periphery: American Women in Science, 1830–1880," *Signs,* 1978, *4:* 81–96; and Margaret Rossiter, *Women Scientists in America: Struggles and Strategies to 1940* (Baltimore: Johns Hopkins University Press, 1982), chap. 1. Normal schools deserve closer analysis with regard to curriculum. The Normal School in Durham, for example, boasted a "considerable Museum and Cabinet of minerals" in its 1856–57 catalogue according to Nora Campbell Chaffin, *Trinity College, 1839–1892: The Beginnings of Duke University* (Durham: University of North Carolina, 1950), p. 199.

Mount Holyoke's Mary Lyon and then Mary Whitman carried on special studies with Edward Hitchcock at neighboring Amherst College. See Arthur C. Cole, *A Hundred Years of Mount Holyoke College: The Evolution of an Educational Ideal* (New Haven: Yale University Press, 1940), pp. 62–63 and 159. This informal education continued at various summer institutes and part-time programs until the end of the century; see Rossiter, *Women Scientists,* pp. 86–88; and Lois Barber Arnold, *Four Lives in Science: Women's Education in Science* (New York: Schocken Books, 1984), pp. 83–85.

In the 1860s the Smithsonian Institution contributed shells and other items to Troy Female Seminary, Pittsburg Female College, Columbia (Indiana) Female Institute, and

Michigan Female College; see volume entitled Specimens Distribution Record Book [1854–1873], RU 120 Registrar, 1853–1920, SIA.

13. The fact that at a number of colleges the former chapel housed natural history collections led one historian to reflect on whether "science [was] driving out religion or could not keep pace with it." See Henry Clyde Hubbart, *Ohio Wesleyan's First Hundred Years* (Delaware, Ohio: Ohio Wesleyan, 1943), p. 251. This tendency was most apparent in the 1870s and 1880s.

14. Board of Trustees, "Rules of the Museum of Natural History" in *Catalogue of the Trustees, Faculty, and Students of the College of Charleston* (Charleston: Walker, Evans, and Co., 1850), pp. 30–33. The rules indicate that students needed consent to use the collections but also that those given permission had designated desks for their use. Lester Stephens kindly shared this citation and his unpublished essay, "Ancient Animals and Other Wondrous Things: The Study of Francis Simmons Holmes, Paleontologist and the First Curator of the Charleston Museum."

15. Thus, Frederick Hall at Washington College (later Trinity) brought minerals in 1825 and took them when he left, as did his successor a Dr. Rogers. Glen Weaver, *The History of Trinity College,* vol. 1 (Hartford, Conn.: Trinity College Press, 1967), 75–76.

16. Walter C. Bronson, *The History of Brown University 1764–1914* (1914; reprint, New York: Arno, 1971), p. 371. A trustee expressed an apparently common hope that scientific study would prepare students to work in "the various industries of Rhode Island"; W. W. Keen to Spencer F. Baird, 19 October 1873, and also see Jenks to Baird, 14 November 1871, both letters in RU 52, Assistant Secretary (Spencer F. Baird), 1850–1877, Incoming Correspondence, SIA.

17. Lewis C. Beck's collection, for example, was purchased and donated by friends. William Demarest, *A History of Rutgers College 1766–1924* (New Brunswick, N.J.: Rutgers, 1924), p. 435.

18. The first president, Landon C. Garland, was reportedly preoccupied with "scientific equipment" and pressed his geology and chemistry faculty to provide him with lists of the specimens and equipment needed, spending thousands of dollars for such materials in the formative years of Vanderbilt University. Paul C. Conklin, *Gone with the Ivy: A Biography of Vanderbilt University* (Knoxville: University of Tennessee Press, 1985), pp. 44–46.

The quotation is from Vivian Lyon Moore, *The First Hundred Years of Hillsdale College* (n.p.: Ann Arbor Press, 1943), p. 391. Josiah Dwight Whitney helped his friend Charles A. Joy identify specimens when the latter took a position at Union College; see Brewster, *Life and Letters,* pp. 153–154.

19. See A. Hunter Dupree, *Asa Gray, 1810–1888* (New York: Athenaeum, 1968), on botanical collections; and Toby Appel, "Jeffries Wyman, Agassiz's Colleague and Counterpart," which points out the scale and size of Wyman's museum of anatomy and physiology in Boylston Hall.

20. "Educational Museums of Vertebrates," *Science,* 1885, *6:* 222–224.

21. Advertisements discussed by Benson, in this volume.

22. Sally Gregory Kohlstedt, "Henry A. Ward: The Merchant Naturalist and American Museum Development," *Journal of the Society for the Bibliography of Natural History,* 1980, *9:* 647–661.

23. Helen Lefkowitz Horowitz, *Alma Mater: Design and Experience in the*

Women's Colleges from their Nineteenth-Century Beginnings to the 1930s (New York: Knopf, 1984).

24. Southern schools were often devastated by fire or collections looted, as at Emory, which had served as a wartime hospital. Henry Morton Bullock, *A History of Emory University* (Nashville, Tenn.: Parthenon Books, 1936), p. 190.

25. Ward to Francis H. Smith at University of Virginia, 13 March 1876, Henry Augustus Ward MSS, Department of Rare Books and Special Collections, Rush Rhees Library, University of Rochester, Rochester, N.Y. (hereafter Ward MSS). Ward to Spencer F. Baird, 22 November 1876, RU 52, SIA.

26. For a discussion of the architecturally controversial structure, see Virginius Dabney, *Mr. Jefferson's University: A History* (Charlottesville: University of Virginia Press, 1981), p. 29.

27. Ward to Cornell, 7 October 1864, Ward MSS.

Morris Bishop, *A History of Cornell* (Ithaca: Cornell University Press, 1962), p. 77. An undated statement [probably 1866] "Fundamental Objects of the Cornell University" suggested that: "Professor of Natural History will cause the students in his class to collect the specimens as far as may be practicable and will see that they prepare them with their own hands; they will stuff animals and birds, and mount insects, and prepare herbariums, and eliminate fossils from the matrix as a daily exercise, until they acquire the requisite degree of expertise. He will also carefully train them in the delicate manipulations required in preparing tissues and cells for examination under the microscope." A. D. White Papers, microfilm copy at Syracuse University.

White to Smith, 1 September 1862, Gerrit Smith Collection, George Arents Research Library, Special Collections, Syracuse University (hereafter SU).

28. Bishop, *History of Cornell,* p. 95, and Philip Dorf, *The Builder: A Biography of Ezra Cornell* (New York: Macmillan, 1952), p. 314.

29. Howard Miller, *Dollars for Research: Science and Its Patrons in Nineteenth-Century America* (Seattle: University of Washington Press, 1970), esp. chap. 8.

30. Laurence R. Veysey, *The Emergence of the American University* (Chicago: University of Chicago, 1965). Uncertainty was evident in the postwar campus mood as well, as reflected in a letter between two professors in 1866 (quoted in Charles Coleman Sellers, *Dickinson College: A History* [Middletown, Conn.: Wesleyan University Press, 1973], p. 258): "The age is groping half blind but conscious of a great want, for a *system of practical education.* We go stumbling and blundering on, mistaking our way, retracing our steps, trying again, dissatisfied with ourselves, and with our doings, yet not disheartened, having strong faith that the right way will be found some time, if not by us, by our successors, and that errors and failures have their uses in achieving ultimate success."

Burke, *American Collegiate Populations,* pp. 13–25, challenges earlier estimates of failures but nonetheless documents the instability in many regions.

31. On this literature, see James Axtell, "The Death of the Liberal Arts College," *Hist. Educ. Qrtly.*, 1971, *2:* 339–352.

32. Many of these were very weak and short-lived; see Bruce, *Launching of American Science,* pp. 327–328.

33. See Jurgen Herbst, "Diversification in American Higher Education," in Jarusch, *The Transformation of Higher Learning,* pp. 196–206. On faculty orientation,

see Burton J. Bledstein, *The Culture of Professionalism: The Middle Class and the Development of Higher Education in America* (New York: Norton, 1976).

34. James Edward Scanlon, *Randolph-Macon College: A Southern History 1825–1967* (Charlottesville: University of Virginia Press, 1983), pp. 153–154.

35. Dana to Spencer F. Baird, 12 August 1863 and 20 December 1866, RU 7002, Spencer F. Baird Collection, 1793–1923, SIA. Later, O. C. Marsh developed the paleontology collection by purchases abroad, field expeditions to the West, and contributions from the Smithsonian; see Marsh to Baird, 4 April 1874, RU 52, SIA.

36. Oberlin College, for example, introduced laboratories after pressure from students. Barnard, *From Evangelicalism to Progressivism,* pp. 51–52.

37. The 1890 act provided for "the most complete endowment and support of the Colleges for the benefit of agriculture and the mechanic arts" by giving $15,000 per state the first year with an increase each year following to a maximum of $25,000. See Merritt Caldwell Fernald, *History of the Maine State College and the University of Maine* (Orono: University of Maine, 1916), pp. 84–86.

38. In 1866 at Yale, for example, George J. Brush built a "mining and metallurgical museum" in the Sheffield Hall for public display, had a mineralogical and geological museum known as the college cabinet on the college square, and kept a private collection available to students; see Chittenden, *Sheffield Scientific School,* vol. 2, pp. 370–371. Ralph V. Chamberlin, *The University of Utah: A History of Its First Hundred Years 1850–1950* (Salt Lake City: University of Utah Press, 1960), p. 238.

Earle Dudley Ross, *The Land Grant Idea at Iowa State College: A Centennial Balance 1858–1958* (Ames: Iowa State College Press, 1958), p. 69.

39. Centre College of Danville, Kentucky, *Catalogue for 1885–1886,* p. 35, and Ernest Cummings Marriner, *The History of Colby College* (Waterville, Maine: Colby College, 1963), p. 196.

40. There is an entire chapter on the museum, a central feature at the college from 1871 to 1885, in Newell Yost Osborne, *A Select School: The History of Mount Union College and An Account of a Unique Educational Experiment, Scio College* (Alliance, Ohio: Mount Union College, 1967), p. 372.

41. Watching her colleagues at Vassar, Maria Mitchell observed that for the individual faculty member, "a large m[useum] means a consumption of physical strength enormously disproportionate to the intellectual advantage." Maria Mitchell Memorabilia, n.d., Item 16, Microfilm of the Maria Mitchell Papers, American Philosophical Society, Philadelphia.

42. Charles Bessey is quoted in Earle D. Ross, *A History of the Iowa State College of Agriculture and Mechanic Arts* (Ames: Iowa State College Press, 1942), p. 159. Bessey with several colleagues, however, went on to collect a "real museum" on his own terms.

43. One historian, aware of a disparity between actual courses and the descriptions in the college catalogs, reported the disgruntled comment of a college president who declared, "America's greatest work of fiction is a college catalogue." Marriner, *History of Colby College,* p. 197.

44. William James Vaughan to Ward, 7 February 1876, Ward MSS.

45. Lawrence Vail Coleman, *College and University Museums: A Message to College and University Presidents* (Washington, D.C.: American Association of Mu-

seums, 1942), p. 40, claims that by 1876 there were seventy-three college museums, but neither defines the term nor gives a source for his information.

46. Requests from colleges are found throughout the incoming correspondence in RU 52 and the "Distribution of Specimens" indicated in the cumulative volumes of RU 120, SIA.

47. James W. Papez, "The Brain of Burton G. Wilder," *Journal of Comparative Neurology,* 1929, *47:* 285–323.

48. Harvard established an exclusive long-term contract for the exquisite glass flowers. Richard Evans Schultes and William A. Davis, *The Glass Flowers at Harvard* (New York: E.P. Dutton, 1982), pp. 1–15. Audrey Davis, "Louis Thomas Jerome Auzoux and the Papier Maché Anatomical Models," *Biblioteea Della,* 1977, *20:* 257–279.

There were instances of such donations from Spencer F. Baird throughout his administration of the collections, from 1850 to 1883. In 1873, faced with a substantial deposit of geological and mineral specimens from the Land Office, he asked for and received an appropriation from Congress to pay for their distribution to colleges and similar institutions. See William J. Rhees, ed., *The Smithsonian Institution: Documents Relative to Its Origin and History 1836–1885,* 2 vols. (Washington, D.C.: Smithsonian Institution, 1901), vol. 1, p. 693.

49. *Preliminary Report upon the Princeton Scientific Expedition of 1882* (Princeton, N.J.: C. S. Robinson, 1882), p. 8. On John Wesley Powell's excursions with students, see Elmo Scott Watson, *The Illinois Wesleyan Story 1850–1950* (Bloomington, Ill.: Illinois Wesleyan University Press, 1950), p. 74.

50. Durward T. Stokes, *Elon College: Its History and Traditions* (Elon College, North Carolina: Alumnae Association, 1982), p. 44. Similarly a Baptist-based college founded in 1827 built a natural history museum in the early 1880s; see Alfred Sandlin Reid, *Furman University: Toward a New Identity, 1925–1975* (Durham, N.C.: Duke University Press, 1976).

51. Dorothy Lloyd Gilbert, *Guilford: A Quaker College* (Greensboro, N.C.: Guilford College, 1937), pp. 206–207. After six years of working his way through Cornell as taxidermist, one student gave up because of the overload; R. W. Corwin to Baird, 10 January 1873, RU 52, SIA.

52. Clarence Luther Herrick was an apparently exceptional teacher who comfortably combined fieldwork and laboratory work. His personal interest derived from participation in a Young Naturalists Society in Minneapolis, which he used as his model for the Scientific Association at Denison. See G. Wallace Chessman, *Denison: The Story of an Ohio College* (Granville, Ohio: Denison University, 1957), p. 129, and Charles Judson Herrick, "Clarence Luther Herrick: Pioneer Naturalist, Teacher, and Psychobiologist," *Transactions of the American Philosophical Society,* 1955, n.s. *45,* pt. 1: 33–34; also Maienschein, in this volume.

53. These early student societies are discussed in Kohlstedt, "Natural History on Campus." Louis Agassiz, for example, declined to attend the students' zoological club, and one student later observed, "I see now that, much concerned for our advancement, his aim was to have us stand alone, or at least to only lean on our mates," in [Sophia Penn Page Shaler], *The Autobiography of Nathaniel Southgate Shaler* (New York: Riverside Press, 1909), p. 103.

Clyde Kenneth Snyder, *Snow of Kansas: The Life of Francis Huntington Snow with Extracts from his Journals and Letters* (Lawrence: University Press of Kansas, 1953), pp. 144–145. Snow had been a member of the active Lyceum of Natural History at Williams College.

54. See Burke, *American Collegiate Populations,* p. 230. A helpful comment by Philip Pauly reminded me of the significance of this important shift in student populations.

55. Pamphlet concerning "Lectures on Physiology" (1874–1875) at Cornell, copy in Wilder Papers, and a folder with an account book and receipts for the Natural History Society, 1887–1897, in the Cornell University Archives. The *Cornellian* yearbook listed twenty-five members in 1870.

56. E. A. Rose, commenting on the teaching of Frederick Starr at Coe College, to S. A. Lattimore, 25 June 1888, in Starr papers, Special Collections, Regenstein Library, University of Chicago. Starr had "sponsored various scientific societies among the students, encouraged individual work and private cabinets, and stimulated attention to the college museum."

57. Page flyer (with internal reference to 1871), University of Pennsylvania Archives, Philadelphia. New types of student societies proliferated in the 1880s and 1890s that were often quite distinct from earlier literary and study groups.

58. *Mark Hopkins and the Log: Williams College, 1836–1872* (New Haven: Yale University Press, 1956), p. 155. Indeed, Ralph S. Bates, *Scientific Societies in the United States* (Cambridge, Mass.: MIT Press, 1965), pp. 28–84, goes out of his way to enumerate societies of the 1830s and 1840s but does not mention those of the last quarter of the century. On Sigma Xi, see Michael Sokal, "Companions in Zealous Research, 1886–1986," *American Scientist,* 1986, *74:* 486–509.

59. Arthur G. Beach, *A Pioneer College: The Story of Marietta* (privately printed, 1935), p. 158; Osborne, *A Select School,* p. 372; and Jonas Villes, *The University of Missouri: A Centennial History* (Columbia: University of Missouri, 1939), pp. 201–203.

60. Arthur Herbert Wilde, *Northwestern University, A History, 1855–1905* (New York: University Publishing Society, 1905), pp. 171–183; and a student paper by Charles Key, "The Northwestern University Museum of Natural History" (1978) provided by Patrick Quinn of the Northwestern University Archives. See clipping file on the Syracuse Natural History Museum, especially items in *Alumni News* (February 1929 and April 1930), SU. Another example was Hillsdale where the class of 1881 presented a wall case and one member sent a box of ores, a bear, a Rocky Mountain wildcat, and a mountain lion, all mounted; see Moore, *The First Hundred Years of Hillsdale College,* p. 391.

61. Chaffin, *Trinity College, 1839–1892,* p. 199.

62. Jordan to Ward, 30 April 1885, Indiana University, Ward MSS. Later that year a major fire destroyed 85,000 specimens, including Jordan's ichthyological collection. Thomas D. Clark, *Indiana University: Midwestern Pioneer* (Bloomington: Indiana University Press, 1970), p. 137.

63. J. W. P. Jenks to Ward, Middleboro, Massachusetts, 22 December 1874, Ward MSS.

64. Cornell University *Register,* 1868–1869 (Ithaca, N.Y., 1969), pp. 34–37.

65. Snyder, *Snow of Kansas,* p. 151.

66. Clifford S. Griffin, *The University of Kansas: A History* (Lawrence: Univer-

sity Press of Kansas, 1974), pp. 33 and 115. Similarly, a new building in 1871 at the University of New Hampshire devoted the top two floors to a museum, including a room on the geology of New Hampshire and Vermont. Anon., *History of the University of New Hampshire, 1866–1941* (Rochester, N.H.: The Record Press, 1941), p. 25.

67. *An Account of the Organization and Progress of the Museum of Comparative Zoology at Harvard College in Cambridge, Massachusetts* (Cambridge, Mass.: Welch, Bigelow, and Co., 1871), pp. 3–7. Eventually, under Alexander Agassiz, the Museum's staff no longer taught Harvard students and simply cared for specimens and published research on them.

68. In 1879, Harvard made clear its rules that incoming faculty who had collections should dispose of them or catalog them clearly so that all work subsequently done by the faculty member would accrue to the college. Clark Elliott generously pointed out these rules, which appear in the *Annual Reports of the President and Treasurer of Harvard College, 1878–1879* (Cambridge, Mass., 1880), pp. 45–46.

69. The *Report of the Commissioner of Education for 1876* (Washington, D.C., 1878), pp. 780–787, indicates that twenty-nine of the forty-three listed museums were at colleges and universities.

70. Thus a local veterinarian and ornithologist G. S. Agersborg collected and organized a collection for his neighboring university. See Cedric Cummins, *The University of South Dakota 1862–1966* (Vermillion, S.D.: Dakota Press, 1975), p. 12. Also see J. W. P. Jenks, mentioned earlier, in Bronson, *History of Brown,* p. 371.

71. Kennicott to Baird, 2 July 1858, RU 7002, SIA.

72. Loss by fire was not uncommon and the reader can only feel sympathy for Joseph Leidy, who had collected and enlarged a museum at Swarthmore College that was destroyed by fire, when he said, "there are ten years of my life gone." Quoted in Edward Hicks, *Sixty-Five Years in the Life of a Teacher 1841–1906* (New York: Houghton Mifflin, 1907), p. 157. Also see Cornelia R. Shaw, *Davidson College* (New York: Revell Press, 1923), pp. 228–229, and George J. Stevenson, *Increase in Excellence: A History of Emory and Henry College* (New York: Appleton-Century-Crofts, 1963), pp. 152–153.

73. Extracts from the Minutes of the Trustee's Meetings Relative to the Museum, bound with Trustees Minutes, 14–15 June 1915. A copy was generously provided by Earle Coleman, Archivist, Princeton University. A series of contributions in 1874 permitted the extensive renovation of Nassau Hall, but within a decade there was a financial crisis when the College refused to take over responsibility for the collections.

74. Francis Holmes oversaw his own collection along with that of the Elliott Natural History Society and the College in Charleston. His lighter teaching load and constant requests for museum appropriations alienated him to some degree from fellow faculty members. See Stephens, "Ancient Animals and Other Wondrous Things."

75. Wilfred Shaw, *The University of Michigan* (New York: Harcourt, Brace, and Howe, 1920), 288. Here the museums were upgraded and turned to uses not unlike those in some urban centers by a herpetologist who later became university president. See Peter E. Van de Water, *Alexander Grant Ruthven: Biography of a University President* (n.p.: William B. Eerdman's Publishing Co., 1977), and Ruthven's own anecdotal, *Naturalist in Two Worlds: Random Recollections of a University President* (Ann Arbor: University of Michigan, 1963).

76. Richard Storr, *Harper's University: The Beginnings* (Chicago: University of

Chicago Press, 1966), p. 335. The efforts are reported in more personal detail in Thomas Wakefield Goodspeed, *A History of the University of Chicago: The First Quarter Century* (Chicago: University of Chicago, 1918), pp. 230–231; some evidence of the debates is in papers of the various faculty members at Regenstein Library, University of Chicago. Coleman, *College and University Museums,* pp. 39–56, takes a stand for consolidation perhaps because so many museums faced direct challenges to their very existence.

77. Frederick Merrill, *Natural History Museums of the United States and Canada,* in *New York State Museum Bulletin,* 1907, *62*: 201–211. Drawing on Merrill's data at the Smithsonian Institution, Alfred Goldsborough Mayer suggested that of 252 institutions with natural history collections, 176 were school, college, or university museums with another forty-four in small colleges. Yet he critically described two-thirds as "mere storehouses for the materials of which museums are made," in *Sci,* 1903, *17:* 843 and 845.

78. Cited in [Horace Coon], *Columbia: Colossus on the Hudson* (New York: Dutton, 1947), p. 209. Also see Douglas Sloan, "Science in New York City, 1867–1907," *Isis,* 1980, *71:* 35–76.

79. Sally Gregory Kohlstedt, "From Learned Society to Public Museum: The Boston Society of Natural History," in Alexandra Oleson and John Voss, eds., *The Organization of Knowledge in Modern America, 1860–1920* (Baltimore: Johns Hopkins University Press, 1979), pp. 386–406.

80. The university had a small collection evaluated earlier by a young visitor as having "nothing of consequence." James Graham Cooper to his father, 17 March 1857, James Graham Cooper MSS (75-4c), Bancroft Library, University of California at Berkeley (hereafter UCB).

Edward Potts Cheyney, *History of the University of Pennsylvania, 1740–1940* (Philadelphia: University of Pennsylvania Press, 1940), p. 428; and Cornell M. Dowlin, *The University of Pennsylvania Today: Its Buildings and Work* (Philadelphia: University of Pennsylvania, 1940), pp. 91–95.

81. *Eleventh Annual Report of the President of The Johns Hopkins University* (Baltimore: Johns Hopkins University Press, 1886).

82. Walter B. Hendrickson and William J. Beecher, "In the Service of Science: The History of the Chicago Academy of Sciences," *Chicago Academy of Sciences Bulletin,* 1972, *11:* 242–243.

83. Sally Gregory Kohlstedt, "International Exchange and National Style: A View of Natural History Museums in the United States, 1860–1900" in Nathan Reingold and Marc Rothenberg, eds., *Scientific Colonialism: A Cross-National Comparison* (Washington, D.C.: Smithsonian Institution Press, 1986), pp. 167–190.

The ambivalence was especially evident in anthropology, as in the varied affiliations of Franz Boas who found it difficult to concentrate on collections in Chicago, New York, and elsewhere before becoming a professor at Columbia. Rebecca H. Welch, "Alice Cunningham Fletcher: Anthropologist and Indian Rights Reformer" (Ph.D. dissertation, George Washington University, 1980), p. 244.

84. George P. Merrill, *Contributions to a History of American State Geological and Natural History Surveys* (New York: Arno Press, 1978 [1920]); and Michele L. Aldrich, "New York Natural History Survey, 1836–1845" (Ph.D. dissertation, University of Texas at Austin, 1974).

85. See Museum of Natural History Papers, 1842–1947, together with a helpful finding aid, at the University of Minnesota Archives, Minneapolis, and Martha Bray, "The Minnesota Academy of Natural Sciences," *Minnesota History,* 1964, *39:* 111–122.

86. Thus, Tulane University took over responsibility for the New Orleans Academy of Science's library and fossil collection, and provided space for meetings in return for free access by students and faculty. Letter, 26 February 1885, New Orleans Academy of Sciences MSS, Tulane University. Lester Stephens kindly shared his notes on the Academy with me. On the absence of a working relationship between the major colleges (Worcester Polytechnic Institute and Clark University) and the local museum in Worcester, Massachusetts, see Sally Gregory Kohlstedt, "Collections, Cabinets, and Summer Camp: Natural History in the Public Life of Nineteenth-century Worcester," *Museum Studies Journal,* 1985, *2:* 10–23, on p. 15.

87. The uncomfortable position of the museum has yet to be detailed, but some clues about the tension between museum directors and the Harvard administration are suggested in Louis Agassiz to Alfred M. Mayer, 30 January 1873, and Alexander Agassiz to Alpheus Hyatt, 4 July 1889, both in Hyatt-Mayer MSS, Firestone Library, Princeton University, Princeton, N.J.

88. Ira Remsen, cited in Hugh Hawkins, *Pioneer: A History of the Johns Hopkins University, 1874–1889* (Ithaca: Cornell University Press, 1960), p. 142.

89. Goode to Baird, 6 March 1877, RU 7002, SIA. Still, he continued, "I do not regret my past work but I think the museum is now better than almost any other college museum in the United States, and meets the present needs of teachers and students."

90. Clipping, 11 May 1883, in Natural History Museum folder, SU. The more substantial challenges came from biologists seeking space to work with living specimens and new equipment. See Donald Fleming, *Science and Technology in Providence, 1760–1914: An Essay in the History of Brown University in the Metropolitan Community* (Providence, R.I.: Brown University, 1952), p. 48.

91. The data are from Philip Pauly, "The Appearance of Academic Biology in Late Nineteenth-Century America," *Journal of the History of Biology,* 1984, *17:* 369–397, on p. 380; their use to show the long-term strength of the natural sciences is my own inference.

92. See, for example, the zoology class that included "laboratory work, assigned readings, and personal investigation into the life histories of our native birds," in Coe College, *Catalogue* (1904), pp. 73–76.

93. An outspoken resistance was offered by C. Hart Merriam who in 1893 published "Biology in Our Colleges: A Plea for a More Liberal Biology," cited in Keir Brooks Sterling, *The Last of the Naturalists: The Career of C. Hart Merriam,* rev. ed. (New York: Arno Press, 1977), p. 200.

94. A precursor of the battles to come in the twentieth century is found in the argument that museums had "very limited utility" when compared with a laboratory; *Sci.* 1884, *3:* 173.

95. Keith R. Benson, "The First Hundred Years: A Century of Natural History at the Burke Museum," *Landmarks,* 1985, *4:* 28–31; and idem, "The Young Naturalists' Society: From Chess to Natural History Collections," *Pacific Northwest Quarterly,* 1986, *77:* 82–93.

96. As late as 1968 a major survey reported, "although universities have tended to

discard their natural history collections because of the present-day emphasis on molecular biology rather than taxonomic biological training, there are more science museums than any other kind on American campuses [today]." *American Museums: The Belmont Report* (Washington, D.C.: Special Committee of the American Association of Museums, 1968), p. 14.

97. See, for example, the autobiographical comments of Hans Zinsser at Columbia in *As I Remember Him: The Biography of R.S.* (Boston: Little, Brown and Company, 1964 [1934]), p. 148.

98. Verne A. Stadtman, *The University of California 1868–1968* (New York: McGraw-Hill, 1970).

99. Hilda Grinnell, *Annie Montague Alexander,* (Berkeley: Grinnell Naturalists Society, 1957), pp. 1–27.

Grinnell actually suggested that the museum be at Stanford (he was an alumnus and would later take a Ph.D. there), but Alexander insisted that Berkeley was closer and the faculty familiar to her. Alexander to Grinnell, 2 November 1907, Annie Montague Alexander MSS, Additions (67/121c), UCB. Grinnell offered to go at his own expense but accepted gratefully when Alexander offered to pay his way to visit the major museums in New York and Chicago. The series of letters between Grinnell and Alexander provides a window on a remarkable patron-director relationship as well as details on the development of the museum. For the discussion of his eastern tour see Grinnell to Alexander, 6 and 11 December 1907, Annie Montague Alexander MSS (C-B 1003), UCB. In 1916 he again went East to a meeting of the American Association of Museums and visited several from Charleston, South Carolina, to Ottawa in Canada (see the letter of 15 March 1916, same series).

100. F. B. Sumner in reviewing Grinnell's *An Account of the Mammals and Birds of the Lower Colorado Valley, with Especial Reference to the Distributional Problems Presented* (1914), for *Sci.*, 1915, *41:* 64–69, emphasized the ongoing relationship between fieldwork and laboratory work: "The evolutionary theories of Darwin and Wallace were largely founded upon personal observations of geographical distribution. The modern student of genetics on the contrary, carries on his studies for the most part in the laboratory and the breeding pen. It is significant, therefore, that Bateson [1913, "Problems of Genetics"], perhaps the foremost living Mendelian, devotes a considerable portion of a recent volume to the problems of geographic variation. One can hardly read that volume attentively without being convinced that the field naturalist holds the key to some of the most important secrets of nature."

Keith R. Benson

2 From Museum Research to Laboratory Research: The Transformation of Natural History into Academic Biology

Providing a reasonable working definition for a "professional zoologist" presents few problems at the present time because the zoological community is highly organized, closely connected with academic and research institutions, and, at least in a general sense, clearly recognized by society as performing a valuable role in science. The American Society of Zoologists (ASZ) reinforces this definitive character by describing itself as an "association of professional zoologists for the presentation, discussion, and public dissemination of new or important facts and concepts in the area of animal biology, and the adoption and support of such measures as shall advance the zoological sciences."[1] One hundred years ago, there was little agreement in defining what was meant by a professional zoologist; indeed, because zoology and biology were both ill defined, no simple definition adequately describes the mélange of amateurs and professionals who were examining the natural world at that time. Nevertheless detailed studies of these investigators and the institutions supporting studies of nature reveal much about the formation of professional science in the United States.

Conventional wisdom, which in this case has considerable historical support, traces the ASZ from the American Morphological Society and the western branch of the American Society of Naturalists.[2] This interpretation justifies celebrating the centennial year 1889–90. However, the chronological history does not illuminate the conditions surrounding the establishment of the organization, nor does it address why any zoological organization was wanted in the 1880s. Discipline-specific institutions were scarce in the natural sciences before the 1880s; natural history, the rarely defined precursor to biology, lacked any cohesive and meaningful institutional framework beyond a few societies and academic institutions that sponsored museums. Certainly one of the first signs of professional development was the growing recognition that natural science needed bolstering. Many early editions of *Science* indicate this need in the 1880s. Noting that the "gravest danger before this branch of learning [natural science] is to be found in the radical imperfection of the methods of science teaching in use in our schools,"[3] the American Association for the Advancement of Science and other organizations called for greater attention to the teaching of science, especially in terms of teaching the methods of science. The Society of Naturalists of the Eastern United States, formed in 1883 and arguably a relative of the ASZ, made this same goal its raison d'être. Instructors in natural history, museum curators, physicians, and any professionally engaged natural historian were invited to meet "for the discussion of methods of investigation and instruction, laboratory technique and museum administration, and other topics of interest to investigators and teachers of natural history, and for the adoption of such measures as shall tend to the advancement and diffusion of the knowledge of natural history in the community."[4]

Curiously, the quotations of the ASZ and the Society of Naturalists, separated by more than one hundred years, have the same tenor in their purpose, namely the "advance of zoological sciences" or the "advancement and diffusion of the knowledge of natural history." But beyond the superficial similarity in tone there are a number of differences. The purpose of the ASZ is directed to professionals who are actively involved in animal biology. Only by inference can one argue that the statement of purpose has an educational focus. Conversely, the aged relative of the ASZ was directed to teach natural history and to disseminate new methods of investigating natural history. Again, only by inference can one argue that this direction was primarily focused to a professional research community. This study will utilize these distinctions to examine the gradual formation of the modern professional community in biology from its roots in the 1870s and 1880s. Specifically, the study will attempt to depart from the traditional linear view that depicts a shift from amateur to professional. Instead this chapter will examine the complexities of the gradual transformation of American biology from its primary location in museum-oriented natural history to its eventual setting within academic and research institutions.

The Popular Appeal of Natural History

An American biology community at the end of the nineteenth century cannot be defined with accuracy. In fact, the work in natural history that characterized aficionados of the natural world in the sixth decade continued to dominate at century's end. Given the extent to which natural history permeated the fabric of American cultural and scientific endeavors in the nineteenth century, its persistence at the beginning of the twentieth century may not be surprising. In general terms, Americans imbibed the alluring wine of natural history in a variety of vintages, including popular culture; government-sponsored collecting and surveying expeditions; public education through lectures, lyceums, and Chautauqua meetings often supported by natural history societies; and specialty education associated with universities and museums. All worked in concert to familiarize the educated American with the goals of the natural historian and the standard methods of collecting, preserving, drawing, identifying, and describing.

Perhaps the most pervasive form of natural history was that obtained through popular culture. Despite the lack of monographs on the American version of popular natural history equivalent to D. A. Allen's and Lynn Barber's work on the British form, a number of studies by Sally Gregory Kohlstedt, Toby Appel, and Margaret W. Rossiter have allowed historians to begin forming a picture of the American scene.[5] What has emerged on the canvas is an image not too dissimilar from the better-understood English tradition.

The popularity of natural history in both countries centered about its perceived moral and practical benefits. "A good naturalist can not be a bad man" was a conviction shared on both shores of the Atlantic.[6] Undoubtedly this feeling was rooted in natural theology, the early nineteenth-century rendition of natural history that extolled nature as the product of a wise and beneficent God. Later in the century, a secular tone was added to the moral and practical benefits of communing with nature as commercial concerns produced books, playing cards, Wardian cases, and aquaria. All enabled the English to conduct their avocation not only in the field but also at home.

The United States, already becoming known after the Civil War for its consumer appetite, witnessed a similar invasion of natural history into the economic arena. Natural history museums, several of which were franchise museums from Charles Willson Peale's Philadelphia museum, became the rage from the first half of the century. Soon the northeastern United States was littered with proprietary museums that not only promoted natural history to the layperson but generated impressive revenues for their owners.[7] Typically, but to the detriment of good science, for every accurate description and educational display there was a sensationalist exhibit: the public demand for the stupendous and the vulgar heightened the popularity of ancient forms of life, many of which were of dubious authenticity, and supernatural life. To underscore the popularity of these exhibits, John Scudder's American Museum, which he founded in New York in 1810, was purchased by Phineas T. Barnum in 1841 who converted the museum into the "Grand Colossal Museum and Greatest Show on Earth."[8]

But the popular-culture aspect of natural history in museums was not completely negative. Many museums, including some of the proprietary ones, were directed either by naturalists or by owners with enough scientific acumen to hire naturalists to supervise the museum work. Other museums, supported impressively by public funds or private donations, served an additional role in creating job opportunities for professional naturalists and, in this manner, moved beyond popular culture. At the Museum of Comparative Zoology (MCZ), for example, Louis Agassiz accepted students to work initially under his guidance and then to function as assistant curators. Salem's Peabody Museum became the employment base for several of Agassiz's assistants. This museum, originally the East India Marine Society until it was purchased by the wealthy industrial financier George Peabody,[9] became well known for its zoological collections organized by Edward S. Morse, Frederick Ward Putnam, and A. E. Verrill.[10] Similarly, at the museum of the Boston Society of Natural History, William Keith Brooks, fresh from his doctoral work at Harvard, was hired by Alpheus Hyatt for his first professional job before becoming part of the original faculty at Johns Hopkins University.

One of the more interesting spin-offs from popular natural history was the formation of businesses that supplied naturalists, museums, public schools, and universities with everything from preservatives to preserved specimens. A major supplier was Ward's Natural History Establishment in Rochester, New York, owned and operated by Henry A. Ward.[11] Located, as his letterhead prominently declared, on College Avenue "opposite [Rochester] University," Ward developed a huge mail order business aimed especially at universities and professional naturalists. Ward was definitely not an apologist for his business and did his utmost to extend his natural history interests. As David Starr Jordan prepared to open the new Stanford University, he received a letter from Ward announcing new casts of *Megatherium* and *Plesiosaurus* that he was constructing for the MCZ and stated that "I hope one of these days to be putting up such big pieces at Menlo Park!"[12] (In addition to his business concerns, Ward noted that he wanted to recommend his son to Jordan as a potential faculty member.)[13] Eventually, the bulk of Ward's zoological material was sold to Marshall Field and formed the base of Chicago's Field Museum.[14]

Ward was not alone in the business of purveying natural history wares. Charles K. Reed of Worcester, Massachusetts, peppered *The Museum,* a journal "devoted to research in natural science," with advertisements promising that "we have the BEST at lowest prices."[15] Interestingly, *The Museum* was edited by Walter F. Webb who also operated his own business, Webb's Natural Science Establishment, and made similar promises of low prices. The *Natural Science Journal* was another journal that dedicated itself "to furnish to universities, schools, and all educational institutions, as well as to individual scientists and collectors, a publication that will at the same time give them the latest news of the progress of events in the Natural Science world, together with instructive and entertaining articles."[16] Like *The Museum,* this journal was filled with advertisements by dealers like G. K. Green ("Corals from the falls of the Ohio a specialty"), L. W. Stilwell of the Black Hills Natural History Establishment ("Your collection will be more valuable if you send, on approval, for some of my 15,000 Stone Relics"), and Robert Burnham ("Dealer in opals, Mexican curios and relics, minerals and shells").[17] Other periodicals, such as *The Oölogist, The Antiquarian, the Linnaean Fern Bulletin, The Naturalist,* and *The Nautilus,* served to popularize natural history, especially through circulating "desiderata," lists of specimens for trade to professional and amateur naturalists.[18] The continued popularity of these publications and economic viability of natural history businesses to the end of the century underscores the widespread nature of popular natural history. Again, the target was the avocational naturalist who engaged in natural history primarily for personal pleasure rather than professional gain.

Even the federal government, in one of its initial adventures into science, became involved in natural history. When the government supported the Lewis and Clark expedition, which, in part, called for collecting plants and animals of the regions through which the expedition traveled, it created, at the very least, a recognition of the value of such collections. The most impressive undertaking, however, was the United States Exploring Expedition, often referred to as the Wilkes Expedition after its leader Charles Wilkes. Originally authorized by President John Quincy Adams in 1828, the expedition did not actually depart until 1838 but, in its four years of labor, gathered natural history specimens from around the world.[19] Importantly, this material was not distributed to private collections and museums as had been the fate of Lewis and Clark's collections. Instead, it became the base of the first federally supported museum, located originally in the National Gallery of the Patent Office.[20] Later, Wilkes's specimens were shifted to the Smithsonian Institution, originally directed by Joseph Henry, a former curator at the Albany Lyceum of Natural History, where they ultimately became part of the National Museum of Natural History.[21]

With the formation first of the U.S. Patent Office Gallery and later the National Museum as the repository for natural history objects, the government began its investment in natural history. But Henry was not interested in forming just another museum: because the Smithsonian was bequeathed to the United States with the expressed purpose "for the increase & diffusion of knowledge among men," he wanted a research collection for the "active expansion of human knowledge." Therefore, the new institution limited its collection "to objects of a special character, or to such as may lead to the discovery of new truths, or which may serve to verify or disprove existing or proposed scientific generalizations."[22] But as the century progressed even this more limited goal amounted to forming a collection so large that Henry could not handle it alone. In 1850 he hired a young assistant, Spencer Fullerton Baird, who during his career at the Smithsonian developed an unparalleled network of professional and amateur naturalists and natural history societies, which helped him establish the impressive holdings of the U.S. National Museum.[23] After 1850 virtually every government survey, including those sent out by the navy and army, was charged with collecting specimens of plants and animals for the Smithsonian. Particularly impressive, but largely overlooked by historians until recently, were the collections from the Pacific railroad surveys, supported by a federal appropriation of $150,000 in 1853.[24]

Similar to the popular-culture version of natural history, amateurs dominated the natural history investigations of the government expeditions and surveys.[25] After all, there were few opportunities to receive professional training and, perhaps more critical, there were even fewer job opportunities that called for the skills of a naturalist with professional training. These amateurs, as Sally Gregory Kohlstedt has clearly demonstrated in her work on the Boston Society of Natural History (BSNH) and the American Association for the Advancement of Science, not only filled the important role of carrying out the actual work in natural history through most of the nineteenth century, but they also formed many of the important institutions that expanded natural science from its amateur roots to a more professional stature.[26] As early as the 1840s and continuing throughout the century, institutions emerged outside of academic science, supported primarily by amateur naturalists. Many of these organizations also included professional naturalists, but their primary orientation was toward public science and not the academic community.

One major reason for establishing new natural history institutions was that the few existing societies did not address many of the needs of the amateurs; in fact, often the interests of amateur-dominated natural history did not harmonize with the interests of established societies.[27] The American Philosophical Society (Philadelphia) and the American Academy of Arts and Sciences (Boston), two of the earliest societies that included science, did not have an explicit natural history orientation despite supporting the work of serious naturalists. In the first three decades of the century, several institutions with an overt natural history orientation were formed, including the Academy of Natural Sciences (Philadelphia), New York Lyceum, Albany Lyceum of Natural History, and three successive Boston societies: New England Society for the Promotion of Natural History, the Linnaean Society, and the Boston Society of Natural History.[28]

Of equal importance to the institutionalization of natural history through these groups was the desire of the societies to promote their activities to the public. These were not societies that specialized in arcane subjects understandable only to scientific literati. Indeed, linked to the progressive attitude of popular culture, the BSNH called for the "improvement, and . . . the cultivation of a taste for natural history in our community."[29] The most common device to accomplish this goal was the popular lyceum or public lecture. Not ostensibly a natural history institution, the Lowell Institute, the descendant of Daniel Webster's Society for the Diffusion of Useful Knowledge, clearly had a natural history orientation in the lecture series.[30] John Lowell Jr.'s will called for "the maintenance and support of public lectures to be delivered in said city of Boston, upon philosophy, natural history, and the arts and sciences, or any of them, . . . and I wish a course of lectures to be established on physics and chemistry with their applications to the arts, also on botany, zoology, geology, and minerology [sic], connected with their particular utility to man."[31] The will was not very restrictive concerning the topics for the lectures, and the lectures in the 1840s had a strong bias toward natural history: lecturers Asa Gray, Jeffries Wyman, Louis Agassiz, Charles Lyell, and J. F. W. Johnston were all naturalists.[32]

The societies encouraged the public to visit the collections of plants and animals on display, in addition to attending lectures. Visits were clearly intended to provide both instruction and pleasure. By mid-century, the BSNH had one of the best collections of specimens in the country: "A large collection has the effect of attracting great attention, and the wandering thousands who are drawn by its exhibition to visit it daily or weekly, enjoy an innocent pleasure that is well worth providing for in all large communities, especially as the influence may often go far beyond gratifying curiosity."[33] The same thrust characterized West Coast societies, although most of these groups dated from later in the century. For example, the Young Naturalists' Society (YNS) in Seattle built a hall to exhibit its specimens and opened the museum to the public every weekend.[34] According to local newspapers, a visit to the "YNS Hall" was a highlight of the weekend for many Seattle residents through the 1890s. The San Diego Marine Biological Association, a natural history group headed by William Emerson Ritter but composed primarily of wealthy San Diego businessmen, formed with the goal of building an aquarium to display natural history objects from the marine world. Support for this effort was due to the conviction that such a venture would be "beneficial" to the people of San Diego.[35]

Some societies also offered courses. By the 1860s and 1870s several museums connected to natural history societies were staffed by professional naturalists, most from the Sheffield School of Science at Yale or Agassiz's museum-based program at the MCZ. Both the Peabody Academy of Sciences and the BSNH offered instruction in natural history and are perhaps the best representatives of this tradition. The Teacher's School of Science begun by the BSNH in 1871 had a particularly popular program, attracting over two thousand students during one year in its two courses in physiology and physical geography.[36] Later in the 1870s and the first half of the 1880s both the Peabody Academy and the BSNH sponsored important marine summer courses for teachers. In all these undertakings, museums and societies were remarkably successful. The new educational programs were either entirely self-supporting through fees or, in a few cases, were underwritten by wealthy patrons.[37]

Natural history societies also contributed to the rapid growth of periodicals. Again, the archetypical example is BSNH.[38] For most of the century there had been a dearth of journals that published articles in the natural sciences. *The American Mineralogical Journal* (1810–14), *American Journal of Science and Arts* (1818), and the journals, proceedings, and transactions from the American Academy of Arts and Sciences accounted for the bulk of the material.[39] The first periodicals to specialize in the zoological or botanical arenas of natural history were the *Journal of the Academy of Natural Sciences at Philadelphia* (1817), the *Journal of the Boston Society of Natural History* (1834), and the *Proceedings of the Boston Society of Natural History* (1841).[40] But these publications represented only the earliest journals and the most widely circulated periodicals. From 1850 through the end of the century periodicals were issued from societies as a matter of course. Examples of the genre include the *Natural Science Journal,* published by the Atlantic Scientific Bureau; *The Nautilus,* a monthly devoted to conchology and published by H. A. Pilsbry at the Academy of Natural Sciences and C. W. Johnson of the Wagner Free Institute; and the *Bulletin of the Brooklyn Conchology Club.* These publications provided a valuable outlet for naturalists to share information, to maintain communication with one another, and, like the journals from commercial institutions, to advertise specimens for exchange.

Natural History in the Academic Setting

The same traditions of collecting, preserving, drawing, identifying, and describing in natural history also characterized the treatment of the natural world in American academic institutions. Although universities and colleges were institutionally distinct from natural history societies, there was a substantial overlap in members between the organizations. However, academic natural history, particularly in its published versions, often made a more overt connection between the study of the wondrous artifacts in nature and the American version of natural theology; that is, one studied nature to observe signs of a beneficent creator who designed the harmonious natural world. At the same time, these aesthetic and religious considerations did not erode the quality of natural history. In fact, Benjamin Silliman, the leading figure in natural science during the first half of the century, demonstrated that good science reinforced religion.[41] The American edition of Robert Bakewell's *Introduction to Geology* (1829) included an appendix in which Silliman modified Bakewell's work to fit the Noachian account of the formation of the world.[42] This tradition in American geology was carried on by Silliman's student Edward Hitchcock through his popular *Religion of Geology* (1851), which served as one of the major geology textbooks at mid-century.[43]

University courses in botany and zoology, which appeared later in the century than geology courses, bore a similar stamp. The connection between religion and botany received its inspiration from Asa Gray, who was America's first adequately supported professional botanist and who wrote one of the earliest textbooks, *Elements of Botany.*[44] Like Silliman and Hitchcock, Gray did not use dogmatic assertions from religion that colored the quality of his science. Instead, he claimed that the design argument, the cornerstone of nineteenth-century natural theology and natural history, represented the boundary layer that joined the philosophy of religion and the philosophy of science.[45] Louis Agassiz adopted a nearly identical position in *Principles of Zoölogy* (1848). This text, in numerous editions, became the textbook for natural history and zoology courses until the century's end.[46] In fact, Joseph LeConte used the work at Berkeley as the only textbook in his zoology course, which he taught until 1892.[47] The text reveals Agassiz's bias in his presentation of the natural world. Nature provided an irrefutable argument for design and the creative agency of God.

> Animals are worthy of our regard, not merely when considered as to the variety and elegance of their forms, or their adaptation to the supply of our wants; but the Animal Kingdom, as a whole, has a still higher signification. It is the exhibition of the divine thought, as carried out in one department of that grand whole which we call Nature; and considered as such, it teaches us most important lessons. . . . Man, in virtue of his twofold constitution, the spiritual and the material, is qualified to comprehend Nature. Being made in the spiritual image of God, he is competent to rise to the conception of His plan and purpose in the works of Creation.[48]

According to Agassiz, the *nommer, classer, et décrire* character of natural history necessarily involved study of the "plan and purpose of God in his creation."

Through Silliman, Hitchcock, Gray, and Agassiz, natural history developed an academic orientation before the Civil War that was distinct from, but parallel to, the popular-culture version of natural history. Some of their work, especially the more esoteric aspects of natural history, was intended to serve only the small but growing scientific community. At the same time, the natural theology component of American biology provided the vehicle to carry natural history from academic institutions to the public. In fact, academicians readily published articles in lay periodicals and spoke before lay audiences to underscore the evidences of divine thought in the natural world. As an example, Gray helped to champion evolution theory after 1859 by arguing that natural selection represented God's efforts to maintain a well-designed natural world. After the Civil War as part of the nation's attempt to rebuild itself, colleges and universities felt pressure to reexamine the role of science in an academic arena. In particular, several American educators and critics of academic institutions wanted to upgrade science in line with new European developments that emphasized a more utilitarian and secular dimension to science.

Many critics worried that there was too little emphasis on research. "America, when compared with other first-class nations, occupies a low position in science. For every research published in our country, at least fifty appear elsewhere. England, France, Germany, Austria, Russia, Italy, and Sweden, outrank us as to producers of knowledge. Our original investigators in any department of learning may almost be counted on the fingers."[49] Part of the problem, of course, was that there were few scientific journals in the United States to publish research. In addition, much of the research in natural history did not necessarily lead to published results; most museum work involved categorizing and cataloging the specimens from the natural world. This work was, of course, research, but rarely was it published. But there was another reason for the paucity of American publication. In the same article quoted above from *Popular Science Monthly* in 1876, the author continued to claim that the sectarian nature of many colleges and universities was pernicious to the development of natural science. The result was that "nearly every American college emphasizes the classic and literary studies, and looks upon natural science as something of minor importance, often as a dangerous accessory, which must be tolerated, but not encouraged."[50] In a large number of academic institutions, science courses were continued in a single class, Natural Philosophy or Natural Science, with a clear natural theology orientation. The sectarian nature obviously left little room for research or publication.

Yet those who desired an expanded role for science, usually after the European model, did not go unchallenged. William P. Atkinson, a professor of engineering at Massachusetts Institute of Technology, commented that any new appreciation of science had to include the spiritual nature that natural science was intended to serve. "The higher uses of science will still be spiritual uses. It has not come into the world merely to carry us faster through space, merely that we may sleep more softly and eat and drink more luxuriously, nor will education become the mere teaching how to do these things. It is with the spiritual educating function alone that we have to deal when we consider it [science] as an element in liberal education."[51] Atkinson perhaps was aware of the implications of European science, especially in the form of evolution theory and German materialist and mechanist ideas, on the heretofore close connection between American science and religion, particularly in the university curriculum. Even the editorial staff of *Popular Science Monthly* considered that scientists should be aware of the public preference for a religious interpretation of science. "That the public is to-day [1875] far more interested in the relations of science to religion than they are in science itself, is because one term of the relation is so thoroughly familiar to the general mind."[52]

To complicate matters even further, other American scientists looked with a jaundiced eye on any strong push for a change in the American community that encouraged research and, by implication, specialization. "The efforts now being made for the endowment of research will, if successful, lead to a still further tendency to limit the fields of scientific labor. A better project would be to keep that connection between inquiry and exposition from which science has had so much profit in by-gone times."[53] Implicit in this statement was the fear that "inquiry" would take a preference over "exposition," especially if research efforts were encouraged. In other words, natural history was in danger of losing the expositional goal of finding evidence of a beneficent and intelligent God.

These critics had legitimate concerns. Despite the research done in natural history within universities and museums, as Sally Gregory Kohlstedt has previously mentioned, the American tradition was clearly impoverished when compared with European studies. The natural theology bias of American natural history alone cannot be blamed for this condition because many first-rate biologists completed studies under its influence. However, the methods of instruction and the placement and role of natural history within the university in the 1870s contributed to the relatively retarded state of American biology.

Before the educational reforms in American universities in the 1870s and 1880s, the usual method of learning natural history involved memorizing a text and reciting the lessons from it on command by the instructor. Andrew D. White, who became the first president of Cornell University and was a leading proponent of the reforms in academic science, made the following remarks concerning his experience in Denison Olmsted's class in natural philosophy at Yale. "The textbook was simply repeated by rote. Not one student in fifty took the least interest in it; and the man who could give the words of the text most glibly secured the best marks."[54] Such courses never involved the active investigation of nature nor did they include any exposure to research in nature. Certainly laboratory exercises were unheard of. In fact, Louis Agassiz's powerful legacy at the MCZ depended largely upon his innovation of adding actual specimens to the study of natural history.

Not only were there flaws in the instructional methods; most schools, especially the smaller colleges and universities, lacked a clear definition of what science encompassed. A professor of natural science might teach "chemistry, physics, astronomy, botany, zoölogy, mineralogy, geology, physiology, and perhaps Paley's evidence on top of all." St. John's had one man who served as "Professor of Natural Philosophy, Chemistry, Mineralogy, and Geology, and Lecturer on Zoölogy and Botany," while Hobart College hired one teacher to be the "Professor of Civil Engineering and Chemistry, and acting-Professor of Mathematics and Modern Languages."[55] Not surprisingly, given the mode of instruction and the broad demands placed on educators, little productive research in the natural sciences took place in academic settings.[56] In fact, as late as 1870 only seventeen institutions had science departments or scientific schools.[57] Furthermore, only the Sheffield School at Yale and the MCZ offered programs of advanced instruction in natural history at this time.

The reform-minded scientists and educators, therefore, not only stressed more research but they vociferously called for a new approach to the teaching of science. The *Popular Science Monthly* in 1876 encouraged new methods. "They could also, perhaps, do something toward breaking up the present vicious and absurd mode of teaching science by mere textbook recitations, and so help forward the adoption of correct methods. . . . Nature must be studied at first hand to be properly understood."[58] Agassiz's teaching methods at the MCZ offered a clear alternative to traditional educational programs. Claiming that his sole interest in studying natural history since coming to the United States "was done with a view to advancing science in my adopted country," Agassiz stressed that students needed to "study nature, not books."[59] What Agassiz emphasized was the close examination of natural specimens and not the dry and sterile recitation of factual information contained in old monographs. His reforms, therefore, were aimed mainly at alleviating the excesses of natural history work, a point that was well received by his students. "He denounced our worldly and bookish education as baseless and unreal, and demanded such a change in our systems of instruction as shall bring the pupils face to face with Nature herself, and call out the mind by direct exercise upon phenomena—the facts, laws, relations, and realities of the world of experience."[60]

Still, Agassiz and his students were primarily concerned with collecting, preserving, drawing, identifying, and describing plants and animals in the MCZ. Implicit in much of this work, especially in the eyes of Agassiz, was the same view of nature that prevailed during much of the nineteenth century. Nature revealed to Agassiz the "plan and purpose in the works of Creation."[61] So while Agassiz's methods were innovative, his view of the natural world was conservative. It was precisely this attitude toward nature and the value of studying nature that gradually changed in major institutions during the last three decades of the century as naturalists turned to study nature at "first hand."

Natural History Research in a New Direction

After the Civil War, many universities and colleges recognized that there was little or no prestige for the college graduate and no clear career path for the college degree.[62] In response, Yale and Harvard, and to a lesser extent Dartmouth, Union, and Columbia, attempted to revive science education by creating distinct science departments with specific curricular paths to separate them from the criticized undergraduate education. But even these attempts did not clearly prosper or point to a new direction of scientific work.[63] Nevertheless, far-sighted educators and naturalists continued to press for reforms in the teaching and methods of natural science. Whether the major forces for reform were internal to the natural sciences or were related to external factors is relatively unimportant for the focus of this study.[64] What is important to stress was the growing recognition of the need for a change in the way science in general and biology in particular was pursued in American colleges and universities.

A case can be made that an important initial impetus for reform came from passage of the Morrill Land Grant Act, first introduced in Congress in 1857 but not passed until 1867. The bill had strong individualistic and utilitarian overtones to it that favored a more important role for academic education. It was framed "to offer an opportunity in every State for a liberal and larger education to large numbers, not merely to those destined to sedentary professions, but to those much needing higher education for the world's business, for the industrial pursuits and professions of life."[65] To accomplish these goals, the bill made provision to set aside federal lands for sale in every state with proceeds to establish at least one college in the state. Contemporary accounts of the bill considered these new institutions as "scientific schools" to deal with the question of "How can the methods and results of modern science be made most conducive to the education of American young men?" The answer followed. "Mathematical, physical, and natural science, the investigation of the laws of nature, are to be the predominant study, rather than language, literature, and history."[66] The explicit assumption was that through the advancement of science, contributions could be made to the national wealth primarily through industrial applications.

Evidence to support this expectation could be found in Germany, where industry flourished as a result of the close link between manufacturing companies, research laboratories, and research institutes. American industry and science could benefit in a similar manner. "We need very much at the present moment an examination of the influence of foreign scientific institutions in promoting the efficiency of industrial undertakings."[67] By the late 1860s and early 1870s, European science was viewed as the exemplar for American institutions. Already, numerous physicians and many scientists had traveled to Europe to study in hospitals, laboratories, and institutes, especially those located in the German states.[68] On their return, these Americans invariably praised their experiences. In an article in *Medical Record* in 1878, an anonymous author urged American medical schools to introduce European-style programs in their training to remedy the defects. "The nearer our medical education approaches to the European models in this direction, the freer shall we be from quackery and humby [sic], and the better able to occupy our proper positions."[69] The trustees at Johns Hopkins University echoed these comments in 1878 when they publicly announced plans for a new hospital. The new model was praised because it had the "same plan as those in the Leipzig Hospital."[70]

Many of these remarks, which have too often been interpreted to support the contention that Americans borrowed their model for modern science wholesale from Europe, were not made without a careful scrutiny of the foreign science. Indeed, coupled with admiration for European laboratories, universities and science, most American reformers cautiously recommended reframing the European models to fit American needs.

> At the same time, we do not believe in copying any foreign institution. The classical colleges of this country are the growth of this country. The technical colleges should be equally our own, adapted to our institutions, our common schools, our modes of life, our national necessities. If they are not American colleges, they will not suit American students. Let us carefully study all that is good in the institutions of other countries, and adapt it so far as possible to our own circumstances and needs.[71]

Such sage advice recognized that European science, especially in the German tradition, was conducted largely through autonomous specialty laboratories that were controlled by one scientist. A good example was Justus von Liebig's laboratory in animal chemistry at Giessen. Clearly this model required modification to fit the American academic scene. In addition, some thought that each European style of science should be evaluated and the best selected for American use. The editors of *Science* claimed German science, which was "predominant over the scientific world," was too philosophical, suffered from poor writing, and was frequently arranged poorly; the English depended too heavily on amateurs; and the French were too provincial.[72] The message to American reformers was to borrow the best and discard the rest.

Despite Agassiz's new model for teaching in the 1860s, critical comments intended to encourage research in the 1870s, and the zeal to reform science teaching in both medical schools and academic institutions during the 1860s and 1870s, change in the character of natural history was not evident until the 1880s. Therefore, it is understandable why the members of the newly formed Society of Naturalists of the Eastern United States charged their organization in 1883 with "the advancement and diffusion of the knowledge of natural history in the community."[73] But many of these same naturalists had another goal that involved the incorporation of graduate instruction in biology and the development of research programs in the natural world. In the same article advocating more attention to the teaching of methods of instruction and observation, the naturalists also noted that Americans returning from Europe were introducing the "latest methods in vogue on the other side of the Atlantic."[74] Because there were no other organizations to discuss these methods, this new society served a critical function in promoting new techniques, most of which involved technical improvements in microscopy.

What is of equal note regarding the interest in European methods is that many of the naturalists were trained completely in the American natural history tradition and had very little exposure to European-style research. But through teachers familiar with these methods, such as Louis Agassiz and Jeffries Wyman at Harvard, young first-rate naturalists came to appreciate fully the innovative research techniques pioneered in the last three decades of the century. Among these men were Alpheus Hyatt and A. S. Packard, both of whom were eager to remedy the valid observation that American science had "never duly fostered research." Soon signs of change appeared among "a little band of men who have before them the model of Germany, and who are working earnestly for the intellectual elevation of their country. Their first object is necessarily to render research more important in public estimation, and so to smooth the way for a corps of professional investigators. Every thoughtful person must wish success to the attempt."[75]

It was precisely in the direction of laboratory investigation, advanced instruction, and research in biology that Johns Hopkins University pioneered in 1876 and, as a result, offered a new direction to the former natural history tradition.[76] By 1883, the university had constructed a new "Biological Laboratory" (the "best to be found in the country"[77]); hired a physiologist and a morphologist, each charged to conduct a graduate program in biology; established a marine research station; and begun to publish scientific journals containing research papers. All were unparalleled developments in biology in this country and all had significant impacts on the new direction in biology. No longer was natural history the sole goal of the biological community. The "little band of men" charged ahead to promote the research ideal in nature, based upon the laboratory model. The thrust was to provide Americans with opportunities in biology that equaled those opportunities in Europe and to ensure that American biology would begin to contribute to science. In a scathing critique of American science, it was claimed that if "we look to the biological laboratories of our colleges, to our medical schools, and to the laboratories connected with our hospitals, we find an almost utter lack of work tended to increase the boundaries of science."[78] Research laboratories were intended to correct the problem. The new physiology laboratory at Harvard was built to "serve primarily as a laboratory of research,"[79] Cornell's first physiology course advertised that the "important feature of the course is the large amount of laboratory practice;"[80] and the Chesapeake Zoological Laboratory (CZL) of Johns Hopkins was praised for its research orientation rather than as a "sort of holiday pic-nic of some weeks," the opinion of one reviewer of earlier American marine stations.[81]

But one must not be easily deceived by the use of words like "laboratory" or "research" as the only signal for a new direction in biology. After all, both words were bandied about from the 1860s with no dramatic change in the character of academic biology. Perhaps this was to be expected; after all, before research programs could be envisioned the rote methods of education had to be replaced by teaching from nature. In an important way, therefore, the laboratories of naturalists in the 1860s and 1870s did represent a notable reform. The new educational experiments were interchangeably described as "summer schools," "summer stations," and "laboratories." In fact, all were educational adjuncts of museums or centers for the collection of marine materials to be used either in museums or schools. With the possible exception of Baird's fish propagation studies and later work at the Fish Commission laboratory, none of these stations had a true research dimension apart from museum-oriented natural history. At the same time, there was a fairly well-articulated program in laboratory work that emphasized original research in the European tradition. This was the approach that slowly trickled into American biology in the late 1870s and early 1880s.

This new direction in biology was pioneered by Alexander Agassiz, Louis Agassiz's son and successor at the MCZ, in his personal laboratory at Newport. According to the younger Agassiz, when his father's marine laboratory at Penikese closed in 1874 he decided to open a smaller laboratory offering opportunities for research for American investigators.

> Ever since the closing of the school at Penikese it has been my hope to replace, at least in a somewhat different direction, the work which might have been carried on there. It was impossible for me to establish a school on so large a scale, but I hope by giving facilities each year to a few advanced students from the Museum [MCZ] and teachers in our public schools, to prepare, little by little, a small number of teachers who will have opportunities for pursuing their studies hitherto unattainable.[82]

Despite the paucity of documentary evidence about all the work at the laboratory, it is clear that the morphologist W. K. Brooks and the physiologist C. E. Brown Séquard were among the investigators who enjoyed Agassiz's hospitality. Furthermore, and more authoritatively, E. Ray Lankester included the Newport laboratory as one of the two American stations where the idea of pure research was understood and practiced.[83]

The other laboratory Lankester referred to was the CZL, established by William K. Brooks and supported financially by Johns Hopkins University. Like Agassiz's Newport facility, the CZL was never intended to serve an educational function for elementary students. In fact, when Brooks initially approached Daniel Coit Gilman, the first president of Johns Hopkins, concerning his plan for the laboratory, he specifically outlined his own research goals investigating molluscan taxonomic problems. Only secondarily did he mention that some advanced students from the department wanted to attend to work on their own research.[84] By the mid 1880s, however, the CZL became the morphology laboratory for the biology department and every graduate student in morphology was required to attend at least one of the long summer sessions (May to October). Brooks was convinced that the summer experience was the critical factor to permit students to develop the necessary skills of an independent investigator.[85] The character of the CZL that set it apart from other summer stations, in addition to the prolific research efforts of Brooks, was that the CZL was intended to enable Johns Hopkins's graduate students either to complete research for their doctoral degree or to conduct research with hopes for publication in scientific journals. Importantly, this was a new direction for marine research and one that led to Lankester's eloquent and enthusiastic endorsement.

Brooks was not the only pioneer, however. True, he may have received the model for the CZL from his experiences at the Newport laboratory because he, unlike several of his colleagues, had never been to a European laboratory. But of equal importance was Brooks's position as one of the original faculty members at the new graduate university in Baltimore. Johns Hopkins University was a novel experiment in education that self-consciously determined to establish research, laboratories, and "seminaries" as characteristics of American higher education. Moreover, the goal was not restricted to improving only educational opportunities; indeed, it also aimed at forming a new research ideal in the United States.

The person who actually orchestrated these new developments was the visionary first president of Hopkins. Gilman was one of a group of reform-minded university presidents at the end of the nineteenth century: the other notable figures were Andrew D. White of Cornell (1868), Charles W. Eliot at Harvard (1869), G. Stanley Hall at Clark University (1889), and later William Rainey Harper of Chicago (1891). What set Gilman apart from his colleagues was that the model he chose to adopt at Johns Hopkins was centered on advanced laboratory-based research. White and Eliot had both made earlier contributions to resolving the conflict between the "classical" university and the "scientific" university program utilizing the graduate school.[86] But neither Cornell nor Harvard emphasized research at the outset, preferring instead to stress practical education consistent with American educational ideals and needs.[87] However, after observing the "Baltimore method" in action after 1876, Cornell, Harvard, Clark, and Chicago began to rely on the model of Johns Hopkins.[88] In fact, President Eliot, who earlier had called for a "slow and natural" formation of the American university instead of a wholesale acceptance of the German model,[89] later looked specifically to Baltimore for the American model. As Gilman boasted in 1880, "JHU is often quoted to Pres. Eliot, & by him; & he has now announced that the chief topic of discussion in the Faculty next year is to be 'Graduate instruction.'"[90] Eliot then commented at the twenty-fifth celebration of the founding of Johns Hopkins in 1902 that Harvard's graduate program began "feebly" in 1870–71 but did not "thrive" until the example of Johns Hopkins convinced the administration and faculty to put their strength in the development of graduate instruction.[91]

Before formulating final plans for Johns Hopkins, Gilman traveled to Europe to observe the best in scientific institutions and to discover firsthand what and how the Europeans had developed. Using this information and the wise counsel of his fellow American administrators, he sought to create a center for research and a training ground for future leaders in American education. Gilman's bias for the direction of the university was clear from his selection of the original faculty. Basil Gildersleeve (philosophy) and Ira Remsen (chemistry) were trained at Göttingen, James J. Sylvester (mathematics) and H. Newell Martin received their education at Cambridge, and their colleague Charles J. Morris (classics) completed his work at Oxford. Only Henry A. Rowland (physics) studied in the United States, at the technically oriented school, Renssalaer Polytechnic Institute.[92] Gilman wanted to bring to Baltimore impressive thinkers and scientists whose major orientation involved teaching advanced courses in graduate programs. After all, Gilman expected that the success of his experiment rested squarely on the shoulders of his chosen few. "It is their researches in the library and the laboratory; their utterances in the classroom and in private; their example as students and investigators, and as champions of the truth; their publications through the journals and the scientific treatises which will make the University in Baltimore an attraction to the best students, and serviceable to the intellectual growth of the land."[93] Gilman's remarks were prophetic and also diagnostic. Johns Hopkins had an incredible impact on the growth of biology precisely because of the model it prescribed for the direction of biological research.

In a real sense, White and Eliot were correct for not encouraging the complete incorporation of the German model because what emerged at Johns Hopkins and then at Harvard, Cornell, Clark, and Chicago was a new American university that was unique. True, as Laurence Veysey comments, the model of German universities opened up new vistas and opportunities. "The German laboratory and seminar offered these future American professors a novel mode of life, a private mode that turned them aside from the everyday world of society, politics, morality, and religion, even from the classroom itself, and removed them during most of their waking hours from their fellow men."[94] But in the hands of the Americans, the laboratory and seminar took on an entirely new look. These were not laboratories and seminars dominated by one established professor: the American interpretation emphasized science departments composed of groups of scientists.

Johns Hopkins University and Laboratory Research

The biology department at Johns Hopkins was an excellent example of the use of the laboratory and the seminar, and also an example of the "new direction" in biology emphasizing research.[95] Easily the largest of the original graduate programs, the department had an early orientation toward physiology.[96] But by the time the department moved into the new biological laboratory in 1883, the morphological side of biology was equally well developed.[97] Furthermore, the morphological and physiological orientation of the department was in a completely new direction compared with the natural history approach in most American colleges and universities in the 1870s and 1880s. The only remnant of the older approach was the "Naturalists' Field Club," an organization of members from the university and local citizens. Martin and Brooks built a well-articulated curriculum of courses arranged from introductory to advanced. For 1884–85, these courses were in general biology, embryology of the chick, human and comparative osteology, plant analysis, mammalian anatomy, animal physiology and histology, elements of zoology, and marine laboratory (summer work at the CZL).[98] Not one of these classes included *nommer, classer, et décrire* of natural history and not one was intended to promote educational opportunities for secondary teachers. The entire curriculum was pointed toward graduate instruction and research in biology: there was a clear orientation toward increased specialization and an appreciation for pure investigation.[99]

But courses alone did not characterize the new direction at Johns Hopkins. The biology department was centralized about the laboratory. After students completed introductory courses, Brooks and Martin directed them to repeat important research projects that had been recently published in order to verify or criticize the published results. This laboratory exercise enabled many students to begin the transition from neophyte to advanced researcher.[100]

Clearly the laboratory was a critical element. Both professors surrounded themselves with laboratory facilities appropriate to the type of research they conducted. When Martin wrote Gilman before boarding an oceanliner to the United States, appropriately named the *Germanic,* he noted he had ordered equipment for the physiology laboratory including a Ludwig's Kymographion (Leipzig), Helmholtz's pendulum myographion (Berlin), a time registering apparatus (Leipzig), recording tuning forks (Paris), and Czermack's rabbit holder (Leipzig).[101] None of this equipment was commercially available in this country. Brooks's task in equipping the morphological laboratory was even more difficult because he lacked the European contacts. But by the time the biological laboratory opened, he had assembled the finest Zeiss microscopes with new oil immersion lenses and the latest rotary microtomes for serial sectioning. The education available in the laboratory in Baltimore and at the CZL, by virtue of the new equipment, was unparalleled in this country and justly allowed Martin to claim proudly that the program was the "best in the country."

Brooks and Martin also contributed to the laboratory emphasis at Johns Hopkins through their own research projects. For his entire career, which lasted from 1876 until 1908, Brooks made lasting contributions to morphology; in particular, he emphasized the need for careful comparative studies of related embryos using the latest microscopical techniques prior to framing any ancestral relationships between the embryos. This placed embryological morphology on a new track from the often fanciful direction of morphology in the 1860s and 1870s.[102] Martin, despite producing less published material, provided American students with the first well-established laboratory approach to the study of the heart and respiration. Most significant, he pioneered the technique to isolate the mammalian heart from the donor organism to determine the reaction of the heart under various experimental conditions, a technique contemporaries considered as the most monumental in American physiology.[103] Unselfishly and pedagogically, both biologists presented their research to the students through departmental seminars, or "seminaries" as they were then called, and journal clubs. In addition, these informal gatherings provided the students with the opportunity to discuss with their mentors the major problems in biology.

Just as the research laboratory and departmental seminar were new to the United States, so was the gradual recognition of the need for periodicals to publish the research results. Until the 1880s, with the exception of specialty publications from societies and museums, the major journals of science were general periodicals like the *American Journal of Science, American Naturalist, Scientific American,* and the *Popular Science Monthly.* Gilman knew this situation and actively encouraged his faculty to begin specialty journals. Of all the academic departments, the biology department excelled at this task. Brooks began *Scientific Results of the Sessions of the Chesapeake Zoological Laboratory,* Brooks and Martin coedited *Studies from the Biological Laboratory of Johns Hopkins University,* Martin served as associate editor of Michael Foster's *Journal of Physiology,* and Brooks later formed *Memoirs from the Biological Laboratory of Johns Hopkins University.* These journals were important for the career development of the new generation of research biologists. Of the thirty-seven Ph.D. degrees awarded by Brooks and Martin in physiology and morphology between 1879 and 1893, one-third of the research for the dissertation was published in a journal originating in Baltimore. But even Martin was not satisfied with these journals. In 1884, he wrote Gilman of the need for an "American Journal of Biology" and the "urgent need of a Journal of Animal Morphology in the United States."[104]

Advanced instruction, laboratory methods, student research, and publication opportunities all contributed to the excellence of Johns Hopkins. Universitywide, 151 doctoral degrees were awarded between 1878 and 1889; by comparison, Harvard awarded forty-three (with an additional twelve Sc.D.'s) in the same period and Yale conferred 101 between 1861 and 1889.[105] But perhaps more impressive than the actual numbers of doctoral degrees is the impact the biology department had on American biology. In 1875, Brooks had received one of the first Ph.D.'s awarded in the United States and the first doctorate in zoology at Harvard. Yet by 1893, Johns Hopkins had awarded thirty-seven degrees in morphology and physiology. Notably, many of these degrees were earned by young biologists who later contributed impressively to the advancement of biology in this country, including Henry Sewall, W. T. Sedgwick, William Henry Howell, H. H. Donaldson, F. S. Lee, S. F. Clarke, E. B. Wilson, James B. McMurrich, E. A. Andrews, F. H. Herrick, Henry V. P.

Wilson, T. H. Morgan, E. G. Conklin, R. P. Bigelow, J. L. Kellogg, M. M. Metcalf, and Ross G. Harrison.

But Johns Hopkins would be less important if it had acted alone. Moreover, the appreciation of the "Baltimore method" was not that it was *the* causal factor for reform throughout the United States. In fact, Gilman's formulation of graduate education and the development of the biology department by Brooks and Martin coincided with a multifaceted movement toward a new view of the natural world and a new approach to study it. Previously, Agassiz had turned the American attention at least partially from the passive glorification of nature to an active participation in and study of nature. In addition, after the Civil War another current moved to the development of "scientific schools" to replace the aged and bankrupt classical model of the first half of the century. Call from other sectors encouraged American universities to serve the industrial needs of the growing country, much as German laboratories served the industrial complex of the German states. Medical reformers urged educators to implement vigorous scientific courses for students prior to entrance into medical school. Other physicians pleaded for new research laboratories to teach physiology, an option not available to most American physicians. Even scientists from outside the United States wrote their American colleagues of the needs for research opportunities in the biological sciences, thereby providing added inducement for new directions.

Not surprisingly, similar educational opportunities and research imperatives appeared in other academic settings. Several institutions also formed biology departments that included a strong research component, often staffed by biologists trained either at Johns Hopkins or in Europe. As Philip Pauly has pointed out in an excellent study, these institutions (Harvard, Clark, Chicago, Columbia, and Pennsylvania) focused originally on the contributions biology could make to medical education, but what ultimately emerged were biology programs largely independent of medicine.[106] In other words, biology came to be recogized as a specific scientific discipline that involved a number of intellectual and organizational elements unique to itself. Pauly lists several of these elements, including the notion of common basic concepts, research problems, shared instruments and techniques, formation of institutions to disseminate information (journals, departments, and laboratories), clearly developed general biology courses, and financial support.[107] Studies like Pauly's point out the dangers inherent in providing an assessment of American biology that emphasizes a linear process from amateurs to professionals following the importation of a European model or a causal relationship between the graduate model at Johns Hopkins and the subsequent proliferation of graduate programs in the United States. Each of the institutions he has examined was unique, which shows the need for historians to paint a larger canvas, especially adding complexity to the simplistic portrayals that have become commonplace in the history of American biology.

The failure of linear or causal explanations in American biology also becomes apparent in evaluating new research directions. Undeniably, a change occurred in the scope of the research, but this change did not involve a simple rejection of natural history in favor of laboratory-based experimental work. Work from American museums, during the second half of the century and most notably at the MCZ, often involved life history studies of new species, clarification of taxonomic debates, or, in the hands of the American neo-Lamarckian community, studies on the mechanisms that produced changes in the paleontological record. These studies did not cease. Indeed, much of this descriptive work was the basis for later studies of a different character or for studies using a new approach.

Whitman's and Brooks's research published in 1878 provides the clearest illustration of the new direction. Whitman's paper, "The Embryology of Clepsine," had a dual impact on American biologists. First, it clearly pointed to the importance of studying early cell events in embryology—the so-called premorphological events. Second, the paper introduced Americans to many research techniques and procedures that had been developed in Europe but not previously imported to the United States.[108] (In fact, this paper represented the only written resource for these tools until Whitman published his book on microscopical methods in 1885.) Brooks's research was related. Using the latest microscopical techniques, most of which were related to the new equipment at Johns Hopkins, Brooks provided a critical evaluation of Haeckel's gastraea-theorie and shed doubt on its universality.[109] Brooks's research may have had a greater immediate impact than Whitman's because of Brooks's academic position and Whitman's relatively obscure positions at the MCZ and the Allis Lake Laboratory. However, Whitman's work later became the basis for the notable cell-lineage studies at the Marine Biological Laboratory in the 1890s.[110]

In addition to new directions for morphology, physiological research offered a new approach to biological investigation. First at Johns Hopkins through the work of Martin beginning in 1876, and then at the Marine Biological Laboratory (MBL) in Woods Hole through the example of Jacques Loeb, physiology provided young American biologists with experimental approaches to investigate questions of animal function. But physiology, with its emphasis upon experimentation, did not supplant any existing American tradition in biology. Instead, Martin's work at Johns Hopkins was the twin to Brooks's morphology. Loeb's course at the MBL was offered alongside marine botany, invertebrate zoology, and embryology. As such, physiology represented a new option for the growing cadre of American biologists; in addition to studies of animal form, questions of animal function could be examined in the new biological laboratories.

Thus, the new research did not replace any former tradition but served to draw attention to an approach that used new techniques to attack existing problems in biology. In appreciating the influence of the work of Brooks and Whitman, one must also note that the research programs developed at Johns Hopkins, the CZL, and the MBL were dependent on recent advances in microscopy. By the end of the 1870s, German microscopists developed differential stains, materials to harden or soften vital materials prior to sectioning, oil immersion lenses to increase resolution, and, perhaps most important, the rotary microtome that allowed sectioning of a continuous ribbon of embryological specimen. In the hands of German biologists, these new techniques permitted insights into the phenomena surrounding cell division and development in the 1880s. Both the techniques and approaches were sought by Americans who migrated to European laboratories in the 1870s and 1880s. Now these could be implemented at home. As E. L. Mark, the zoologist at Harvard, remarked, the combination of microscopical innovations and the questions surrounding cell division "made possible investigations which could never have been achieved without its aid."[111]

Laboratory Research and the Definition of Biology

What slowly emerged from the combination of new departments, new research approaches, new techniques, and a new enthusiasm for biology was an effort to define with clarity what biology encompassed. Philip Pauly is certainly correct to point out the inability to define biology with precision prior to the twentieth century.[112] But the recourse to polemics in the debate over exactly what did define biology was a result of the change in emphasis away from natural history. Whereas many naturalists had previously specialized in "departments" of conchology, ichthyology, ornithology, geology, and so forth, these labels very rarely provided an academic designation. Typically, a college or university offered courses in natural science or natural history that were taught by professors of natural history, usually with poor academic distinction.[113] In contrast, the new departments established in the 1880s and 1890s adopted a different orientation. Brooks and Martin separated their courses not on the basis of taxonomic units, but in terms of questions of animal form or function. Martin taught physiology and anatomy, while Brooks taught zoology, embryology, and histology. Martin later wrote a letter to Gilman in which he included a desiderata of the major components of a research department, including a professor of biology and associates in comparative anatomy and zoology, physiology, human anatomy, and botany.[114] Whitman's ideal department included appointments in paleontology, comparative anatomy, histology, embryology, taxonomy, cytology, and physiology.[115] At institutions with well-established traditions in natural history, the change was not always so rapid. Harvard used the designation "Natural History" until 1890 when it disappeared from the catalog, and this area of study was separated into botany, zoology, and geology. Interestingly, this change occurred many years after Harvard offered courses in these separate subjects.[116] Similarly, natural history remained at the University of Washington until 1899 when separate departments of zoology and botany were created.

But again, the new specialization and definition of biology cannot be separated from the growing emphasis on research in nature and from novel attitudes about investigations in the natural world. By the 1880s, no respectable biologist would submit a serious paper to a research journal that contained the adulatory tone toward natural history typical of mid-century articles. Now many publications were analytic expositions of attempts to frame questions about the natural world with a method to extract an answer from nature. Thus, naturalists lost their contentment with merely examining nature for evidence of design or to provide recreational enjoyment. The new biologist acted to tease apart the natural world in an effort to explore the underlying themes and mechanisms.

It was this approach that soon characterized the biology departments in major universities. The biological laboratory at Johns Hopkins never exposed its students to a museum of representative species; Martin and Brooks used laboratory specimens solely to illustrate the major biological principles. Even Brooks's handbook on laboratory approaches in invertebrate zoology, the first of its kind in the United States, had this theme.[117] General biology texts also turned from the design argument common in Agassiz's *Principles of Zoölogy,* which was still in print in the 1880s. The first textbook with the fresh focus was written by W. T. Sedgwick and E. B. Wilson, students of Martin and Brooks. Perusing the text indicates how far removed it was from the earlier traditions. It offered a detailed anatomical and physiological treatment of one plant and one animal to provide an overview of the "properties of matter in the living state."[118]

Not surprisingly, these changes were not effected without self-conscious reflection and, in some cases, animosity. When Brooks established the CZL, his real concern was not in the value of the summer experience but in justifying the existence of the CZL to the trustees when the demand for a research institution was small compared with the demand on marine stations with practical instruction for elementary students.[119] Fortunately, the success of the research-oriented CZL was compelling enough to the trustees to ensure its continuous funding, at least until financial troubles besieged Johns Hopkins in the 1890s. But Brooks was clear on the distinction between what he was attempting to do with the CZL and what the goals of other marine laboratories were. In a letter to Gilman, Brooks thanked his president for supporting his efforts to promote "pure research" through the CZL.[120] Indeed, Brooks's view of the research orientation of the CZL was a factor in his attempt to discourage the effort to establish a permanent marine station at Woods Hole. Brooks wrote to S. F. Baird in 1883 that he had "often thought of Woods Hole as a possible locality" for his own laboratory, which moved quite frequently.[121] After Baird built a laboratory of the U.S. Fish Commission at Woods Hole in 1885 and gave Brooks and his students research access to the facility, Brooks understandably backed away from moving his own itinerant station there. After all, with the CZL enjoying a southern setting and the Fish Commission laboratory located to the north, Brooks and his students could draw on two diverse habitats. When Brooks heard of the plans for a second laboratory at Woods Hole that apparently was to stress educational opportunities, he was reluctant to support the venture. Brooks's reasons were multiple; he did not favor elementary instruction, he already had research facilities in Woods Hole, and if a permanent laboratory were to be built, it should be built on the southern coast to complement the Fish Commission facility.

At the same time, other naturalists lamented the increasing specialization in biology as a sign that the well-educated biologist soon would become extinct. George W. Peckham, writing to D. S. Jordan concerning his views on the "organization of the biological side of a great university," cautioned his friend not to turn aside from systematic work in natural history at the new university in Palo Alto.

> I should be the last one to underestimate the importance of General Biology, since for many years I have obtained my crackers and cheese by teaching it. But it seems to me that Morphology and Embryology have usurped too much of the attention of the workers in the universities of America. I really believe that there has been more bad cell-making than bad species making. The new Clark University under my friend Dr. Whitman will turn out numerous young morphologists, but not a man with any sympathy for general Natural History work.[122]

A similar complaint was also made to Jordan by William R. Dudley, a botanist at Cornell who opposed the economic orientation of the specialized experimental station at Cornell. Dudley argued forcefully to have some natural history in the new station, but "no one on the purely Natural History side was appointed."[123]

The criticisms by Peckham, Dudley, and others were accurate. By the 1890s, graduate programs in biology no longer produced naturalists for museum work or natural historians who occupied positions that emphasized a wide faunal or floral scope. Moreover, the mark of prominence in biology gradually shifted toward the research ideal as practiced in the graduate programs and laboratories of colleges and universities. Perhaps the best example of the changing emphasis in American biology is in the development of marine laboratories.

Such laboratories at Penikese, Salem, and Annisquam were originally exercises directed to produce teachers with better preparation in natural history. Archival records and manuscript sources do not indicate any appreciable component of advanced research in biology. Concurrently, Alexander Agassiz and W. K. Brooks pioneered another model for a marine research laboratory that pointed to a new direction for marine studies.[124] Brooks's rendition of a marine station, the CZL, had a notable impact upon the character of American biology because many of his forty-three doctoral candidates were exposed to research methods for the first time at the CZL and then continued their careers with research programs frequently centered at marine laboratories. The paradigmatic example of the research at a new-style laboratory was the work at the MBL at Woods Hole.

The MBL, however, did not have a clear focus for research from its inception. There were two major groups with interests in the idea of an independent, permanent marine station, but each group had different goals for the station. One group, spearheaded by the Woman's Education Association (WEA) of Boston, wanted to establish a permanent summer school for teachers on the model of Alpheus Hyatt's Annisquam. In fact, this lobby pushed the candidacy of B. H. Van Vleck as director of the MBL. Van Vleck was the naturalist who taught the invertebrate zoology course for Hyatt at Annisquam. The other group, composed of scientists with ties to Harvard and Johns Hopkins, was both smaller and less defined but was more focused; the members wanted a new direction for the laboratory and a director with experience in marine research, apparently with the CZL serving as the model. The latter group preferred a station combining research and instruction, and to implement their model, they favored the selection of Brooks. Professor Brooks declined the opportunity, presumably due to his opposition for a second station at Woods Hole, and the offer was extended to Whitman, another American with vast laboratory experience.

Despite the selection of Whitman, the MBL still lacked a clearly charted future. The initial course offered in 1888 was Van Vleck's invertebrate zoology course; botany was added in 1889. These two courses differed little from those offered at the MBL's forerunners. Whitman, however, had a more defined course for the new institution. After returning from Europe and visiting Naples, he wrote an article in which he called for greater opportunities in American laboratories to teach the European methods as the best way to improve biology in this country. "Whatever improves our facilities for study will tend to increase the general interest in biology, and to augment the number of naturalists who will seek the best that the world affords in the way of methods."[125] These remarks harmonized with the goals of the Society of Naturalists of the Eastern United States: Americans lacked educational opportunities in the natural sciences, especially in learning new methods. Whitman, however, did not initially oppose the natural history orientation of the MBL. But most certainly by 1893, if not before, Whitman made his intent clear. In an article published in *Popular Science Monthly,* he complained bitterly of the "purulent education" available at most colleges with no attention to "scientific culture"; "the best antitoxine [*sic*] would be the creation of research laboratories."[126] Relying on letters from major European biologists for support, Whitman called for a "marine biological observatory" devoted exclusively to research, clearly the model Whitman preferred for the MBL.

> The aim from the outset has been to provide for both investigation and instruction, but for the latter as subsidiary to the former. The problem has been to combine the two in such relations that each would contribute most to the same end—the advancement of science. We have always kept in view the necessity of providing as early as possible a separate building or the exclusive use of investigators.[127]

That the MBL eventually did champion research was clear by the end of the century. General physiology (1892) and embryology (1893) were added to the curriculum; the Friday evening lecture series featured the latest in scientific work; studies in cell lineage, artificial fertilization, and cytology were standard; research tables became available in 1893; and the *Biological Bulletin* was published to carry the results of the research to the larger biological community. In contrast, undergraduate teaching became a secondary focus and, eventually, the WEA withdrew its support.

The MBL is a fine example of an American institution in which research objectives came to dominate the biological programs. The maturation of this institution and the entire American biological community was due to a number of formative influences from different directions. Naturalists, medical reformers, university administrators, American biologists trained in Europe, European biologists who immigrated to the United States, industrialists and philanthropists, professional societies, new scientific organizations, new academic institutions, and novel research approaches accompanied by unique methods acted in concert to create a fresh mélange—research biology. It was this new version of biology in the United States that provided the hybrid vigor to vault the American community into a prominent position. As late as 1884, an article in *Science* bemoaned the paucity of research in biology that advanced the science in this country.[128] But the changes toward research work and advanced instruction helped to create a biological community with such prominent zoologists that by 1922 Maurice Caullery, the French zoologist, doubted that "at the present time, many other countries could find an equivalent" list of prominent zoologists.[129]

Concurrently, American biologists adopted a new attitude toward nature that was reflected in the kind of research they did. Heavily influenced by the pragmatism of Charles Saunders Peirce and William James, they no longer exhibited the same adulation of nature by looking for the ideas of a Designer-God. By the 1890s biologists expanded the research goals of biology in a different direction. Often using the word "experimental" in an uncritical fashion, for this became the umbrella term also to include instrumental approaches for research, biologists borrowed Peirce's view that nature needed to be probed and manipulated to reveal its secrets. Research was no longer passive; now research tools allowed biologists to pursue nature actively.

As a result, biology for some scientists became a part of the social enterprise to better the standard of living for Americans. Nowhere was this attitude more clearly enunciated than in J. McKeen Cattell's survey of the state of American science, *American Men of Science*. In the second edition he noted the importance of science to the American social fabric by stating that the "applications of science have quadrupled the wealth which each individual produces and have doubled the length of human life." Moreover, Cattell's "science" was no mere buzzword. He referred specifically to science as it was practiced in the new research orientation. "The most important recent development of science has been the establishment of endowed institutions for research. . . . In all our leading universities there are professors whose attention is devoted to advanced students and investigation, and their laboratories may be regarded as research institutions."[130] These research institutions created, by the beginning of the twentieth century, a totally new way of conducting biological work that more than rivaled European laboratories. Soon the tide of Americans flowing to Europe for advanced training slackened and, by the early twentieth century, Europeans looked to the United States. In addition, while Brooks, Edward A. Birge, J. W. Fewkes, and Walter Faxon represented the bulk of American-granted Ph.D.'s in biology in 1880, between 1898 and 1915 American institutions produced almost one thousand Ph.D.'s in zoology, botany, biology, or physiology. As C. M. Coulter at the University of Chicago stated, "research is the *nervous system* of the university."[131]

Conclusion

The new application of research within an academic setting involving graduate studies was not universal, however. In fact, for most of the nineteenth century academic institutions preferred the "classical" model for education that emphasized studies of language, history, and philosophy within a theological context; not until the twentieth century did scientific ability come to rival theological soundness in most American academic institutions.[132] Nevertheless, as the twentieth century flowered so did the ideal of biological research, usually on a similar model to that of the reforming universities like Harvard, Johns Hopkins, Chicago, Michigan, and Cornell. For example, William Emerson Ritter implemented laboratory instruction with a research orientation in the zoology program at Berkeley in 1892, finally replacing the former tradition based on Agassiz's *Principles of Zoölogy*.[133] A similar orientation characterized the natural science program at the University of Washington and its marine station at Friday Harbor after T. C. Frye, from the research-oriented University of Chicago, was appointed to the faculty in 1903. Without exaggeration, the new role of research in biology contributed importantly to resculpting the visage of the American biological community.

But the face-lifting did not mark a rejection of the former natural history tradition nor did it mark a simple move of biology from an amateur setting to a professional setting. As Sally Gregory Kohlstedt has demonstrated (in this volume), the *nommer, classer, et décrire* methods of natural history were refined along the lines of research objectives in museums located on American university campuses. These institutions certainly had an important educational function. However, beginning in the 1860s and continuing through the end of the century, biology moved beyond the museum. First into nature, most notably at Penikese and its successor marine stations, and then into the university laboratory, biological investigations expanded. While the museum work continued and proliferated, biology became academic and developed specialized fields of research. Embryology, physiology, ecology, animal behavior, and genetics represent several of the specialty areas of biology that became clearly demarcated by the early twentieth century. Natural history remained alive and well, primarily within museums. Academic biology emerged as a byproduct of the new research objectives and research methods in natural history; the new mission for graduate education in science; the importation of European ideals, techniques, and research questions to budding research institutions; and the more secular and utilitarian attitude toward science at century's end.

At the same time, the American emergence of biology was not accomplished by restyling amateurs into professionals nor was it accomplished by replacing amateurs with professionals. In fact, it was only after American naturalists formed societies, regardless of whether the individuals were professionals or amateurs, that a community of biologists could be recognized. As Toby Appel clearly illustrates in her study of professional societies (in this volume), these groups were important in defining the character of a scientific discipline as it matured. The gradual move of biology into academic centers and research institutions was accomplished by a gradual change in the character of the social structure of biology. But here again the debt of American biology to its natural history forebear is clear; as stated in the opening of this chapter, the "professional zoologists" of today's ASZ are related to the instructors in natural history, officers of museums and scientific institutions, physicians, and other persons who are professionally engaged in some branch of natural history of the Society of Naturalists of the Eastern United States.[134] By century's end, the increasing importance of research centers in biology, especially in American colleges and universities, led to the gradual formation of a professional group of biologists defined by specialized research interests.

Notes

1. Mary Adams-Wiley, executive officer, *Handbook for Officers and Committees* (Thousand Oaks, Calif.: American Society of Zoologists), p. 18.

2. Brother C. Edward Quinn, "Ancestry and Beginnings: The Early History of the American Society of Zoologists," *American Zoologist,* 1982, *22*: 735–748. Brother Edward's research has had the fortuitous result of adding substantially to the archives of the ASZ, now housed in the Smithsonian. See also Appel, in this volume.

3. Anonymous, "The American Association for the Advancement of Science," *Science,* 1883, *2*: 153.

4. Anonymous, "The Society of Naturalists of the Eastern United States," *Sci.,* 1883, *1*: 412.

5. The two English works are David E. Allen, *The Naturalist in Britain* (London: Allen Lane, 1976) and Lynn Barber, *The Heyday of Natural History, 1820–1870* (London: Jonathan Cape, 1981). Three recent contributions to our understanding of American natural history at mid-century are Toby A. Appel, "Science, Popular Culture, and Profit: Peale's Philadelphia Museum," *Journal of the Society for a Bibliography of Natural History,* 1980, *9*: 619–634; Sally Gregory Kohlstedt, "Henry A. Ward: The Merchant Naturalist and American Museum Development," *J. Soc. Bibliog. Nat. Hist.,* 1980, *9*: 647–661; and Margaret W. Rossiter, "Benjamin Silliman and the Lowell Institute: The Popularization of Science in Nineteenth-Century America," *New England Quarterly,* 1971, *44*: 602–626.

6. Barber, *The Heyday of Natural History,* p. 17.

7. Appel, "Science, Popular Culture, and Profit," pp. 630, 633.

8. Barber, *The Heyday of Natural History,* p. 159.

9. Ernest S. Dodge, "The Peabody Museum of Salem, Mass.," *Mariner's Mirror,* 1961, *47*: 90–100.

10. Dorothy G. Wayman, *Edward Sylvester Morse, A Biography* (Cambridge, Mass.: Harvard University Press, 1942), pp. 66, 69–70.

11. Kohlstedt, "Henry A. Ward."

12. Henry A. Ward to D. S. Jordan, 17 July 1891, Jordan Papers, Stanford University, Archives, Stanford, Calif. (hereafter, Jordan Papers).

13. Ibid.

14. Kohlstedt, "Henry A. Ward," p. 655.

15. Advertisement in *The Museum,* 1899, *5*: n.p.

16. Anonymous, "Editorial," *The Natural Science Journal,* 1897, *1*: 24.

17. Ibid., pp. 28–29.

18. It is curious indeed, as Professor Kohlstedt points out in her article on Henry Ward, that despite the material that is available concerning these dealers in natural history, there has been no systematic study on them.

19. Herman J. Viola and Carolyn Margolis, eds., *Magnificent Voyagers: The U.S. Exploring Expedition, 1838–1842* (Washington, D.C.: Smithsonian Institution Press, 1985).

20. Douglas E. Evelyn, "The National Gallery at the Patent Office," in Viola and Margolis, *Magnificent Voyagers,* p. 227.

21. Nathan Reingold and Marc Rothenberg, "The Exploring Expedition and the Smithsonian Institution," in Viola and Margolis, *Magnificent Voyagers,* p. 243.

22. Ibid., pp. 245–247.

23. William A. Deiss, "Spencer F. Baird and His Collectors," *J. Soc. Bibliog. Nat. Hist.*, 1980, *9*: 635–645.

24. Max Meisel, *A Bibliography of American Natural History: The Pioneer Century, 1769–1865*, vol. 3 (New York: Hafner Publishing Company, 1967), p. 189. These surveys have been largely neglected by historians. A recent examination of the importance of the railway surveys for zoology is John A. Moore, "Zoology of the Pacific Railroad Surveys," *Am. Zool.*, 1986, *26*: 311–341.

25. As a definition of "amateur," I accept Sally Gregory Kohlstedt's description as someone between the general public and a practicing scientist who has an active interest in science, is an avid collector or data gatherer, lacks specialized training and college degree(s), has no scientific publication, and has few contacts with the professional world. See Sally Gregory Kohlstedt, "The Nineteenth-Century Amateur Tradition: The Case of the Boston Society of Natural History," in Gerald Holton and W. A. Blanpied, eds., *Science and Its Public* (Dordrecht: D. Reidel Publishing Company, 1976), p. 173.

26. Sally Gregory Kohlstedt, *The Formation of the American Scientific Community* (Urbana: University of Illinois Press, 1976); Kohlstedt, "The Nineteenth-Century Amateur Tradition"; and Kohlstedt, "From Learned Society to Public Museum: The Boston Society of Natural History," in Alexandra Oleson and John Voss, eds., *The Organization of Knowledge in Modern America, 1860–1920* (Baltimore: Johns Hopkins University Press, 1979).

27. Kohlstedt, *The Formation of the American Scientific Community,* p. 33.

28. Ralph S. Bates, *Scientific Societies in the U.S.* (Cambridge, Mass.: MIT Press, 1965), p. 39, and Kohlstedt, "The Nineteenth-Century Amateur Tradition," p. 178.

29. Kohlstedt, *The Formation of the American Scientific Community,* p. 23, and Kohlstedt, "The Nineteenth-Century Amateur Tradition," p. 178.

30. Rossiter, "Benjamin Silliman and the Lowell Institute, p. 603.

31. Ibid., p. 606.

32. Ibid., p. 609.

33. As quoted in, Kohlstedt, "The Nineteenth-Century Amateur Tradition," pp. 180–181.

34. Keith R. Benson, "The Young Naturalists' Society and Natural History in the Northwest," *Am. Zool.*, 1986, *26*: 356.

35. Helen Raitt and Beatrice Moulton, *Scripps Institution of Oceanography* (La Jolla, Calif.: The Ward Ritchie Press, 1967), pp. 14–24.

36. Anonymous, "Intelligence from American Scientific Stations. Public and Private Institutions. Boston Society of Natural History. Teacher's School of Science," *Sci.*, 1883, *1*: 55–56.

37. The BSNH enjoyed wide-scale support from wealthy Bostonians for educational programs. The "Teacher's school of science" was underwritten by several prominent locals including Elizabeth Agassiz. See Kohlstedt, "From Learned Society to Public Museum," p. 398. The Woman's Education Association, an organization of wealthy Boston-area women, also supported the educational aims of the society: unfor-

tunately, despite repeated references to the group in the literature, there is no careful study of the crucial role it played in education, the development of science, or women's issues at the end of the century.

38. Kohlstedt, "From Learned Society to Public Museum," p. 388.

39. Bates, *Scientific Societies in the U.S.*, p. 35.

40. Kohlstedt, "From Learned Society to Public Museum," p. 388.

41. John C. Greene, "Silliman, Benjamin," *Dictionary of Scientific Biography*, vol. 12 (New York: Charles Scribner's Sons, 1975), p. 433.

42. Herbert Hovencamp, *Science and Religion in America, 1800–1860* (Philadelphia: University of Pennsylvania Press, 1978), p. 125.

43. Ibid., p. 140.

44. A. Hunter Dupree, "Gray, Asa," *Dict. Sci. Biog.*, vol. 5, p. 512.

45. A. Hunter Dupree, "Christianity and the Scientific Community in the Age of Darwin," in David C. Lindberg and Ronald L. Numbers, eds., *God and Nature: Historical Essays in the Encounter between Christianity and Science* (Berkeley: University of California Press, 1986), p. 361.

46. Keith R. Benson, "William Keith Brooks (1848–1908): A Case Study in Morphology and the Development of American Biology" (Ph.D. dissertation, Oregon State University, 1979), p. 16.

47. Richard M. Eakin, "History of Zoology at the University of California, Berkeley," *BIOS*, 1956, *27*: 69.

48. Louis Agassiz and A. A. Gould, *Principles of Zoölogy: Touching the Structure, Development, Distribution and Natural Arrangement of the Races of Animals, Living and Extinct*, 1866 edition (Boston: Gould and Lincoln, 1866), pp. 25–26.

49. F. W. Clarke, "American Colleges versus American Science," *Popular Science Monthly*, 1876, *9*: 467.

50. Ibid., p. 470.

51. William P. Atkinson, "Liberal Education of the Nineteenth Century," *Pop. Sci. Mthly.*, 1873, *4*: 24.

52. Editors, "Professor Newcomb on American Science," *Pop. Sci. Mthly.*, 1876, *6*: 244.

53. Anonymous, "Specialization in Scientific Study," *Sci.*, 1884, *4*: 36.

54. Walter P. Rogers, *Andrew D. White and the Modern University* (Ithaca: Cornell University Press, 1942), pp. 28, 34.

55. Clarke, "American Colleges Versus American Science," p. 469.

56. Edward Shils, "The Order of Learning in the United States: The Ascendancy of the University," in Oleson and Voss, *The Organization of Knowledge in Modern America, 1860–1920*, p. 20.

57. Rogers, *Andrew D. White and the Modern University*, p. 114.

58. Clarke, "American Colleges Versus American Science," p. 477.

59. Elmer Charles Herber, *Correspondence between Spencer Fullerton Baird and Louis Agassiz—Two Pioneer American Naturalists* (Washington, D.C.: Smithsonian Institution Press, 1963), p. 225 (Agassiz to Baird, 15 October 1873). Latter quotation is from Agassiz's sign at Penikese.

60. Anonymous, "Editor's Table—Agassiz," *Pop. Sci. Mthly.*, 1874, *4*: 497.

61. Agassiz and Gould, *Principles of Zoölogy*, p. 26.

62. Laurence R. Veysey, *The Emergence of the American University* (Chicago: University of Chicago Press, 1965), p. 4.

63. R. J. Storr, *The Beginning of Graduate Education in America* (Chicago: University of Chicago Press, 1955), p. 46.

64. See Veysey, *The Emergence of the American University,* p. 2, and Hugh Hawkins, ed., *The Emerging University and Industrial America* (Lexington, Mass.: D. C. Heath, 1970), p. viii, for discussions of the reform pressure.

65. William Warren Ferrier, *Origin and Development of the University of California* (Berkeley, Calif.: The Sather Gate Book Shop, 1930), p. 43.

66. Anonymous, "Our National Schools of Science," *The North American Review,* 1867, *105*: 499, 508.

67. Ibid., p. 515.

68. See especially Thomas Neville Bonner, *American Doctors and German Universities* (Lincoln, Neb.: University of Nebraska Press, 1963). This is a persistent theme in the development of American science at the end of the nineteenth century, but a theme in need of critical evaluation to determine the exact nature of the German influence on the United States.

69. Anonymous, "The Preliminary Medical Course at Johns Hopkins," *The Medical Record,* 1878, *14*: 152.

70. Anonymous, "The Johns Hopkins University," *The Boston Medical and Surgical Journal,* 1878, *49*: 605.

71. Anonymous, "Our National Schools of Science," p. 516.

72. Anonymous, "National Traits in Science," *Sci.,* 1883, *2*: 455–457.

73. Anonymous, "The Society of Naturalists of the Eastern United States," p. 412.

74. Ibid., p. 411.

75. Anonymous, "National Traits of Science," p. 457.

76. See the series of articles on the contributions of the Biology Department at Johns Hopkins University in *Am. Zool.,* 1987, *27:* 745–817.

77. Anonymous, "Biological Study at the Johns Hopkins University," *Med. Rec.,* 1881, *19*: 409.

78. Anonymous, "The Importance of Chemistry in Biology and Medicine," *Sci.,* 1884, *4*: 455.

79. Anonymous, "The Harvard Physiological Laboratory," *Sci.,* 1884, *4*: 130.

80. B. G. W. [Burt G. Wilder], "Preliminary Medical Education," *Bost. Med. Surg. J.,* 1875, *92*: 759.

81. E. Ray Lankester, "An American Sea-side Lab," *Nature,* 1880, *21*: 498.

82. Anonymous, "A Zoological Laboratory," *Nat.,* 1879, *19*: 318–319.

83. E. Ray Lankester, "The Endowment of Biological Research," *Sci.,* 1883, *2*: 514.

84. W. K. Brooks to D. C. Gilman, [early] 1878, Gilman Papers, Eisenhower Library Manuscript Room, Johns Hopkins University, Baltimore (hereafter Gilman Papers).

85. "William Keith Brooks," *Journal of Experimental Zoology,* 1910, *9*: 19.

86. Josiah Royce, "Present Ideals of American University Life" [from *Scribner's Magazine,* 1891], in Hawkins, *The Emerging University,* p. 18.

87. See Veysey, *The Emergence of the American University,* p. 95 and Rogers, *Andrew D. White and the Modern University,* p. 197.

88. Royce, "Present Ideals of American University Life," p. 19.

89. Rogers, *Andrew D. White and the Modern University,* p. 197.

90. Veysey, *The Emergence of the American University,* p. 197 (D. C. Gilman to B. L. Gildersleeve, 21 July 1880).

91. Johns Hopkins University, *Celebration of the Twenty-Fifth Anniversary* (Baltimore: Johns Hopkins University Press, 1902), p. 105.

92. J. C. French, *A History of Johns Hopkins University* (Baltimore: Johns Hopkins University Press, 1946), p. 35.

93. Gilman as quoted in Hugh Hawkins, *Pioneer: A History of the Johns Hopkins University, 1874–1889* (Ithaca: Cornell University Press, 1960), p. 65.

94. Veysey, *The Emergence of the American University,* pp. 132–133.

95. Larry Owens has published an essay on the "new direction" in science education based on the laboratory: "Pure and Sound Government: Laboratories, Gymnasia, and Playing Fields in Nineteenth-Century America," *Isis,* 1985, *76*: 182–194.

96. Hawkins, *Pioneer: A History of the Johns Hopkins University,* p. 90. Of the initial fifty-four graduate fellows and nonfellows, nineteen were in biology. The only other fields with large numbers of graduate students were language (twelve) and chemistry (nine).

97. Benson, "William Keith Brooks (1848–1908)," pp. 75–76.

98. Johns Hopkins University, *University Circulars,* 1884–85, pp. 8–9.

99. Veysey, *The Emergence of the American University,* p. 142.

100. Hawkins, *Pioneer, A History of the Johns Hopkins University,* p. 224.

101. H. Newell Martin to D. C. Gilman, 21 June 1876, Gilman Papers.

102. Benson, "William Keith Brooks (1848–1908)," chap. 4.

103. David Frick, "The Life and Career of Henry Newell Martin: Pioneer of American Physiological Education and Research " (M.A. thesis, University of Washington, 1986), pp. 12–19.

104. H. Newell Martin to D. C. Gilman [n.d.] 1884 and 7 March 1884, Gilman Papers.

105. Hawkins, *Pioneer: A History of the Johns Hopkins University,* p. 122.

106. Philip J. Pauly, "The Appearance of Academic Biology in Late Nineteenth-Century America," *Journal of the History of Biology,* 1984, *17*: 369–397, on p. 373.

107. Ibid., p. 371.

108. C. O. Whitman, "The Embryology of Clepsine," *Quarterly Journal of Microscopical Science,* 1878, *18*: 215–314.

109. W. K. Brooks, "Preliminary Observations upon the Development of the Marine Prosobranchiate Gasteropods," *Johns Hopkins University, Studies from the Biological Laboratory,* 1878, *1*: 13.

110. Maienschein, "Agassiz, Hyatt, Whitman," pp. 31–32.

111. E. L. Mark, "Zoology," in S. E. Morison, ed., *The Development of Harvard* (Cambridge, Mass.: Harvard University Press, 1930), p. 384. The first American manual with these techniques was C. O. Whitman, *Methods of Research in Microscopical Anatomy and Embryology* (Boston: S. E. Cassino, 1885).

112. Pauly, "The Appearance of Academic Biology," p. 371.

113. See earlier remarks in this paper.

114. H. Newell Martin to D. C. Gilman, 1 April 1878, Gilman Papers.

115. C. O. Whitman, "Biological Instruction in Universities," *American Naturalist,* 1887, *21*: 507–519.

116. Mark, "Zoology," pp. 387–389.

117. W. K. Brooks, *Handbook of Invertebrate Zoology* (Boston: Bradlee Whiddon, 1880).

118. W. T. Sedgwick and E. B. Wilson, *General Biology* (New York: Henry Holt, 1886). Special thanks to Philip Pauly for pointing out this reference to me.

119. W. K. Brooks to Walter Faxon, 15 December 1880, Gilman Papers.

120. W. K. Brooks to D. C. Gilman, 21 August 1888, Gilman Papers.

121. W. K. Brooks to S. F. Baird, 16 November 1883, Archives, Office of the Secretary (1865–1873, 1883–1912), Smithsonian Institution, Washington, D.C.

122. George W. Peckham to D. S. Jordan, 28 April 1891, Jordan Papers.

123. W. R. Dudley to D. S. Jordan, 10 May 1892, Jordan Papers.

124. Keith R. Benson, "Laboratories on the New England Shore: The 'Somewhat Different Direction' of American Marine Biology," *New Eng. Quart.*, 1988 (in press).

125. C. O. Whitman, "The Advantages of Study at the Naples Zoölogical Station," *Sci.,* 1883, *2*: 97.

126. C. O. Whitman, "A Marine Biological Observatory," *Pop. Sci. Mthly.,* 1893, *42*: 466.

127. Ibid., p. 464.

128. Anonymous, "The Importance of Chemistry and Biology in Medicine," *Sci.,* 1884, *4*: 455.

129. Maurice Caullery, *Universities and Scientific Life in the United States* (Cambridge, Mass.: Harvard University Press, 1922), p. 163.

130. J. McKeen Cattell, *American Men of Science. A Bibliographical Directory,* 2d ed. (New York: The Science Press, 1910), p. 564.

131. Caullery, *Universities and Scientific Life in the United States,* p. 97 and p. 155. Italics in the original.

132. Rogers, *Andrew D. White and the Modern University,* pp. 98, 105.

133. Eakin, "The History of Zoology at the University of California, Berkeley," p. 70.

134. Anonymous, "The Society of Naturalists of the Eastern United States," p. 412.

Part Two

Centers of Cooperation

Toby A. Appel

3 Organizing Biology: The American Society of Naturalists and its "Affiliated Societies," 1883–1923

In 1946 Wallace Fenn, secretary of the American Physiological Society, received a letter inviting him to be a charter member of the Society of General Physiologists. Unhappy at the news of yet another biological society, Fenn commented on the sad state of biological societies in contrast to the perceived unity of other disciplines. His model was the American Chemical Society: "The chemists in this country have a single strong and influential society and profit accordingly. The curse of the organization of biological societies in this country is that they refuse to pull together."[1] Even with the creation in 1947 of the American Institute of Biological Sciences, an umbrella organization that its founders hoped would unite the biological sciences, the "curse" of disunity remained. Fenn's concern with the fragmentation of biology and his wistful comparison of biologists and chemists was not at all new; these themes go back to nearly the beginning of the organization of the biological sciences in the 1880s and 1890s.

The contrast between the unity of chemistry and the disunity of biology was apparent to observers as early as at the turn of the century. Because of fundamental differences between chemistry and biology, it was impossible then, as it was later, to form an American biological society in any way comparable with the American Chemical Society. Almost all chemists at the turn of the century shared a similar basic training, usually in an undergraduate department of chemistry. "Biologists" shared no such common ground. Even at the undergraduate level, a student quickly specialized in botany or zoology. While some schools formed departments of biology, the greater number had departments of botany and zoology. Practical fields such as forestry or agriculture were often separated institutionally from the more theoretical fields. These differences were reflected and augmented by very different forms of professional organization. The American Chemical Society, founded in 1876, became by the first decade of the twentieth century large and prosperous because it gathered together chemists of all levels of training and all occupations. Although biologists had looked with envy on the chemists as early as 1908, they had not chosen to organize their own societies on a similar broad basis and remained unwilling to liberalize their membership policies. Separate societies for the biological sciences were formed very early. The first predated national societies in mathematics, physics, and astronomy. Even the American Chemical Society, though founded before the biological societies, was primarily a local New York society before its reorganization in the early 1890s.[2] From the beginning, the disciplinary societies in the biological sciences were set up as exclusive organizations, limited not only to active investigators with appropriate degrees, but to researchers doing certain kinds of work in certain kinds of places. It is not surprising that such specialized research-oriented societies were unable to unite under the aegis of "biology."

We are now about to witness a rash of centennials of national biological societies. The American Physiological Society and the American Association of Anatomists have held centennial meetings in Washington, D.C., in 1987 and the American Society of Zoologists is planning its centennial meeting for 1989. We can soon expect to celebrate a hundred years of botany, of psychology, and of microbiology. The origin of most of the major national disciplinary societies in the biological sciences—those that corresponded to university and medical school departments—dates to a fifteen or twenty year period beginning in the late 1880s.[3]

The 1880s and 1890s were in retrospect a remarkable period of expansion in American science. Graduate programs were being instituted; new departments were being formed and old ones enlarged; and research was becoming expected of faculty members in the better schools (see Benson, in this volume). It was a time characterized by the emergence of a "culture of professionalism" not only in scholarly disciplines but in many other walks of life.[4] Just as the formation of the modern university in the late nineteenth century was a product of forces unique to America, so too was the rapid creation of professional disciplinary societies a peculiarly American phenomenon.[5] Much has been written of confrontations of amateurs and professionals in nineteenth-century American science.[6] Although one can find evidence of such conflicts throughout the century, the 1880s and 1890s saw the strains between researchers and cultivators of science at their height, in both medicine and biology. These conflicts not only divided amateurs from professionals—those who held paid positions—but also divided the so-called professionals. It was a time when the tensions between the new elite, concentrated in the better schools in the East, and the rank and file of teachers and workers all over the country were at their sharpest.

This paper examines the emergence of national disciplinary societies in the biological sciences, their membership and role in establishing disciplinary boundaries, and their efforts, ultimately unsuccessful, to unite together in some form of an American biological society. The starting point is 1883, the year that the American Society of Naturalists was founded. This society, an offshoot of the American Association for the Advancement of Science (AAAS), was not itself a disciplinary society but played a key role in the organization of the biological sciences. It gave rise to most of the biological disciplinary societies under discussion, and enabled them to remain viable through the 1890s as very small, financially poor, and elitist groups. In particular, we shall focus on the Naturalists, the American Physiological Society, the American Association of Anatomists, the American Morphological Society (later the American Society of Zoologists), the Botanical Society of America, and the Society for Plant Morphology and Physiology. The ending point of this paper, some forty years later, corresponds to the inauguration of the Union of American Biological Societies. Its foundation was the apparent culmination of over thirty years of effort to found a federation of biological societies to speak for all biologists. But the union was a general federation of biological societies in name only. Its sole function was that of sponsoring *Biological Abstracts,* founded in 1926.

The history of professional societies in America, and in particular the history of the biological societies, is an area practically untouched by historians of science. Until recently, the history of such societies had been written almost entirely by past officers at times of anniversary celebrations. But for the historian of biology, the study of societies complements the internal history of biology by revealing in an accessible manner how the study of the phenomena of living beings was divided into specialties and what the real lines of cleavage were. One discovers that, in terms of societies, there was no radical "revolt from morphology."[7] By a gradual transformation, the American Morphological Society incorporated experimental studies in its program and, without serious conflict, eventually recognized the shift by changing its name to the American Society of Zoologists. However, there were clearly discernible cleavages between the biomedical sciences, based in medical schools, and those biological sciences primarily based in universities;[8] between the practitioners of traditional natural history employed in colleges, museums, and government agencies, who were often self-taught, and those well versed in morphology and physiology through training in scientific schools, study abroad, or doctoral programs; and, related to this, between academic and applied biology. Societies for areas of practical biology—such as forestry, animal breeding, fisheries, and horticulture—grew up entirely outside the orbit of the disciplinary societies. The competition among the biological societies, and between the biological societies and older organizations such as AAAS, and the failure of the various federation plans, give us another perspective from which to examine the question of what biology meant in the American context.

Founding of the American Society of Naturalists

The first biological offshoot of the AAAS and the mother of many of the disciplinary societies that began at the end of the 1880s was the American Society of Naturalists, founded in 1883.[9] A hybrid sort of society, the Naturalists was neither a large general group such as the AAAS nor a true disciplinary society. The American Society of Naturalists was a halfway house that, by its existence, encouraged the formation of disciplinary societies and enabled them to survive their early years when there were still few prospective members with the desired training and attainments. At least through the 1890s, the biological societies that it spawned continued to meet with it in a loose form of affiliation and to participate in various joint activities. The American Society of Naturalists and its affiliates in the 1890s, more than any other organization, came closest to resembling an American biological society that would unite all biologists.

By 1880, the predominant national scientific society, the AAAS, no longer fulfilled the needs of the new discipline builders. It was a large society encompassing all sciences and open to anyone who would pay dues, whether amateur or professional, well trained or not.[10] Its annual national meetings were held in the summer in different locations each year, often in inconvenient places for the eastern-based elite. Before 1882, there were only two informal sections that held simultaneous sessions and therefore there was very little time on the program for any one discipline. In 1882, responding to the demand for more program space and specialization, the AAAS established nine sections identified by letter, each to hold separate sessions presided over by a vice-president. Among them was Section F, Biology, which, after 1893, was divided into Section F, Zoology and Section G, Botany.[11] The creation of sections, far from satisfying special interests, seems to have precipitated discussion leading to the formation of disciplinary societies with restricted membership.

The Society of Naturalists of the Eastern United States, known after 1886 as the American Society of Naturalists, was founded largely on the initiative of Alpheus Hyatt (who became its first president) and Samuel F. Clarke (who became its first secretary). Hyatt, curator of the Boston Society of Natural History and sometime instructor at the Massachusetts Institute of Technology, represented the generation of students trained in the 1860s by Louis Agassiz and Jeffries Wyman at the Lawrence Scientific School of Harvard. Clarke, professor of natural history at Williams College, represented the new generation of Ph.D.'s from Johns Hopkins. The invitation, emanating from Clarke at Williams, to attend the initial meeting to be held in Springfield, Massachusetts, on 10 April 1883 began: "A number of American workers in biology desiring to have established an Association of American Naturalists for business purposes, extend to you a cordial invitation to join in a movement looking toward that end." Although "biology" was not in the title of the new organization, for it was to include geologists also, the society from the beginning identified itself with biology as well as with traditional natural history.[12]

Probably to avoid competition with Section F of the AAAS, it was decided to hold annual meetings not for presenting research papers but rather for the discussion of educational and research techniques or, what Hyatt called in his presidential address of December 1883, "the business of the naturalist."[13] Among the topics proposed in the letter of invitation were methods of museum work; methods of exhibition; laboratory techniques; "new and valuable points in staining, mounting, cutting and preserving of specimens"; "systems of instruction"; and "the position which the observational sciences should hold in the college curriculum" and the amount of preparatory work required. A list of sixteen persons said to be interested in the enterprise followed. All taught at northeastern schools or were employed by the Boston Society of Natural History.

At the organization meeting, attended by Hyatt (chairman), Clarke, Alpheus Packard (Brown), Charles Sedgwick Minot (Harvard Medical School), J. S. Kingsley (a graduate student at Princeton), Henry Fairfield Osborn (Princeton), William Berryman Scott (Princeton), William North Rice (Wesleyan), J. M. Tylor (Amherst), and J. H. Pillsbury (high school instructor in Springfield, later professor of biology at Smith), the justification of a new society entirely separate from AAAS was discussed. In response to Packard's inquiry of the feasibility of uniting the new organization to AAAS, Minot gave arguments for why a geographically restricted and selective society would be more effective. He maintained that the AAAS "comprises a very large and varied membership" and that "it would be very difficult to pursue any definite line of professional research without interruption within that body." In addition, the AAAS met each year in widely different localities, making it difficult for the same persons to attend each year. He concluded, "If a few men, who are thoroughly in earnest, meet together at an appointed place and time, for a single definite purpose, the chances are in favor of their accomplishing something worth the doing."[14]

According to Minot, who soon replaced Clarke as secretary, the American Society of Naturalists was the first society to restrict its membership to professionals through its constitution.[15] At the organizational meeting, Packard proposed that the society "be restricted to professional naturalists." Tylor argued for the addition of "medical men," presumably anatomists, physiologists and physicians who contributed to natural history.[16] The membership clause as finally adopted read: "Membership in this society shall be limited to Instructors in Natural History, Officers of Museums and other Scientific Institutions, Physicians and other persons professionally engaged in some branch of Natural History." [17] In contrast to later societies, the Naturalists were relatively liberal. A "professional" was taken to be someone with a position related to natural history and not necessarily an active researcher. Prospective members were to be nominated by current members and then approved by the executive committee and a majority of members present at a meeting. Dues, as in most of the early societies, were modest, in this case two dollars (later reduced to one dollar). As the meeting locations were limited by the constitution to the northeastern seaboard—Maine to the District of Columbia—membership was also effectively limited geographically as well. According to Minot, the society also pioneered in the choice of the Christmas recess—the week between Christmas and New Year's Day—as the time to hold meetings.[18] The society's first annual meeting was held in New York, at Columbia School of Mines, 27 and 28 December 1883.

One hundred nine individuals accepted the invitation to become charter members—a nearly inclusive list of all the workers in biology and geology living in the East, academic scientists as well as those employed in museums and geological surveys. There were scientists of Agassiz's generation such as James Hall, James Dwight Dana, Joseph Leidy, and Spencer F. Baird as well as many recent recipients of Ph.D.'s from schools such as Hopkins, Harvard, Yale, Princeton, and the German universities. Even the Rochester taxidermist and museum supplier, Henry A. Ward, was included. In the first membership list, published in 1884, the charter members as well as some elected members were listed with their fields of interest, presumably supplied by themselves. Of the 109 charter members, thirty-eight were categorized as geologists and paleontologists. Another thirty-five were zoologists or animal morphologists (seven thus styled themselves). There were ten botanists; handfuls of anatomists, physiologists, anthropologists; one psychologist (G. Stanley Hall); several miscellaneous; and six who chose to call themselves biologists. Less than a majority held a Ph.D. (twenty-four) or M.D. (thirty); four held both.[19]

The society's first annual meetings were well attended and largely devoted to the aims set forth in the constitution—the presentation of papers on techniques of research and education. At the first annual meeting in December 1883, attended by thirty-eight members, for example, the program included such papers as "Applications of photography to the production of natural history figures and lecture-room charts," by Simon Henry Gage; "Methods of preparing rock sections," by James Hall; "On some methods of pursuing teratological research," by Harrison Allen; "On the classification of tissues and organs with regard to the arrangement of collections," by Minot; and "Methods of section cutting," by E. B. Wilson.[20] But this phase of the society's history lasted only through 1888.

Emergence of the Disciplinary Societies

Beginning in 1887, new disciplinary societies began to form that cut into the Naturalists' membership. They were from the outset national in intent, although their membership was concentrated in the East and all early meetings were held in the East. The earliest of these societies were the American Physiological Society and the Association of American Anatomists (changed in 1909 to the American Association of Anatomists). These two societies were first to be formed not because physiology and anatomy were more mature or better organized than other sciences, but because their formation was precipitated by the impending first meeting of a federation of specialist medical societies in Washington in 1888, known as the Congress of American Physicians and Surgeons. This congress reflected the same sorts of tensions between the elite and the rank-and-file in the medical community as were evident in the scientific community. The congress was conceived as a rival to the International Medical Congress of 1887, sponsored by the American Medical Association (AMA), the control of which was wrested from the eastern elite and turned over to the hands of representatives of state medical societies. The concern of the latter was to make the congress representative of the nationwide membership of the AMA. As a result of the battles over control of the congress, much of the eastern elite boycotted the congress, which had become to them a supreme embarrassment. The alternative Congress of American Physicians and Surgeons was structured as a federation of medical specialty societies with limited membership and was to emphasize medical research to the exclusion of all medical politics. If the physiologists and anatomists wanted to be part of this new medical organization, as they had been part of the AMA and the original organization of the International Medical Congress, they had to form national societies.[21]

The Physiologists and Anatomists form an interesting contrast; the Physiologists defined their membership relatively narrowly, leaving out many who might have wanted to join, while the Anatomists initially included all those who had an interest in anatomy. How each of the disciplinary societies defined their membership depended on a number of factors—in particular, the composition of the field (whether it contained many or few amateurs), the geographical distribution of the field (whether an eastern elite could readily be separated), the research orientation (whether a more modern-style research could readily be separated from older forms of research), and finally the perspective of the founders (who got there first).

The American Physiological Society was founded at the initiative of Henry Pickering Bowditch, Henry Newell Martin (both charter members of the Naturalists), and S. Weir Mitchell in New York on 30 December 1887, just after the Naturalists' meeting in New Haven. Seventeen attended the meeting and twenty-eight were named charter members. The Physiologists might be considered the first offshoot of the Naturalists, although by 1887 just fourteen of the charter members were members of the Naturalists.

The constitution adopted in 1887, apparently written by Martin, was in part modeled on that of the Naturalists, but the membership provision was much more restrictive. The physiologists were the first to have a requirement in their constitution that prospective members must have some work published. The membership clause limited membership to persons who had "conducted and published an original research" in the area represented by the society. Physiology was a very generally used term in the nineteenth century, in the popular mind often nearly equated with hygiene. It was taught in most medical schools and encompassed an older nonexperimental clinical methodology as well as the newer instrumentation-oriented animal physiology imported from Germany, France, and Britain. The charter members were mostly young men, all well known to the founders of the society, and more than two-thirds had spent time working in foreign laboratories. Some were technically amateurs (practicing physicians not employed as physiologists), as was Mitchell himself, but this mattered less than the kind of training they had received and the kind of research they had published. Deliberately excluded from the society were a great many teachers of physiology in medical schools throughout the country, some of whom published what they considered to be physiological research. Although the charter members represented a wide range of interests, including plant physiology (George Lincoln Goodale), the society soon became identified largely with the use of animal experimentation to investigate human physiology. Anyone too clinical or too zoological in orientation was unwelcome. The new society held its first "special" meeting with the Congress of American Physicians and Surgeons in September 1888 and its first annual meeting in December in Philadelphia just after the American Society of Naturalists' meeting in Baltimore. In 1889 and thereafter through the 1890s, it met with the Naturalists.[22]

The Association of American Anatomists differed in its initial approach to membership selection because it was less clear what direction modern anatomy was taking (there was considerable concern that anatomy was a completed and, hence, dead science), and because the "wrong" people organized the society. It might be remembered that for anatomy, there was no Johns Hopkins until the medical school opened in 1893 and Franklin P. Mall was appointed professor. In 1892, the society had no members at all from the state of Maryland. The Anatomists were organized, apparently without much forethought, on 17 September 1888 during the meeting of the Congress of American Physicians and Surgeons by "anatomists who were attending the Congress or who were already resident in the city." Immediately thereafter, the association held its first sessions for the presentation of papers as an unofficial part of the congress. The person who took the initiative in forming the association was a shadowy figure, Alexander H. P. Leuf, Director of Physical Education at the University of Pennsylvania, who was elected first secretary but by 1890 ceased being a member. The elderly Joseph Leidy, who was not present, was elected first president. Although some anatomists of distinction attended the organizational meeting—Harrison Allen, Horace Jayne, and D. S. Lamb—there were also a number of obscure practitioner/teachers and U.S. Army surgeons; major centers of medical research such as Harvard were not represented.[23] In the early years of the society, a deliberate effort was made to bring into membership all teachers of anatomy. At the organizational meeting, the executive committee "was instructed to inform the professors and demonstrators of anatomy in the regular medical schools [as opposed to the homeopathic schools] of the United States and Canada, as well as all others interested in the subject, of the object of the Association and invite them to become members."[24] There were eighty-five names on the first membership list published in 1891; none of the members had a Ph.D.[25] Beginning the following year a table of the geographical distribution of members was published, a recognized sign of a concern for scientific democracy. Probably because research was not a criterion for membership in the Anatomists, the Congress of American Physicians and Surgeons was reluctant to accept the society into its federation, although it eventually did so.[26] It is noteworthy that Mall was a charter member of the American Morphological Society but did not become a member of the Anatomists until 1898.[27]

Not until about 1900, when an entirely different group had achieved control of the association, was the membership policy reconsidered, and membership limited, as in other biological disciplinary societies, to well-trained investigators. By 1906, when Mall was president of the society, the revised constitution stated: "Candidates for membership must be persons engaged in the investigation of anatomical or cognate sciences."[28] Even then, the Anatomists were more lenient than other societies and, unlike the Physiologists or Zoologists, were willing to take in occasional students who had not yet completed their degrees. The Anatomists met with the Naturalists for most of their annual meetings in the 1890s, but with less regularity than the Physiologists or Morphologists.

In 1888, also, the geologists formed their own society. Geology was a well-developed science, with representatives all over the country, many of them employed by the state and federal geological surveys. An eastern leadership was not readily separable. The Geological Society of America, formed as an offshoot of the AAAS, was initially relatively liberal in its membership policy. Charter membership was open to members of Section E, Geology, of the AAAS, but nongeologists were deliberately discouraged by a very high annual dues of ten dollars. By the end of 1889, the membership list of 191 fellows included almost all the geological professionals in the country, although the society later became increasingly selective.[29] The Geological Society occasionally met with the Naturalists in the 1890s but was otherwise not closely tied.

The formation of these three new societies made serious inroads into the Naturalists' meetings. The Naturalists responded by inviting the new societies to meet with them in New York in December 1889.[30] With that meeting, the Naturalists adopted a new role that it carried out for the next two decades, as coordinator of meetings of biological disciplinary societies. The Naturalists no longer held a program of its own except for a symposium on a topic of general interest to all the societies and a presidential address. This arrangement made it exceedingly easy to form new societies; all that was needed was for a small group to meet, adopt a constitution, and select charter members. The Naturalists took care of all the work of arranging the annual meetings, and all the secretary of the fledgling disciplinary society was required to do was arrange for a separate session of papers in the area of the society. Societies thus could remain viable even if as few as ten members attended the meetings. Proliferation of new societies soon resulted.[31]

The American Morphological Society was founded at the Naturalists' meeting in Boston in December 1890 at the initiative of Charles Otis Whitman, who became its first president. The letter of invitation to the organizational meeting was signed by Whitman, Henry Fairfield Osborn, E. B. Wilson, Edward G. Gardiner, and J. Playfair McMurrich. In 1884 Whitman and E. L. Mark had initiated a discussion about forming a highly exclusive "embryological club," with an admission fee of $100 to support a journal, but the plan threatened to alienate too many people to be carried out.[32] The Morphological Society was broader in scope, yet the membership could still be readily circumscribed. A knowledge of morphology, like the new experimental physiology, could only be obtained through an academic training in a limited number of places.[33] Of the twenty-six charter members, seventeen held Ph.D.'s and only one, Mall, held an M.D. The Morphologists met every year through the 1890s with the Naturalists.[34]

The Morphologists were soon followed by the American Psychological Association, founded at the initiative of G. Stanley Hall who invited a number of psychologists to meet at Clark University in July 1892. Hall became its first president and Joseph Jastrow its first secretary, and thirty-one charter members were named. Patterned on the model of the American Physiological Society of which Hall and Jastrow were charter members, the society limited membership in its constitution to those engaged in "the advancement of Psychology as a science." The members were academics and all had Ph.D.'s or M.D.'s. In 1895 and afterwards, the Psychologists met annually with the Naturalists.[35]

The Botanists presented a somewhat different story. Botany was well established and its leaders were not limited to eastern institutions. Taxonomy was not held in quite the same low esteem as it was by leading zoologists. But unlike the case of the geologists, the leaders wished to set up a relatively exclusive society but could not agree on how to do so. As a result, competing societies were created, which led to considerable dissension. At the AAAS meeting in 1892, on the initiative of Liberty Hyde Bailey, a number of leading botanists discussed the possibility of forming a society. Bailey was made chairman of a committee to report the following year. The majority of the committee decided that the time was not yet right. A minority decided to go ahead and form a highly exclusive society for the promotion of research, the Botanical Society of America, with twenty-five specially selected charter members. Members were to contribute a twenty-five dollar initiation fee and ten dollar annual dues, the money to be later used to fund a publication. Despite its ambitious name, the Botanical Society of America was a club, like the embryological club envisioned by Whitman in 1884, to which several prominent biologists did not care to belong. Five of the original twenty-five asked to join refused, among them William G. Farlow, professor of cryptogamic botany at Harvard. William Trelease, Director of the Missouri Botanical Garden, was elected its first president in 1894. Beginning in 1895, annual meetings of the society were held with the AAAS.[36]

Meanwhile, following discussions in 1895 and 1896, another botanical society was formally established at the Naturalists meeting in 1897 to fill the botanical void at the Christmastime gatherings. To avoid taking in systematic botanists, the new society was called the Society for Plant Morphology and Physiology. Farlow became its first president. It was decided in 1897 that "Candidates for Membership shall show special ability in original research as indicated by published papers or work in progress." As with the other societies meeting with the Naturalists, dues were kept low, in this case just enough to cover expenses of stationery, postage, and printing—about a dollar a year. In 1898, Bailey and Roland Thaxter, active members of the new society, resigned from the Botanical Society of America. When the two societies finally united in 1906, the Botanical Society of America had fifty-two members and eight associate members, and the Society for Plant Morphology and Physiology had seventy-one members.[37]

By the 1890s much of the original membership of the Naturalists was splintered into separate societies held together by a common meeting place. But not all the members of the Naturalists were welcome in the national disciplinary societies. Left out were the systematic botanists and zoologists, the museum curators, and the teachers of the various branches of natural history who did not conduct research and publish. Although some remained members of the American Society of Naturalists, there was no place for them at the combined meetings.

The Naturalists and Biological Unity in the 1890s

The decade of the 1890s was a "golden age" for biological unity. The American Society of Naturalists acted very nearly like an American biological society, coordinating meetings and providing the means for joint programs and projects. The *American Naturalist,* official organ of the Naturalists (though privately owned) published abstracts of the meetings and carried sections on advances in each of the biological sciences. The presidency of Naturalists was held in turn by members of the various disciplines. Among the presidents in the 1890s were the physiologists Martin (1890) and Bowditch (1898), the geologist Rice (1891), the morphologist and paleontologist Osborn (1892), the physiological chemist Russell Chittenden (1893), the anatomist and embryologist Minot (1894), the morphologist Scott (1896), the biologist and bacteriologist W. T. Sedgwick (1898), and the botanist Farlow (1899). The meetings were remembered years later as exceedingly congenial. They took place on northeastern college campuses—Harvard, Yale, Johns Hopkins, Princeton, Columbia, University of Pennsylvania, Cornell—and featured an invited symposium, receptions, a presidential address, and a banquet all organized by the Naturalists. Enough people attended to allow for reduced fares on railroads, an important consideration for all who attended scientific meetings. The number of societies meeting with the Naturalists increased through the 1890s. In 1897, Section H of the AAAS (Anthropology) began to hold meetings with the Naturalists (the American Anthropological Association was founded in 1902), and in 1899, the Society of American Bacteriologists was founded at the New Haven meeting.[38] By 1901, the last year the Naturalists met on their own, there were nine societies meeting with them.

The annual symposium and the presidential address explored themes of general interest to all biological discipline builders. The symposia sometimes dealt with biological topics such as "Inheritance of Acquired Characters" (1890, 1896) or "Recent Discoveries Regarding the Cell" (1893), sometimes with institutional or educational topics such as "Marine Biological Laboratories" (1892) or "Advances in Methods of Teaching" (1898), and sometimes with issues of science policy such as "Position of the Universities Regarding Investigation" (1899) and "Attitude of the States Toward Scientific Investigation" (1900). Usually the topic would be discussed from the perspective of the various specialty areas represented at the meeting. The Naturalists also formed committees on which representatives of the various societies participated. An active committee was that on "Scientific Teaching in the Schools," which in 1891 sent out a circular letter to nearly four hundred American universities and colleges. Another committee worked for the repeal of the duty on scientific and physiological apparatus. A committee was even appointed to combat the antivivisectionists' attempts to have a law passed outlawing animal experimentation in the District of Columbia.[39]

In the mid-1890s, the Naturalists made a concerted but ultimately unsuccessful effort to get their informally affiliated societies to affiliate on a more formal basis. A committee appointed to consider a closer union of the Naturalists with the Anatomists, Morphologists, Physiologists, and Geologists recommended in 1892 that the Society of Naturalists invite the other societies to unite under a common general secretary and form a union at the next meeting on terms to be determined by a conference committee. The conference or "Affiliation Committee"—which consisted of representatives of the Naturalists, Morphologists, and Physiologists—suggested in 1893 a union of these societies and the Anatomists under the following terms: that the affiliated societies elect their own members, "but none but 'professionals' shall be eligible to any one of them"; that election to any of the societies carried with it election to the Naturalists; and that meetings be held at a common time and place. Perhaps the most difficult provision to accept was that a common fund be held by the Naturalists, although special levies could also be made by the societies. The fund would be used to organize the meetings and to publish the constitutions, bylaws, lists of members, and a report of the proceedings of the meetings. Minot proposed that the plan also be submitted to the Geological Society.[40]

The societies discussed and voted on the plan of affiliation in 1893 and again in 1894. None was willing to give the proposal an unconditional acceptance. Ultimately, no group wished to surrender any of its independence as to where to meet, or to force all of its members to become members of the American Society of Naturalists. Many members of the various organizations were in fact not members of the Naturalists. Unofficial membership lists published by the Naturalists in 1897 identified as members of the Naturalists, sixty-two of eighty-six members of the American Morphological Society, twenty-four of fifty-nine members of the American Physiological Society, eleven of eighty-seven members of the American Psychological Association, and nineteen of 119 members of the Association of American Anatomists. All of the societies, however, expressed their desire to continue meeting with the Naturalists. The plan of formal affiliation thus fell through, and informal affiliation continued as before.[41]

Meetings and Publications

Before proceeding further, it is worth pausing to consider the chief activities of the various specialist societies. Their main purpose was to encourage research, and this was accomplished primarily through the holding of scientific meetings for the presentation of research results and for the demonstration of specimens, preparations, and new instruments and techniques. Membership in a society generally conferred the right to present a paper at the meetings.

One gained a place on the program by writing to the secretary a few weeks before the meeting and providing a title. An important aspect of any society was that members could also sponsor papers of nonmembers. This enabled graduate students to participate in the program and thus make themselves and their work known to leaders in the field. Each society held only one session at a time so that all members could hear all papers. Although societies sooner or later faced the problem of a too-crowded program, they clung as long as they could, and longer than they might have, to the notion of the unity of the society. Program pressure did mean that societies became increasingly jealous of the time available to them to hold scientific sessions. Societies were less willing to break for the Naturalists' symposium in the first decade of the twentieth century than they were in the 1890s, especially if the symposium seemed of peripheral interest. As with societies today, members recalled that the most important aspect of the meetings involved the personalities and the informal discussions in the halls.[42]

The availability of opportunities for publication concerned all of the societies, but most could do little about the problems for their resources were minimal, the memberships small, and their dues low. The immediate need for publication of abstracts (which everyone seems to have expected by the mid-1890s) was met for the time by *Science,* which published abstracts from all the societies. Publications of the societies (in particular the Naturalists, Morphologists, Physiologists, Anatomists, and Botanists) were limited in the 1890s to small brochures containing the constitution, officers, and members and sometimes including reports of the meetings and abstracts.

Research journals were beyond the capacities of professional biological societies, at least until about World War I. Journals before that time were started not by societies but by one or a few individuals who put their personal funds and a great amount of dedication into them. They were sometimes founded with a society in mind whose members, it was hoped, would subscribe and contribute articles. Usually the board of editors would consist of prominent members of the appropriate society, but there was no official connection between society and journal. Sometimes societies established an informal relationship with a privately owned journal, naming it as their "official organ." This meant that the journal agreed to publish news of the society and abstracts. It was not until later that some societies (but not all) were able to acquire journals, often by donation or purchase rather than by initiating them.

The *American Naturalist,* founded in 1867 by E. S. Morse, A. S. Packard, Alpheus Hyatt, and F. W. Putnam, served as the official organ of the Naturalists.[43] The *Journal of Morphology,* founded by Charles Otis Whitman in 1887, had a close but unofficial relationship to the American Morphological Society. The Psychologists published their early proceedings in *Psychological Review,* founded in 1894 under the auspices of the Psychological Review Company organized the previous year by James McKeen Cattell and James M.

Baldwin. Between 1925 and 1927, the journals of the company were acquired by the American Psychological Association.[44] The Physiologists were fortunate to have obtained at an early date a closer relation to their associated journal than most other societies; William T. Porter agreed to found a journal and assume financial and editorial responsibility for it under a contract with the society. The *American Journal of Physiology* was begun in 1898, and in 1914, when Porter agreed to relinquish his claim, the society acquired full ownership. But even when the society owned the journal, subscription was not made mandatory for members, because the additional dues required to purchase three volumes a year would have decimated the society.[45] The Botanical Society of America had an informal relationship to the *Botanical Gazette,* founded and edited since 1875 by John Merle Coulter. Because the original Botanical Society of America accumulated a considerable treasury, it was able to establish, with the help of the Brooklyn Botanic Garden, the *American Journal of Botany* in 1914. *The American Journal of Anatomy* was founded in 1901 by establishing an "Anatomic Journal Trust," the original trustees of which were Huntington, Mall, and Minot.[46]

The economic prospects for journal publishing in the 1890s were not good. Few professional journals were financially stable. Porter had to subsidize the *American Journal of Physiology.* The *Journal of Morphology* was said to be published at a considerable loss, and could stay afloat only with the assistance of "munificent gifts" from Edward Phelps Allis;[47] Whitman was forced to suspend publication temporarily in 1903.

For the Anatomists and Zoologists, the Wistar Institute in Philadelphia, formed in 1905, took on the role of journal publisher. Through the efforts of the anatomist Horace Jayne, in charge of the Department of Publications, Wistar deliberately sought to acquire existing journals and by 1908 was publishing five of them. Its first acquisition was *The Journal of Morphology.* Whitman accepted the proposal of the Wistar Institute and assigned management of the journal to the Institute, which set up an editorial board and resumed publication in 1908. The same year, *The American Journal of Anatomy* and its offshoot, the *Anatomical Record,* begun in 1906, were leased to the Wistar Institute. Wistar also acquired the *Journal of Comparative Neurology and Psychology,* begun in 1891, and the *Journal of Experimental Zoology,* begun in 1904 under the editorship of Ross Harrison. In this period, anatomy was revived by its association with morphology and embryology. The ties between the morphologists and anatomists were further strengthened by the common links to the Wistar Institute, which made arrangements with both societies to supply all members with journals at a low cost. Although this proposition meant raising the dues from a nominal dollar or so to five dollars or more, the terms were sufficiently favorable that the societies agreed. After 1908, for dues of five dollars (all but fifty cents went to Wistar), members of the Anatomists received the *American Journal of Anatomy* and the *Anatomi-*

cal Record. A similar arrangement provided all members of the Zoologists with a choice of journals. *Anatomical Record* published proceedings and abstracts of both societies.[48]

The AAAS and the Establishment of Convocation Week

In 1901, a new factor was thrust upon the biological societies by the decision of the AAAS to change its meeting time from summer to the Christmas vacation time used by the Naturalists and to take over the Naturalists' successful role as coordinator of specialist societies. The AAAS invited all societies to affiliate with it. A constitutional amendment provided that affiliated societies would be represented on the Council of the AAAS. The AAAS lobbied universities so that all faculty would be relieved from regular work during this period. The week was to become known as Convocation Week, the time set aside for meetings of scientific and learned societies. Convocation Week was inaugurated at the 1902 meeting of the AAAS in Washington, D.C.[49]

The Naturalists anticipated the change by devoting the annual symposium of their 1901 meeting to "The Relation of the American Society of Naturalists to Other Scientific Societies." The meeting was the first held outside the East, at the University of Chicago, and attested to the rising importance of research centers in the Midwest and West.[50] That same year a society of naturalists for the central states was founded. The chief issue of the symposium of 1901 was the future role of the Naturalists. Minot, who had been involved in implementing the decision of the AAAS, thought that the Naturalists could safely entrust to the AAAS their role as "the organ of affiliation for societies which are concerned with the various branches of natural history." The step of forming "a wide-reaching organization of science, national in extent and power," he believed, was "destined to rank among the great achievements of the century upon which we are just entering." Minot and the other participants in the symposium were united in the belief that the American Society of Naturalists, now numbering 250 members, should be continued but there was no clear sense of what its role ought to be.[51] In practice, the Naturalists continued to invite affiliated societies to join with them in the Naturalists' smoker, banquet and presidential address, and annual symposium.

The new meeting time for the AAAS brought together a large and heterogeneous gathering of academic disciplinary societies from physics to anthropology, as well as an increasing number of societies representing the applied sciences. Depending on the location of the meeting, the total attendance could reach several thousand. There was constant competition for meeting places (which societies would get the best hotels for smokers, etc.) and for time to hold scientific programs. The gathering brought together a number of groups that had hitherto deliberately been kept separate. It brought together the Botanical Society of America and its rival, the Society for Plant Morphology and Physiology, and created pressure for their merger, which came about in 1906. The new organization, also called the Botanical Society of America, met each year with AAAS and was unwilling to support the Naturalists' schemes to secede. The meetings after 1902 also brought together the various sections of AAAS that had no membership requirement and the corresponding specialized societies that were highly selective about who was admitted. Although the officers of the sections were usually members of the corresponding specialty societies, and many of the speakers on the sectional programs were also members, there were always members of the AAAS sections who were not eligible for the specialty societies. This led to conflicts in programming that were especially disturbing for the zoologists.

About the same time the AAAS changed its meeting time, the American Morphological Society became the American Society of Zoologists. In part, the change was made so that the Central Zoologists, organized in 1901, might become the central branch of the larger society. Another factor was the desire to invite the International Congress of Zoologists to meet in America. Morphology did not correspond to university departments or to the structure of foreign societies, whereas zoology did. But in large part, the change represented a recognition that the society was no longer limited to morphology. There had been no revolt from morphology; rather, there had been a gradual introduction of papers into the program of the American Morphological Society that were outside the traditional bounds of morphology. The name change therefore did not signal a radical change in membership policy.

The name change was first suggested during the 1901 meeting in the course of a general debate over means to alleviate the crowding of the scientific program (holding more than a single session was considered inappropriate). E. L. Mark moved "That the Executive Committee be empowered to classify the papers to be presented before the Society, and to reject as unsuitable such as from their titles are clearly outside the scope of the Society." C. B. Davenport suggested that the name of the society be changed to the American Zoological Society "so as to allow in the program papers other than strictly morphological." The matter was referred to a committee consisting of Davenport, Henry Baldwin Ward, E. B. Wilson, Maynard Mayo Metcalf, and Herman Cary Bumpus.[52]

At the 1902 meeting, the committee reported with respect to the society's name: "It is believed that the present terminology is misleading, since some of our most valuable members are not morphologists, and some of the most interesting papers deal with other phases of biological science. In conformity with other organizations in this country, and especially with cognate societies in France and Germany, it is recommended that the name should be changed to the Zoological Society of America, recognizing at once the breadth and strength of this term in its proper use."[53]

In fact, the American Society of Zoologists was no more hospitable to systematic zoology than was the American Morphological Society. It is noteworthy that none of the curators of the U.S. National Museum was a member of the American Society of Zoologists, including Richard Rathbun, in charge of the U.S. National Museum, and Frederick True, head curator of the department of biology, who were identified by stars as eminent zoologists in *American Men of Science*. The officers of the society were up in arms when, at the Baltimore meeting of AAAS in 1908, Section F ran a program at the same time as the society with many of the same participants. The following year, the vice-president for Section F suggested that the secretary of the society take charge of the entire zoology program; however, the society did not want to allow into its own program any papers by nonmembers that were not introduced by members and, therefore, after much arguing back and forth, rejected the offer. The sectional chair did the best he could but still had to schedule papers conflicting with the society's program. C. Judson Herrick, representing the AAAS, complained to H. S. Jennings about the narrow-mindedness of the society and its unwillingness to cooperate with the AAAS:

> It is fully charged in some quarters that the present policy of the administration of the zoological society is hostile to any interests save those of experimental zoology and allied subjects and that faunistic, systematic and other phases of zoology are not given a fair hearing. This may be an unjust criticism, but if the zoological society is to be fully representative of zoology in America and not merely of a single cult, some provision must be made for systematic and similar work that is just as truly zoological as the study of chromosomes.[54]

When papers in the scientific program were listed by category in the 1910s, the categories did not include mammalogy, ornithology, herpetology, and so forth but were rather comparative anatomy, embryology, cytology, genetics, evolution, comparative and general physiology, ecology, parasitology, and demonstrations.

By 1908 there was a general feeling among the officers of the Naturalists and Zoologists that desperate measures needed to be taken to bring the biologists together and extract them from the AAAS. In 1907, the symposium of the American Society of Naturalists was devoted to "cooperation in the Biological Sciences." J. Playfair McMurrich took as the theme of his presidential address the question: "What is the use of continuing the existence of the Society of Naturalists?" He saw the difficulties faced by the Naturalists as due in large part to

> the irruption of the American Association for the Advancement of Science into the quiet and sociable serenity of convocation week, and the conse-

> quent desire on the part of some that the Association should assume responsibility for all the fostering which the different scientific societies may require. . . . [In this situation,] the solidarity of the biological sciences would, by the absorption, be lost, and they would become, if I may be permitted to misquote a celebrated definition, members of an indefinite, incoherent heterogeneity, instead of, as now, parts of a definite, coherent homogeneity.

He argued for the need of an intermediary organization to foster cooperation among biologists in research matters as well as in scientific education in the colleges and high schools.[55]

The following year the need to take action became still more urgent. A. P. Mathews's "Plan for an American Biological Society," circulated at the 1907 meeting, suggested that the name of the Naturalists was unfortunate and that the society should have no role to fill in the future coordination of biological societies. To make matters worse, the AAAS organized a Darwin Symposium to honor the fiftieth anniversary of the *Origin of Species* for the 1908 meeting without involving the Naturalists in any way. H. McE. Knower, Hopkins anatomist and secretary of the Naturalists, wrote in *Science,* "Here arrangements, peculiarly the province of the naturalists, have been perfected without consulting their official representative!" He argued for a plan to give the American Society of Naturalists the dignity it deserved as "a primary natural division of the American Association." In Knower's plan, all members of the affiliated societies automatically would be made members of the Naturalists, which would then cease taking in new members on its own. The affiliated societies would collect as part of their dues a small annual subsidy to be devoted to the purposes of the Naturalists. The new organization would then "secure a proper adjustment" with the AAAS.[56]

At the 1908 meeting, the Naturalists adopted a "New Temporary Plan for the Society," worked out by the president of the society, T. H. Morgan. The society would organize its own program of contributed research papers devoted to "the study of evolution in all of its many-sided aspects (historical, environmental, experimental, etc.). There is no subject which would so well hold together all the present members (including botanists, zoologists, physiologists, anatomists, paleontologists, anthropologists, etc.)." It was hoped that this redefinition of the function of the society would attract a number of new members.[57] Although the programs were originally supposed to approach evolution from varied perspectives, in practice they were largely limited to papers on genetics.

Instead of attempting an adjustment with the AAAS as Knower had recommended, most of the leaders of the Naturalists were ready in 1908 to plan for a "bolt." H. S. Jennings wrote to Lorande Loss Woodruff, secretary of the Zoologists, before the 1908 meeting, "There seems to be considerable feeling among many that our society and others originally affiliated with the Naturalists are making a mistake in following the American Association about." He urged the executive committee of the Zoologists to give the Naturalists' plan of meeting elsewhere "a trial for a year or two at least. . . . It seems to me that if we are to follow the A.A.A.S., we ought logically to merge ourselves in Section F and go out of existence as a separate society."[58] The botanist D. P. Penhallow in his presidential address before the Naturalists in 1908 saw the alternatives not as a revision of relations with AAAS or extinction, but "the real issue should be stated in terms of continued companionship." The purpose of AAAS as he saw it was "to popularize scientific knowledge and effect its widest distribution" by methods that are "not in harmony with the purely scientific spirit." "The issue is a clear one and should be won or lost on the simple question as to whether we shall continue to meet with the American Association or choose our own time and place."[59]

As the AAAS was meeting in 1909 in Boston, a favorite meeting place of the biological societies, it was reluctantly decided to put off the experiment until 1910 when the AAAS would be meeting in Minneapolis.[60] Unfortunately, the Naturalists chose a location for the 1910 meeting—Ithaca, New York—that had not been previously (1897) satisfactory to the Physiologists because there was no medical school (and the professors at the veterinary school were not members of the society). The Physiologists and their recently formed offshoots, the American Society of Biological Chemists (1906) and the American Society for Pharmacology and Experimental Therapeutics (1908) decided instead to meet in New Haven. The Naturalists did succeed in bringing with them the Anatomists, the Eastern Branch of the American Society of Zoologists, and the Society of American Bacteriologists. The next year the experiment was tried again, but again the choice of location, Princeton, was a poor one to attract the Physiologists, Pharmacologists, and Biochemists. Raymond Pearl, secretary of the Eastern Branch of the American Society of Zoologists saw this meeting as crucial to bringing about a "closer coordination of all the distinctly *biological* societies about the American Society of Naturalists as a center,"[61] but the Naturalists managed to bring with them to Princeton only the Anatomists and Zoologists. The move for secession was at an end, and the Naturalists went back to meeting with the AAAS in Cleveland in 1912.

Mathews's Plan for an American Biological Society

At the same time as discussion was underway for forming a federation of existing societies around the Naturalists, an alternative plan to form a popular American Biological Society was proposed in 1908 by Albert Prescott Mathews, professor of physiological chemistry at the University of Chicago.[62] Though well situated academically, Mathews was a controversial figure who was somewhat outside the mainstream of society leadership. Besides being inherently unworkable, Mathews's plan was also at variance with the entire history of biological societies.

At its 1907 meeting, Mathews proposed to the council and membership of the American Physiological Society in general terms a plan for a large, popular biological society that would unite biologists.[63] The following year, he distributed a more formal proposal in the form of a "multigraphed" circular entitled "A Plan for the Organization of the American Biological Society." It began: "The present condition of the biological interests of the country may be called chaotic. There is no general organization and little cooperation between various subdivisions of the science; there are a multitude of small societies and a large number of journals, few with any permanent support. . . . The time has come to effect some kind of cooperation of all biologists and to secure the advantages which come from cooperation."[64] The proposed society was to be modeled on the visibly successful American Chemical Society, which was organized by divisions and local sections and boasted by 1908 more than four thousand members. It had a year past established *Chemical Abstracts* and for dues of ten dollars supplied its large membership with this publication as well as the *Journal of the American Chemical Society* and the *Journal of Industrial Chemistry.* Mathews envisioned a parallel society for biology, open to all who would pay the dues. For the finances of his plan to work, the society would require two thousand members or nearly all the biologists in the country. Like the Chemical Society, there would be regional meetings as well as national meetings. The existing societies would act as sections of the new society.

The heart of the proposal was a giant journal-publishing scheme. The proposed society would coordinate or take over the publication of existing biological journals and in addition undertake the publication of a new biological abstracting journal. Mathews believed that the abstracting journal "would do more to unify and stimulate biology than any move we could make."[65] The Wistar Institute might be called on to manage all of the journals. He listed thirteen journals, all of which, along with the abstract journal, were to be provided to the members for an annual dues of twenty-five dollars. These included not only the journals associated with the American Physiological Society, The American Society of Zoologists, the Biological Society of America, and the Anatomists (including all the Wistar journals), but also three medical journals, the *Journal of Biological Chemistry, Psychological Review,* and *Biological Bulletin.*[66]

The Physiologists responded to the proposal at the 1907 meeting to the extent of appointing a committee, chaired by their president, William H. Howell, to consider whether the society should liberalize its membership policy and thus achieve a membership of a thousand or more physiologists with a correspondingly greater impact on the public, or whether the society should remain as it was, a small organization of active investigators. The committee reported in 1908 that they liked things the way they were. Those who were not eligible to belong to the society could join Section K, Physiology and Medicine, of AAAS; attend society meetings; and, if they had a worthwhile paper to present, could get a member to sponsor them. Although there was no need for an enlarged society to include all physiologists, the society was willing to consider cooperation with other biological societies and appointed Mathews a committee of one to pursue the matter.[67] Mathews's plan was most fully aired not in 1908 but between 1912 and 1914.

There remained considerable hope from 1912 to 1914 among the leadership of the Naturalists and Zoologists that a federation of biological societies would be formed. Ross Harrison wrote to John H. Gerould, secretary of the American Society of Zoologists, in 1912, "I am in favor of meeting in Cleveland next winter and trust that by so doing the relations between the Eastern and Central Branches may be finally straightened out. We ought also to form a federation of biological societies like the old affiliation with the naturalists and be quite independent of the A.A.A.S."[68] A federation was formed in Cleveland in 1912, but not the general federation of biological societies that Harrison had in mind. The Federation of American Societies for Experimental Biology was limited to the biomedical societies, and served, in effect, to sever relations of these societies not only from AAAS but from the Naturalists and Zoologists as well.

The federation was called into being on 31 December 1912 by the representatives of the American Physiological Society, the American Society of Biological Chemists, and the American Society for Pharmacology and Experimental Therapeutics. This organization was a simple expedient for arranging the time and place of the annual meeting and for rationalizing the scientific programs of the societies. The executive committee of the federation consisted of the presidents and secretaries of the societies, with the chairmanship rotating from society to society in order of seniority. The societies agreed to hold their meetings in common at places to be determined by the executive committee and to publish a common program. With permission of the secretaries, papers could be shifted from the program of one society to another. In matters of membership, dues, and publications, the societies were left autonomous. As the federation had no funds of its own, the societies divided the cost of the annual meeting according to number of members. The three original members were joined in 1913 by the newly formed American Society for Experimental Pathology. Included in the provisions for the federation, adopted in 1912, was the statement "that a common meeting place of the Federation with the anatomists, zoologists and naturalists is desirable but not mandatory."[69] The next day, 1 January 1913, there was to be a meeting of representatives of the Naturalists and of the various biological societies, but nothing appears to have been decided.

At this time, discussion of Mathews's plan was reopened by publication of the plan in the *Biochemical Bulletin,* the organ of the Columbia University Biochemical Association. The editor of the *Bulletin* stated that the journal was publishing the plan in its original 1908 form because the recent formation of the federation had given new interest to it.[70] Mathews was well known to the Columbia biochemists, for he had received his doctorate from Columbia in 1898. To promote the plan, the editor distributed a reprint of it to all the members of "four of the leading biological societies" and requested expressions of opinion for publication. During the next year, the *Biochemical Bulletin* published seventy-three responses, mostly from members of the biomedical societies but also from a few members of the Zoologists.[71] Despite the fact that by 1914 the plan was already outdated, for the number of biological journals had continued to grow, many respondents were uncritical of it and announced themselves ready to pay twenty-five dollars in annual dues for a shelf-full of journals. Many others, however, were skeptical of whether the number of members could be obtained, whether anyone would want or need that many journals, or whether the estimates for the abstract journal were at all realistic. A number questioned the need for the abstract journal because there were several German journals that covered the ground. Some of the more thoughtful respondents, however, recognized that biology and chemistry were very different fields, and that the plan for an all-inclusive American biological society was doomed from the start.

A.J. Carlson, secretary of the Physiologists, first secretary of the Federation of American Societies for Experimental Biology, and Mathews's colleague at Chicago, was among those who favored the current division of biological interests into separate, research-oriented societies.

> I think we must face the fact that the *special societies* have not only come to stay, but that specialization will increase as the years go by. There is strength in smaller organizations of men of similar training and aim not found in larger and more heterogeneous societies. There would be little gained and much lost by converting the present special societies into sections of a general organization. . . . [The proposed American biological society] is contrary to the best tradition of all the present societies, with the possible exception of the anatomists. These societies are primarily organizations of research men. The qualification for membership is not willingness or ability to pay the dues, but scientific attainments, I think a reorganization involving the abandoning of this ideal would be fatal.[72]

He and several others, including E. G. Conklin, advocated instead a federation plan that "would leave the present societies intact."[73] One respondent summed it up: "The whole matter is to be decided by whether exclusive membership or strength in numbers is desired. Each has its advantage. Probably the limited membership is more especially advantageous to the individuals; and strength in organization, to the science as a whole, since it promotes dissemination."[74]

Others pointed to a fundamental difference between chemistry and biology: biology simply encompassed too much diversity for this or any federation plan to succeed. Bradley M. Davis of the University of Pennsylvania wrote,

> It seems to me that the biologists are far more diversified in their interests than are the chemists and that it would be correspondingly more difficult to organize them satisfactorily into a single society. Many men would not care to pay the heavy dues when their interests are chiefly centered in one journal or at most a small group of journals. It would be very difficult to hold the interests together and what is now well done by the enthusiasm of each group separately would be poorly done when brought under an organization in common.[75]

The endocrinologist Roy G. Hoskins thought the plan unfeasible because it proposed "to amalgamate interests too widely diverse—for instance, paleontology and pharmacology."[76]

There were serious reasons to question whether the success of the American Chemical Society could be applied to biology. One respondent pointed out that this organization was so successful because the pure and applied chemists worked together, with the higher-paid practical chemists serving principally to maintain the organization. In biology, "the 'pure' biologists will not cooperate with the workers in applied lines."[77] Another respondent attributed the success of the Chemical Society to the fact that when it was organized "and for a great many years afterwards, there were no strong societies occupying any part of the field." The same was not true for biology.[78] Raymond Pearl, then at the Maine Agricultural Experiment Station, argued that biologists, unlike chemists, did not share a fundamental base of knowledge: "To be a chemist of whatever sort of degree, or even to be interested in chemistry, implies some technical knowledge and experience with the fundamentals of the science. Interest in biology carries no such implications. There are a great many people who are, or think they are, interested in biology who have not the slightest real knowledge of the fundamentals (speaking in a technical sense) of any one of the biological sciences."[79] Yet another writer questioned whether "the unity of investigative method and the viewpoint of data and the breadth of interest of those who represent widely different subjects . . . are . . . sufficiently developed today to hold such an organization together? . . . Are the divisions of biology as now pursued naturally articulated so that the analogy with chemistry holds or would they be conglomerate units so as to be analogous to the A.A.A.S.?"[80] It became clear from the discussion that a general biological society either with or without a publications program would never succeed, and the only feasible plan was to form a federation of biological societies on the model of the Federation of American Societies for Experimental Biology. Some respondents, like the biochemist William Gies, saw the federation as "in effect an embryonic American Biological Society."[81]

In 1913, there was some hope that the federation might be expanded, but the naturalists and zoologists were soon to be disappointed. The AAAS was meeting that year in Atlanta. Only after considerable cajoling, the federation agreed to meet with the Naturalists, Zoologists, and Anatomists in Philadelphia in 1913. Representatives of the various societies participated in the annual symposium, which was devoted to "The Scope of Biological Teaching in Relation to New Fields of Discovery." There was considerable discussion of Mathews's plan and of the forming of a federation, but, as the *Biochemical Bulletin* reported, "there seemed to be a general disposition to 'await developments in the Federation,' before attempting to proceed further with a central organization."[82]

But in 1914 hopes for a federation of societies centering about the Naturalists collapsed with the secession of the biomedical societies. While the AAAS, the Naturalists, Psychologists, Zoologists, Botanists, and Bacteriologists met in Philadelphia, the federation went west to St. Louis, where they had received an invitation from the Washington University School of Medicine. The Anatomists, faced with the unpleasant choice between the Zoologists and Physiologists, went with the federation. Ross Harrison, then president of the Anatomists, was especially unhappy with the state of affairs.[83] The federation proved large enough to hold viable meetings on its own, apart from the AAAS and apart from the Naturalists and Zoologists as well. From then on, with only a few exceptions, the federation chose meeting places of its own, thus severing the biomedical sciences from the more purely biological sciences.

By the 1920s, the affiliated societies of the Naturalists were reduced to the Botanists and Zoologists. These three societies met every year with AAAS. In 1927, the federation, probably to avoid further competition with AAAS, moved to a spring meeting, a practice it has followed every since. About the same time, the Anatomists also moved to a spring meeting, choosing different dates and places from the federation.

In 1923, when a federation of biological societies was finally formed, it was too late to be effective in coordinating meetings of biological societies or in acting as a forum for general biological concerns. The movement to form the Union of American Biological Societies was initiated by the Naturalists, Botanists, and Zoologists working with the recently formed Division of Biology and Agriculture of the National Research Council. Preliminary meetings were held in 1920 and 1921, and a more formal meeting in 1922 took place under the auspices of the National Research Council in Washington, D.C., at which a constitution was approved. After ratification of the constitution by a number of biological societies, the union was formally inaugurated in

April 1923 by a meeting of a council composed of representatives of fifteen societies.[84]

The union was originally intended to serve biology in a wide variety of ways, but in fact, it served biology in only a single way, through the publication of *Biological Abstracts*. By 1923, the hope of getting biological societies to meet together each year had been abandoned. A central agency to manage biological journals was no longer needed as many of the biological societies were profitably managing their own journals. So much solicitation had been given to retaining the autonomy of each of the societies that the union was left with no power at all. The societies were all willing enough to join, for there were no obligations (except that of naming representatives), not even the obligation to pay dues. Still, some hoped that the union would unite efforts in the areas of biological education, "biological propaganda," and research support. In practice, the only function of the union was to obtain the cooperation of biologists in the project of establishing a biological abstract journal. An abstract journal had been the keystone of Mathews's plan of 1908, but now in the aftermath of World War I and the disruption of German science, it had become more urgently needed. *Biological Abstracts* began publication in 1926 with the aid of a large ten-year grant from the Rockefeller Foundation. C. W. Greene, the representative of the Physiologists to the union, reported to his society in 1929 that "aside from the establishment and management of Biological Abstracts very little has been undertaken by the Union."[85]

Conclusion

Despite many efforts from 1889 to 1923 to establish a central biological organization, the proposition was one with little hope for success. Biology, unlike most other academic departmental entities, was not a discipline. This paper raises but leaves largely unanswered the question of what biology was and what, if anything, did unite biologists. The history of professional biological societies lends support to the notion that biology achieved a certain identity in the 1890s through a core program of research in embryology, heredity, and evolution. This program was best represented by the memberships of the American Society of Naturalists and the American Society of Zoologists.[86] For others—including many who favored an American biological society—biology represented a broader ideal of unification of all the life sciences. In practice, however, biology was a conglomeration of diverse interests. Although many at the time did not seem to realize it, biology was not at all the equivalent of chemistry, despite the fact that there might be parallel departments of chemistry and biology in universities. Biologists, unlike chemists, did not share a fundamental body of knowledge, nor did they share a common basic training.[87] Although national biological societies organized early, from the beginning they were characterized by specialization and elitism, and were jealous of their autonomy. Even when fields such as anthropology and psychology are set aside, the cleavages between the biomedical sciences and the more purely biological sciences as well as between biology in academia and biology in museums, government agencies, and industry, were too great for any lasting common effort. Numerous biological sciences were established in America but no unified science of biology.

Notes

1. Wallace Fenn to L. V. Heilbrunn, 11 September 1945, Archives, American Physiological Society Minutes, American Physiological Society, Bethesda, Md., (hereafter APS Minutes), vol. 12, 1945–46.

2. On the American Chemical Society, see Charles A. Browne, ed. "A Half-Century of Chemistry in America, 1876–1926: A Historical Review Commemorating the Fiftieth Anniversary of the American Chemical Society," *Journal of the American Chemical Society,* 1926, *48,* pt. 2, no. 8A; and Herman Skolnick and Kenneth M. Reese, eds. *A Century of Chemistry: The Role of Chemists and the American Chemical Society* (Washington, D.C.: American Chemical Society, 1976).

3. Ralph H. Bates, *Scientific Societies in the United States,* 3d ed. (Cambridge, Mass.: MIT Press, 1965), pp. 85–136.

4. See Laurence R. Veysey, *The Emergence of the American University* (Chicago: University of Chicago Press, 1965); and Burton J. Bledstein, *The Culture of Professionalism: The Middle Class and the Development of Higher Education in America* (New York: W. W. Norton, 1976).

5. Much more work needs to be done on comparing the rise of professional scientific societies in different national settings. Although in several cases disciplinary societies were created earlier in Britain than in America (as for example, the Zoological Society of London founded in 1826), the societies were London-based organizations in which professionals and knowledgeable gentlemen-amateurs mixed rather than organizations that served the needs of academic discipline builders. Paul Farber makes a useful distinction between a profession and a discipline in his *The Emergence of Ornithology as a Scientific Discipline: 1760–1850,* (Dordrecht: D. Reidel, 1982), pp. 130–132. In the period under consideration in this paper, the process of discipline formation and professionalization were occurring together.

6. On professionals and amateurs in American science, see Sally Gregory Kohlstedt, *The Formation of the American Scientific Community: The American Association for the Advancement of Science, 1848–1860* (Urbana: University of Illinois Press, 1976); and Nathan Reingold, "Definitions and Speculations: The Professionalization of Science in America in the Nineteenth Century," in Alexandra Oleson and Sanborn C. Brown, eds. *The Pursuit of Knowledge in the Early American Republic: American Scientific and Learned Societies from Colonial Times to the Civil War* (Baltimore: Johns Hopkins University Press, 1976), pp. 33–69.

7. Garland Allen, *Life Science in the Twentieth Century* (New York: Wiley, 1975), esp. pp. 21–72; Jane Maienschein, Ronald Rainger, Keith R. Benson, Garland

Allen, and Frederick B. Churchill, "Special Section on American Morphology at the Turn of the Century," *Journal of the History of Biology,* 1981, *14*: 83–191.

8. This tension has been noted in Philip J. Pauly, "The Appearance of Academic Biology in Late Nineteenth-Century America," *J. Hist. Biol.*, 1984, *17*: 369–397; and in Robert E. Kohler, *From Medical Chemistry to Biochemistry: The Making of a Biomedical Discipline* (Cambridge: Cambridge University Press, 1982).

9. On the American Society of Naturalists, see Edwin G. Conklin, "Fifty Years of the American Society of Naturalists," *American Naturalist,* 1934, *68*: 385–401. Minutes of meetings and lists of members were published in American Society of Naturalists, *Records.* Its archives are located in the American Philosophical Society, Philadelphia.

10. See Kohlstedt, *Formation of the American Scientific Community.*

11. Herman L. Fairchild, "The History of the American Association for the Advancement of Science," *Science,* 1924, *59*: 365–369, 385–390, 410–415, esp. pp. 368, 414.

12. American Society of Naturalists, *Records,* 1884, *1*: pt. 1: 22.

13. Alpheus Hyatt, "The Business of the Naturalist," *Sci.*, 1884, *3*: 44–46.

14. American Society of Naturalists, *Records,* 1884, *1*, pt. 1: 23–24. Those who attended the organizational meeting are listed in a note in American Society of Naturalists, "Vol. 1. Annual Meetings, Constitution & By-Laws: 1883–1941," American Society of Naturalists Archives, American Philosophical Society, Philadelphia. Biographical data here and elsewhere in this paper are derived from the various editions of *American Men of Science* and from Clark A. Elliott, *Biographical Dictionary of American Science: The Seventeenth Through the Nineteenth Centuries* (Westport, Conn.: Greenwood Press, 1979).

15. Charles Sedgwick Minot, et al., "The Relation of the American Society of Naturalists to Other Scientific Societies," *Sci.*, 1902, *15*: 241–255, esp. p. 242.

16. American Society of Naturalists, *Records,* 1884, *1*, pt. 1: 24.

17. "Constitution and By-Laws of the Society of Naturalists of the Eastern United States, 1883," American Society of Naturalists, "Vol. 1." The clause was reworded and tightened in December 1883 to "Membership in this society shall be limited to persons professionally engaged in some branch of Natural History as Instructors in Natural History, Officers of Museums and Other Scientific Institutions, Physicians, and others." The definition of a professional naturalist was elaborated upon in a bylaw to include "those who have published investigations in pure science of acknowledged merit" and "teachers of natural history, officers of museums of natural history, physicians, and others who have essentially promoted natural-history sciences by original contributions of any kind." See American Society of Naturalists, *Records,* 1884, *1*, pt. 1: 5, 31, 32, 39.

18. Minot, et al., "Relation of the American Society of Naturalists to Other Scientific Societies," p. 243.

19. The 109 original members as well as several more recently elected members are listed in American Society of Naturalists, *Records,* 1884, *1*, pt. 1: 10–21.

20. American Society of Naturalists, *Records,* 1884, *1*, pt. 1: 29–32.

21. Toby A. Appel, "Biological and Medical Societies and the Founding of the American Physiological Society," in Gerald L. Geison, ed. *Physiology in the American*

Context, 1850–1940 (Bethesda, Md.: American Physiological Society, 1987), pp. 155–176, esp. pp. 158–162.

22. Ibid., pp. 162–166. On the American Physiological Society, see also, John R. Brobeck, Orr E. Reynolds, and Toby A. Appel, eds. *History of the American Physiological Society: The First Century, 1887–1987* (Bethesda, Md.: American Physiological Society, 1987); and William Henry Howell and Charles Wilson Greene, *History of the American Physiological Society Semicentennial, 1887–1937* (Baltimore: American Physiological Society, 1938). Manuscript records of the society are located at the American Physiological Society, Bethesda, Md., and are described in Toby A. Appel, "The American Physiological Society Archives," *Physiologist,* 1984, *27*: 131–132.

23. *The Association of American Anatomists organized at Washington, D.C., September 17, 1888. Its History, Constitution, Membership, and the Titles and Abstracts of Papers. For the Years 1888, 1889, 1890.* (Washington, D.C.: Association of American Anatomists, 1891), p. 3. See also Nicholas A. Michels, "The American Association of Anatomists: A Sketch of its Origin, Aims and Meetings," *Anatomical Record,* 1955, *122*: 685–714. Many of those who attended were either from Philadelphia or Washington, D.C. No early manuscript minutes for the Anatomists exist. The National Library of Medicine, Bethesda, Md., has a full set of the published minutes and membership lists. After 1906, minutes and membership lists were published in *Anat. Rec.*

24. *Association of American Anatomists, History,* p. 3.

25. Ibid., pp. 7–12.

26. *Proceedings of the Fourth Annual Session of the Association of American Anatomists, Held at Washington, D.C., September 23, 24 and 25, 1891, to which is Appended a List of Members.* (Washington, D.C.: Association of American Anatomists, 1892), p. 3–4, 19.

27. *Proceedings of the Eleventh Annual Session of the Association of American Anatomists, Held in New York City, December 28 to 30, 1898,* . . . (Washington, D.C.: Association of American Anatomists, 1899), p. 5.

28. *Anat. Rec.*, 1906–1908, *1*: 94. On the reformation of the American Association of Anatomists, see Florence Rena Sabin, *Franklin Paine Mall: The Story of a Mind* (Baltimore: Johns Hopkins University Press, 1934), pp. 223–231.

29. Edwin B. Eckel, *The Geological Society of America: Life History of a Learned Society,* Memoir 155 (Boulder, Colo.: Geological Society of America, 1982), pp. 7–11.

30. E. D. Cope and J. S. Kingsley, "Editors' Table," *Am. Nat.*, 1889, *23*: 32–34.

31. For example, Edwin O. Jordan recalled that the idea to form the Society of American Bacteriologists (now the American Society for Microbiology) originated when the anatomist F. P. Mall suggested to him in casual conversation at the meeting of the American Society of Naturalists in 1898, "Why don't you bacteriologists have a society of your own? I just met Abbott also wandering around like a lost soul." Jordan, A. C. Abbott, and H. W. Conn's circular letter sent to some forty bacteriologists specifically noted that the proposed society was to be formed on the models of the American Physiological Society and the American Morphological Society and that it would meet with the Naturalists and other biological societies. See E. O. Jordan to Barnett Cohen, 24 June 1935, American Society for Microbiology Archives, Cantonsville, Md., Uni-

versity of Maryland Baltimore County, and Barnett Cohen, *Chronicles of the Society of American Bacteriologists, 1899–1950* (Baltimore: Society of American Bacteriologists, 1950), pp. 1–3; and L. S. McClung, "The American Society for Microbiology/Society of American Bacteriologists: A Brief History," *ASM News,* 1978, *44*: 446–451, esp. p. 1.

32. E. L. Mark to Henry Fairfield Osborn, 24 March 1884, 4 May 1884, 17 May 1884, 26 May 1884; Benjamin Sharp to Scott and Osborn, 11 May 1884, Henry Fairfield Osborn Papers, New-York Historical Society, New York. Sharp argued that the basis of the club should be broadened to morphology rather than limited to embryology. I thank Ronald Rainger for information about these letters.

33. Whitman came to envision biology as composed of the two complementary sciences of morphology and physiology, rather than of the more traditional botany and zoology. It was thus appropriate to have an American Morphological Society to complement the American Physiological Society. See Jane Maienschein, "Physiology, Biology, and the Advent of Physiological Morphology," in Geison, *Physiology in the American Context,* pp. 177–193.

34. On the American Morphological Society, see Brother C. Edward Quinn, "Ancestry and Beginnings: The Early History of the American Society of Zoologists," *American Zoologist,* 1982, *22*: 735–748; idem, "The Beginnings of the American Society of Zoologists," *Am. Zoo.,* 1979, *19*: 1247–1249; "Historical Review," *Anat. Rec.,* 1917, *11*: 546–554; and "Minutes and Proceedings of the American Society of Zoologists and its Forerunners. Book I. 1890–1918," American Society of Zoologists Archives, Smithsonian Institution Archives, Washington, D.C. (hereafter ASZ Minutes).

35. On the American Psychological Association, see Samuel W. Fernberger, "The American Psychological Association, 1892–1942," *Psychological Review,* 1943, *50*: 33–60; idem, "The American Psychological Association: A Historical Summary, 1892–1930," *Psychological Bulletin,* 1932, *29*: 1–89, esp. p. 22; Wayne Dennis and Edwin G. Boring, "The Founding of the APA," *American Psychologist,* 1952, *7*: 95–97; Dorothy Ross, *G. Stanley Hall: The Psychologist as Prophet* (Chicago: University of Chicago Press, 1972), pp. 169–185; and Thomas M. Camfield, "The Professionalization of American Psychology, 1870–1917," *Journal of the History of the Behavioral Sciences,* 1973, *9*: 66–75.

36. On the Botanical Society of America, see Oswald Tippo, "The Early History of the Botanical Society of America," in William Campbell Steere, ed., *Fifty Years of Botany: Golden Jubilee Volume of the Botanical Society of America,* (New York: McGraw-Hill, 1958), pp. 1–13, esp. pp. 1–4; and *Brief Historical Sketch of the Society,* Publication 11 (Ithaca, N.Y.: Botanical Society of America, 1899). Archival material includes "Minutes, 1894–1926," and "Minutes of the Council, 1895–1923," Botanical Society of America Archives, University of Texas, Austin. I am very grateful to Jeffrey L. Meikle, American Studies, University of Texas, for researching this material for me. The first four presidents of the society were Trelease, Charles E. Bessey, John Merle Coulter, and Nathaniel Lord Britton. Requirements for membership included not only publication but participation in meetings; membership was forfeited if one did not attend or present papers for three years. The associate membership category was suggested in 1898 as a means of including those who did not wish to pay the high fees; associate members could not vote or hold office. Beginning in 1903, some of the accumulated funds were used for grants in aid of research.

37. On this society, see "Society for Plant Morphology and Physiology. The Secretary-Treasurer's Book, 1897–1906," Botanical Society of America Archives, and Botanical Society of America, *Report of the Fourth Annual Meeting, a List of Officers and Members of the Society, as well as of the Affiliating Societies, together with a Brief Historical Sketch of the Botanical Society of America, and of the Affiliating Societies,* Publication 45 (St. Louis: Botanical Society of America, 1910), pp. 18–28, esp. 22–26. The first four presidents were Farlow, John Muirhead McFarlane, David Pearce Penhallow, and Edwin Frink Smith. William Francis Ganong served as Secretary-Treasurer through the merger in 1906.

38. On meetings of the Naturalists, see Conklin, "Fifty Years," esp. pp. 388–389; and the American Society of Naturalists, *Records,* vol. 1 (1884–1895) and vol. 2 (1896–1911). A collection of programs is in the American Society of Naturalists Archives.

39. American Society of Naturalists, *Records,* 1893, *1,* pt. 10: 281–282; 1894, pt. 11: 304–305, 306–307, 309; 1895, *1,* pt. 12: 335–337; 1896, *2,* pt. 1: 35.

40. American Society of Naturalists, *Records,* 1897, *2,* pt. 2: 37–47. Responses of the various societies are noted in ASZ Minutes, 26 December 1895.

41. On anthropology, see George W. Stocking, Jr., "Franz Boas and the Founding of the American Anthropological Association," *American Anthropologist,* 1960, *62*: 1–17; and [W. J. McGee] "The American Anthropological Association," *Am. Anthrop.,* 1903, 5: 178–192. Because anthropology relied heavily on wealthy amateurs for support and because McGee rather than Boas founded the society, the American Anthropological Society did not limit its membership to professional researchers. On the bacteriologists, see n. 31.

42. Conklin, "Fifty Years," pp. 399–400; Howell and Greene, *American Physiological Society Semicentennial,* pp. 63–64.

43. Lynn Keller Nyhart, "The *American Naturalist,* 1867–1886: A Case Study of the Relationship Between Amateur and Professional Naturalists in Nineteenth-Century America" (Senior thesis, Princeton University, 1979). The American Society of Naturalists assumed editorial control of the journal in 1951.

44. Camfield, "Professionalization of American Psychology," pp. 68–69; Fernberger, "American Psychological Association: A Historical Summary, 1892–1930," pp. 72–78. The *American Journal of Psychology,* founded by Hall in 1887, was not acquired by the American Psychological Association; it survives today as a small quarterly.

45. Toby A. Appel, "The First Quarter Century, 1887–1912" and "The Second Quarter Century, 1913–1937" in *History of the American Physiological Society,* pp. 31–87; and Howell and Greene, *American Physiological Society Semicentennial,* pp. 78–83.

46. Michels, "American Association of Anatomists," pp. 706, 709–710.

47. ASZ Minutes, 28 December 1898.

48. "Five Biological Journals. The Growth of Scientific Literature at the Wistar Institute," *Old Penn. Weekly Review of the University of Pennsylvania* 7 (24 October 1908), copy in envelope, "ASZ, 1908–1922," ASZ archives, Smithsonian Institution, Archives, Washington, D.C.; Michels, "American Association of Anatomists," pp. 703–711; *Anat. Rec.,* 1917, *11*: 468–470. The arrangement with the Anatomists

lasted until 1923, after which journal subscriptions to members at reduced rates were optional.

49. *Proceedings of the American Association for the Advancement of Science,* 1901, p. 386; Fairchild, "History," p. 387.

50. See Maienschein, this volume.

51. Minot, et al., "Relation of the American Society of Naturalists to Other Scientific Societies," quotations on pp. 241, 244.

52. ASZ Minutes, 2 January 1902.

53. ASZ Minutes, 31 December 1902. The committee also recommended a program committee with the function of "passing upon abstracts and determining the character of each program." This would "regulate definitively the objection which might be presented with regard to the greater latitude offered in the broader name which has been recommended for this organization." When the committee's report was discussed section by section by members of the society, the name was changed to American Society of Zoologists and the provision for a program commitee dropped. After adopting the revised report, the society voted to invite the International Zoological Congress to meet in America in 1907.

54. See C. Judson Herrick to H. S. Jennings, 11 December 1909 and other correspondence in, "The American Society of Zoologists Eastern Branch, 1908, 1909, 1910, 1911, 1912," Smithsonian Institution Archives, ASZ archives, (hereafter Records of the Eastern Branch of the ASZ, 1908–1912). An example of a systematic zoologist refused membership is J. Chester Bradley, A.M., author of more than twenty entomological papers and Secretary of the Entomological Society of America, whose application was rejected in 1908. Jennings, in discussing the candidates for membership, wrote to Woodruff, 10 December 1908, "if his papers are merely on systematic entomology, it would seem that the Entomological Society is his proper place."

55. Frank R. Lillie, et al., "Cooperation in Biological Research," *Sci.,* 1908, *17*: 369–386; J. Playfair McMurrich, "The American Society of Naturalists Presidential Address," *Sci.,* 1908, *27*: 361–369, quotations on pp. 361, 363–364.

56. Albert P. Mathews, "A Plan for the Organization of the American Biological Society," *Biochemical Bulletin,* 1912–13, *2*: 261–268, esp. p. 263; "The Darwin Anniversary Meeting of the American Association," *Sci.,* 1908, *28*: 602–603; H. McE. Knower, "Discussion and Correspondence: The American Society of Naturalists," *Sci.,* 1908, *28*: 18–19. The symposium was published as *Fifty years of Darwinism, Modern Aspects of Evolution: Centennial Addresses in Honor of Charles Darwin Before the American Association for the Advancement of Science, Baltimore, Friday, January 1, 1909* (New York: Holt, 1909).

57. "The American Society of Naturalists," *Sci.,* 1908, *28*: 879–880; H. McE. Knower, "The American Society of Naturalists," *Sci.,* 1909, *29*: 707; idem, "The American Society of Naturalists," *Sci.,* 1910, *31*: 234–235.

58. H. S. Jennings to Woodruff, 2 December 1908, Records of the Eastern Branch of the ASZ, 1908–1912. The term "bolt" is used by Edward L. Rice to Woodruff, 25 December 1909.

59. D. P. Penhallow, "The Functions and Organization of the American Society of Naturalists," *Sci.,* 1909, *29*: 679–690, esp. pp. 682, 683.

60. See correspondence in Records of the Eastern Branch of the ASZ, 1908–1912.

61. Raymond Pearl to Executive Committee, 11 May 1911, Records of the Eastern Branch of the ASZ, 1908–1912.

62. Seymour S. Cohen, "Some Struggles of Jacques Loeb, Albert Mathews, and Ernest Just at the Marine Biological Laboratory," *Biological Bulletin,* 1985, *168* (supplement): 127–136.

63. APS Society Minutes, vol. 1, 1 January 1908; Mathews, "Plan for the Organization of the American Biological Society," p. 261n.

64. Mathews, "Plan for the Organization of the American Biological Society," p. 261.

65. Ibid., p. 263.

66. Ibid., p. 264. The medical journals were *Journal of Infectious Diseases, Journal of Experimental Medicine,* and *Journal of Medical Research.*

67. APS Minutes, vol. 1, 1 January 1907 and 30 December 1908; Reid Hunt to Joseph Erlanger, 21 January 1908 (and enclosure), Erlanger Papers, School of Medicine Archives, Washington University, St. Louis. See also A. J. Carlson, "The Federation of American Societies for Experimental Biology," *Sci.,* 1914, *39:* 217–218.

68. Ross G. Harrison to John H. Gerould, 24 August 1912, Records of the Eastern Branch of the ASZ, 1908–1912.

69. Appel, "Biological and Medical Societies," pp. 171–172; idem, "First Quarter Century," in *History of the American Physiological Society,* pp. 48–51.

70. Mathews, "Plan for the Organization of the American Biological Society," p. 261n.

71. "The Mathews Plan for an American Biological Society," *Biochem. Bull.*, 1912–13, *2*: 491–508, 582–588, and *Biochem. Bull.*, 1913–14, *3*: 134–142, quotation on p. 490; Walter H. Eddy, "The Mathews Plan: A Summary of Published Opinions," *Biochem. Bull.*, 1913–14, *3*: 142–147.

72. "The Mathews Plan," p. 492.

73. Ibid., pp. 492, 493.

74. Ibid., p. 141.

75. Ibid., p. 493.

76. Ibid., p. 495.

77. Ibid., pp. 500–501.

78. Ibid., p. 504.

79. Ibid., pp. 503–504.

80. Ibid., pp. 138–139.

81. Ibid., p. 135.

82. *Biochem. Bull.*, 1913–14, 3: 344.

83. See Samuel J. Meltzer to Ross G. Harrison, 15 February 1913, Box 65, folder 30, and Harrison to Charles R. Stockard, 5 January 1914, Box 65, folder 31, Harrison Papers, Sterling Library, Yale University, New Haven, Conn. I thank Jane Maienschein for these references. Harrison had felt strongly that anatomy should be closely related to both physiology and zoology. See Ross G. Harrison, "Anatomy: Its Scope, Methods and Relations with Other Biological Sciences," *Anat. Rec.*, 1913, *7*: 401–410.

84. William Campbell Steere, *Biological Abstracts/BIOSIS, The First Fifty Years: The Evolution of a Major Science Information Service* (New York: Plenum Press, 1976), pp. 1–52.

85. APS Council Minutes, 1922, 1923, 1929.

86. "Introduction," in this volume; Pauly, in this volume.

87. Even the basic courses in the life sciences given in colleges varied widely in content. Only a small proportion of first courses, even those labeled "biology," included both plants and animals. Botanists felt rightly that the "biology course" was often a means of excluding botany from the curriculum. See A. J. Goldfarb, "The Teaching of College Biology," *Sci.*, 1913, *38*: 430–436; McMurrich, "American Society of Naturalists Presidential Address," p. 362; Penhallow, "Functions and Organization of the American Society of Naturalists," pp. 684–685.

Philip J. Pauly

4 Summer Resort and Scientific Discipline: Woods Hole and the Structure of American Biology, 1882–1925

By the middle of the 1890s an important group of American scientists and academic administrators, centered around such leading universities and colleges as Johns Hopkins, the University of Chicago, Columbia, Bryn Mawr, and Harvard, believed in the existence of a science of biology. Biology was, in large part, an umbrella term representing all the different sciences of life. At the same time, however, it embodied a much stronger claim: that within the broad domain of the life sciences there existed a core of concepts, problems, and techniques that was, or would be, basic to all more specialized and applied fields. The belief in biology sharply distinguished American conceptions of the structure of the life sciences from those prominent, for example, in Germany; as the sciences evolved in the twentieth century, that faith had, as it turned out, important consequences for research, scientific organization, and education.[1]

The strength of the American belief in biology during the late nineteenth century is retrospectively rather perplexing, for a number of reasons. First, it was a sudden phenomenon historically. While the term "biology" had existed since the beginning of the century, it emerged as a principle of real significance in the United States only in the last two decades of the century. A second problem was the intellectual ambiguity of biology. The potential domain of biology included all the life sciences, from the experimental embryology of invertebrates to the geographic distribution of large mammals and the pathology of the human brain. Even with the principle of evolution there was no lasting consensus regarding the conceptual centrality of any particular body of research. Finally, the organization of biology was extremely indefinite. Although a number of biology departments were established in the 1880s and early 1890s, most (including those at Hopkins, Chicago, and Columbia) had fragmented by the turn of the century. Furthermore, Toby Appel's paper in this volume shows that American biologists never established a single unified disciplinary society comparable to the American Chemical Society.[2]

This apparent paradox can be resolved by granting that in science, as in other social spheres, neither clear intellectual definition nor explicit social organization is necessary for the maintenance of certain fundamental cultural structures. Since the beginning of this century, many American social theorists, themselves influenced by the first generation of biologists, have argued that the strongest beliefs and bonds are those characterizing "primary groups" or intimate communities. From the standpoint of such commentators, community bonds are powerful in part because such ties are neither articulated fully (which would open them to disputation) nor expressed in formal organizations (which are visible targets for attack). Defining and organizing are essential only in "secondary" societal groups that often develop in the face of the weakness of community.[3]

Biology was an unusually powerful ideal, at least during the period from 1890 to 1915, because it expressed the life of a community of scientists. Unlike most investigators, who experienced the "scientific community" largely as an abstraction that referred to the variety of contacts they made in schools, journals, and professional meetings, biologists structured their professional lives around one place—the Marine Biological Laboratory (MBL), in the village of Woods Hole, Massachusetts. At Woods Hole, experimentation and scientific discussion were part of a larger long-term framework of friendships and families. This unique environment crucially shaped the concept of biology.

Woods Hole, however, was not the "traditional" village of social theory, based on permanent residency, functional diversity, and subsistence. The American village, as well as its academic analogue, the college, were under great pressure in the late nineteenth century as a result of the development of a national economy and a national network of professional specialization. Woods Hole was an example of a peculiarly modern form of American community—that of the genteel summer resort. Like villages in the Berkshires and on eastern Long Island, it was rapidly becoming a place limited to the upper and upper-middle classes; at the same time, however, it was potentially open to national participation. The seasonal rhythm of recreation was replacing the routines of earning a living. Furthermore—and this was a matter of considerable importance for life scientists—it was close to nature, yet sheltered in such a way that it was accessible to children, women, and men not interested in enduring a strenuous life. The plumbing supply magnate Charles R. Crane, a leading summer resident and the major patron of the Woods Hole scientific colony, captured the nature of the community when he noted to John D.

Rockefeller, Jr., that the MBL was "more than a Laboratory and had many of the elements of a Biologist's Club." As a resort club, the MBL was extremely effective in structuring the experience that defined biology, and also in creating and maintaining an American biological elite.[4]

Beginning with Frank R. Lillie's comprehensive but decorous history, a number of writers have explored the origins of the MBL. They have focused on its relations to previous marine laboratories (most notably the Naples Zoological Station), on late nineteenth-century interest in the study of marine invertebrates, and on the emergence of an American scientific profession devoted to experimental research.[5] This paper, by contrast, examines the development of Woods Hole from the 1880s to the 1920s as a scientific resort. It considers the origins of scientific activity at Woods Hole, describes the nature of biologists' experiences there, evaluates the village's significance for science, and argues that the centrality of Woods Hole diminished substantially after 1915 as biology became a more formal entity. The basic terms of analysis are those central to consideration of genteel American resorts: community, relaxation, and nature. My central claims are, on the one hand, that the Woods Hole experience generated and maintained a "core discipline" of biology and, on the other, that the nature of American biology has been substantially influenced by its background as a "resort science."

Genteel Resorts

Most nineteenth-century American cities were particularly unpleasant places to live during the summer months. The heat and humidity were notoriously oppressive, especially in contrast to northern Europe. Garbage and manure produced overpowering smells, and flies and mosquitos were unavoidable. Residents feared malaria and other epidemic diseases. A further problem during the summer, from the viewpoint of the well-to-do, was the increased visibility of the immigrant poor, who thronged streets and parks in efforts to find alternatives to poorly ventilated tenements.[6]

Most economically comfortable urban American families left the cities during July and August. In the early part of the century they would, like the young Henry Adams, travel a few miles to live with nonurban relatives, or they would board informally with nearby farmers. After the Civil War, however, many closer areas became urbanized, and the lower classes began to frequent many of the most convenient spots, such as Coney Island and Atlantic City. New places, further away but otherwise more desirable, became geographically specialized as genteel resorts—places where wealthier families could return annually for extended summer stays. Most were in economically depressed areas, including such mountain regions as the Berkshires and Catskills, or various shore points on eastern Long Island, Cape Ann, or Cape

Cod. Vacationers could rent rooms or houses for low rates, and in some places they gradually began to buy the property of local farmers and build summer homes.[7]

While genteel resort areas developed primarily as the result of climate and accessibility, a number of other elements were also of considerable importance. The most obvious of these was closeness to nature. Resort hotels and homes opened out onto landscape and seascape and were surrounded by trees, flowers, and wildlife. The emphasis was on the picturesque and exotic. Summer residents who bought subsistence farms gradually reestablished forested panoramas. They planted unusual flowers and shrubs and admired the presence of nondomestic animals—birds especially, but also small mammals in upland areas and marine fauna on the coast. Yet with the possible exception of the "camps" built in the Adirondacks, nature was never overwhelming at eastern resorts. In comparison to the raw power, monotony, and vermin of the western frontier, the experience of nature in these places was comfortable and easy, suitable for the elderly, women, and children; they offered an enlarged, Americanized version of the ideal English garden. Although limited in grandeur, resort nature made possible an experience of the living world that was intimate and sometimes intense.[8]

A second aspect of resorts was their relaxed, playful atmosphere. Vacationers were away, for the most part, from work and school; they could experience the pleasures of living in unhurried, unhierarchical surroundings. Moreover, because most resorts were segregated by class and ethnicity, residents were free from the social tensions that formed a constant element of life in the cities. Lastly, social conventions, particularly regarding gender roles, were simpler in such places set temporarily apart; women could walk freely without fearing derogatory comments from teamsters, and men could cultivate roses without losing status as leaders in business competition.[9]

Resorts were also important as places where the upper classes could experience community of a sort that approached the ideals of contemporary social theorists. Genteel resorts were centered around old, often quaint villages such as Stockbridge, Marblehead, or East Hampton. Residents constructed large houses on small lots in order to maintain a walking environment. These villages usually had two sharply defined social strata—the wealthy who worked elsewhere, and a nonmobile lower class that combined service roles with some traditional employment, such as fishing. Yet because the lower-class residents were limited in numbers, were dependent on their wealthier seasonal neighbors, and were often old-stock Americans, the summer people usually perceived the community as socially and economically homogeneous in comparison to the cities, crowded with unassimilated immigrants. With physical closeness and time on their hands, summer residents were able to develop personal ties that were often difficult in the cities. Because vacationers were independent of the local economic hierarchies, they perceived community primarily as the result of voluntary activities. Furthermore, at resorts it was possible to establish close relations on a regional, if not national, scale, transcending the limitations of municipal social networks; wealthy residents of smaller cities in particular could go to resorts to establish social ties with metropolitan elites.[10]

The basic problem facing genteel resorts was that of development. Railroad access was fundamental, as was a certain amount of land subdivision and residential construction. Development also often included restoration of picturesque historic structures, removal of industrial eyesores, and return of land to a more natural state. But once begun, development was very difficult to control. Market pressures were such that a resort could easily be "overrun" by Jews, Irish, or other "alien" elements and perhaps even degenerate into a Coney Island.[11]

The Problems of Biology

Biology began as an urban science. It developed between 1870 and 1895 primarily at major metropolitan institutions such as the Johns Hopkins University, Columbia University, the University of Chicago, and the University of Pennsylvania. Academic administrators promoted it as a form of cooperation between zoologists and practitioners of the basic medical sciences of anatomy and physiology, in order to broaden the scientific culture of their premedical students. The geographic center of the new discipline was the biological laboratory of the Johns Hopkins University, completed in 1884. Designed primarily for the practice of physiology, it combined the features of a city schoolhouse with those of a slaughterhouse. Ethan Allen Andrews, a student and later professor, was struck by the fact that the laboratory seemed "even more in the midst of the city than the other buildings of the University." The unadorned main entrance faced a major street, and a small side door served for vendors of dogs and other experimental animals. Chutes for sanitary disposal of sacrificed material ran from the workrooms to a basement incinerator, and the building's rear overlooked an alley where, the racist Andrews noted, "nothing white lived . . . except a freak sparrow with white feathers."[12]

The seasonal problems that faced wealthier urbanites, including academics, were doubly pressing for biologists. The heat and insects that made city life unpleasant during the summer months rendered work with organisms—living or dead—unbearable. Biologists were also inhibited by the artificiality of cities. They had little opportunity to study animals in their natural surroundings, and access to material for laboratory studies was limited largely to a few insects and domestic animals. These difficulties could theoretically be counteracted by field trips each summer, but such a routine raised its own problems regarding both work and social roles. While most academics relaxed with their families each summer, biologists such as Hopkins professor William Keith Brooks and his students sweltered at various isolated sites on the Chesapeake Bay and then on Pamlico Sound in North Carolina. The "Chesapeake Zoological Laboratory" was a deceptively concrete name for Brooks's annual struggle, beginning in 1878, to negotiate with suspicious locals for a place to set up simple equipment brought from Baltimore, and then to find material, complete research, and provide some guidance to graduate students. If material was scarce or students undisciplined, a summer field trip could amount to nothing more than a "picnic."[13] Yet in their interest to gain significant results, Brooks and his students gradually moved further afield. By 1886 they sailed a charter schooner from Baltimore to Grand Turtle Cay in the Bahamas. The real difficulty was that such work was limited largely to energetic young men. Brooks himself seldom joined in field trips after he reached the age of forty-five; older and less aggressive scientists could be more productive working as best they could, in a laboratory or a study, on the routine of preserving, staining, drawing, and writing that was necessary for transforming biological material into scientific text.[14]

The division of labor between the field and the naturalist's "closet," combined with the increasing emphasis on painstaking microscopic observation, put urban-dwelling life scientists in the forefront of one of the major identity problems among well-to-do American men in the late nineteenth century—the crisis of masculinity.[15] Personality characteristics that led to success in biology, including patience, delicacy, and artistic ability, were increasingly being perceived as effeminate. At least one group of the Harvard zoologist Louis Agassiz's male students was described as "schoolma'amy" (i.e., sexually indeterminate). Both the "shy and gentle" Brooks and his Harvard counterpart, Edward L. Mark, had difficulties at their universities (though not within their discipline) because they were not sufficiently aggressive. The Harvard botanist William Gilson Farlow, as described in his National Academy of Sciences obituary, was a stereotype of waspish delicacy.[16] Theodore Roosevelt, obsessed with his and the nation's masculinity, gave up his ambition to become a life scientist after study at Harvard in the early 1880s because he perceived that biologists preferred fussiness to immersion in the grander forms of nature.[17]

This "problem" was reinforced as women took on increasing amounts of the routine work in taxonomy, and as female school teachers sought to learn about nature study. By the late 1880s, as Margaret Rossiter has noted, natural history was increasingly perceived to be "women's work." The issue was of such importance that *Science* published an article combating the apparently widespread opinion that botany, for example, was "not a manly study"—that it was "suitable enough for young ladies and effeminate youths, but not adapted for able-bodied young men who wish to make the best use of their powers."[18]

Biologists' anxieties related not only to the pressures of work and to gender roles, but also to their disciplinary fragmentation. Concern about lack of order was widespread in late nineteenth-century America, but biologists were in a particularly uncertain position. Botany, zoology, and anatomy were relatively firmly established disciplines in the United States; physiology, while less developed, had strong models in both England and Germany. The claim of biologists that these and other sciences could be united meaningfully was a tenuous one. Self-defined biologists were limited to a few universities, and although most of these were elite institutions, there was no guarantee that the rubric would spread. Biology had relatively little input from or impact on botanists in particular, who, by the early 1890s, were actively resisting the "sham biology" that ignored what they considered half the content of its science. Biologists also had minimal contact with amateur naturalists or with scientists at the new agricultural experiment stations; although these groups included few research leaders, they could be influential in the long run, especially regarding the growing state universities.[19]

The Society of Naturalists of the Eastern United States, soon renamed the American Society of Naturalists (ASN), was created in 1883 to unify and organize the Eastern biological elite. It successfully resolved the question of status by limiting membership to those actively participating in academic research; it was thus able to exclude most women and most federal employees. The society was much less satisfactory, however, as a way to unify the different kinds of university-based life sciences. Its only activity was an annual convention. In order to fill the meeting with subjects of general interest, society leaders focused on problems of teaching and technique; but they were unable to expand their program to include research reports. These were soon being presented at "affiliated" societies organized around more specific research interests and disciplinary identities. By 1890 the ASN was an umbrella organization loosely linking a number of specialist societies. Although biology shared in the prestige of European elite science and was potentially important in undergraduate and high school curricula, there was no assurance that it would mean anything more than the sum of the life sciences.[20]

The Beginning of Biology at Woods Hole

Woods Hole did not start as a radical break in scientific practice. Zoologists had been studying marine organisms from the beginning of the nineteenth century. The sea fauna represented a new frontier for taxonomists who had explored most European terrestrial forms, and it took on increased intellectual significance as evolutionism became respectable after 1860. Marine organisms could provide valuable insights into the evolutionarily crucial early stages of embryological development. The idea that permanent observatories were important for marine zoology was well established by the 1880s, most notably through Anton Dohrn's creation and promotion of an international research center, the Naples Zoological Station. Yet these intellectual and organizational developments did not determine the particular form and significance of Woods Hole. For that we need to understand how it developed as a biological resort, oriented, like other American resorts, around community, relaxation, and nature.[21]

The first American institution for marine zoology was established in 1873 by Louis Agassiz on Penikese, a small and desolate island in Buzzards Bay fifteen miles south of New Bedford, Massachusetts. This summer school for teachers closed after two seasons, partly because of Agassiz's death in December 1873, and partly because his son Alexander, who took over for the second year, was unwilling to spend precious summers on what was one of the most isolated spots in Massachusetts. A few years later, after securing a fortune from copper mining, the younger Agassiz opened a second marine laboratory, this time in a more congenial location—the grounds of his new summer home near the plutocratic resort town of Newport, Rhode Island. He shared this private and exclusive facility, however, only with friends and a few Harvard students.[22] In 1879 Alpheus Hyatt of the Boston Society of Natural History established a smaller, more informal copy of the Penikese summer school at the Cape Ann artists' colony of Annisquam. Supported by the Woman's Education Association, it catered largely to women teachers involved in natural history.[23] None of these seasonal "stations," however, was as significant for research and graduate training as the peripatetic Hopkins laboratory, which, by the mid 1880s, was moving to the tropics.

The village of Woods Hole, located at the southwest corner of Cape Cod in the town of Falmouth, boasted a pleasant climate, beautiful views, and abundant wildlife; it was also only two hours from Boston by train.[24] Yet it was an inappropriate location for a resort until it lost its unsavory industrial base, the Pacific Guano Works, in 1880.[25] A year later, Spencer Fullerton Baird, Smithsonian Secretary and U.S. Fish Commissioner, used his influence in congressional back rooms to expand the Woods Hole fish-culture station into a facility with a harbor, a research laboratory, and a summer residence for the fish commissioner and his family. To finance the laboratory, lessen the distance between government and academic scientists, and gain interesting neighbors, Baird invited universities to help purchase the land for the laboratory in return for the right to send workers there on a regular basis. Hopkins, Princeton, and Williams College invested from $500 to $1,000 in Baird's enterprise, as did Alexander Agassiz on behalf of Harvard's Museum of Comparative Zoology.[26]

The Marine Biological Laboratory originated in the desire of the Woman's Education Association to establish a more substantial and less segregated center than Annisquam for women seeking to study natural history. In 1887 the association successfully approached a number of Boston academic scientists who had been unable or unwilling to commit the funds necessary to join Baird's scheme.[27] Following a year of fundraising, this alliance raised $10,000 for a private institution at Woods Hole that would complement the male Fish Commission laboratory by providing elementary instruction and by being open to women. They erected an unfinished two-story building on a plot of land near the Fish Commission's much larger and more ornate facility, arranged to use the commission's water supply and collecting facilities, and persuaded Charles Otis Whitman to act as unpaid supervisor of a six-week summer course in elementary zoology and of facilities for scientific investigation.[28]

The plan to maintain the MBL as a private annex to the Fish Commission laboratory, based at least in part on the sexual division of labor well established in natural history, soon foundered. The Fish Commission declined as a center for federal science in the years following Baird's death in 1887. By 1890 it was the subject of a Senate investigation for favoritism, waste, and possible corruption. With such pressures, new biology departments—including those established at Clark University (1889), Columbia University (1891), and the University of Chicago (1892)—could not gain access to the government's laboratory; in fact, only Agassiz and Brooks had the power to retain space at its facilities for their students. Between 1889 and 1892, therefore, participants in a number of graduate biology programs became more interested in the MBL.[29] As a number of scholars have noted, the leading role in transforming the MBL into a national biological center was taken by Whitman, its first director.

Whitman was an extreme case of the difficult experiences that progressive Americans endured in the Gilded Age. Born in 1843, he grew up in the poor, hilly woodlands of western Maine. After attending an academy in Norway, Maine, and teaching for a number of years in local schools, he worked his way through nearby Bowdoin College. He left Maine on graduation in 1868 to teach in an academy a few miles west of Lowell, Massachusetts, and three years later he obtained a position at the Boston English High School. As Whitman gradually moved from the country to the city, he underwent a typical Victorian crisis of faith, largely as a result of his growing familiarity with English evolutionary positivists such as Herbert Spencer, Alexander Bain, and T. H. Huxley. He first abandoned his father's apocalyptic evangelicalism for a Universalist celebration of death as the door to another, better, world; by 1871 he embraced agnosticism and relativism, arguing that the idea of immortality required "scientific demonstration" and that evolutionary "progress" was a directionless process by which organisms ceaselessly sought to adapt to fluctuating environments. According to Whitman, "non-adaptation is the only evil."[30]

Whitman's reading of evolutionism steered him toward the study of zoology. Following attendance at Penikese in 1873 and 1874, he decided, at the age of thirty-one, to study with the German zoologist Rudolf Leuckart. This decision led to nearly two decades of wandering from one short-term job to another. In his pursuit of a career in the life sciences, Whitman worked on three continents, always in cities; these included Leipzig, Naples, Tokyo, Boston, Milwaukee, and Worcester. In 1892 he finally gained stability as head of the biology program at the new University of Chicago.

While ultimately very successful, Whitman had a difficult time adapting as an urban scientific professional. His German ideals and Ph.D. isolated him from the well-established networks of life scientists in the United States: he was unable to convince Spencer Baird that his research would be relevant to the concerns of the Fish Commission, and, with his frequent moves, he made few close friends among scientists of his generation.[31] The positions he did obtain were tense; for seven years following his return to America in 1882 he was a personal employee, first of Alexander Agassiz at Harvard and then of the wealthy young amateur zoologist, E. P. Allis, Jr., in Milwaukee. Also, Whitman was separated from nature. Living in cities, usually in rented quarters, he was unable to keep and observe live animals; he was repeatedly in the anxious position of scheduling short field trips to obtain material, relying on unfamiliar local guides who would sometimes have fun at the expense of this city dweller looking for worthless animals. Whitman's willingness to take the unpaid position of MBL director was strongly influenced by his desire for interaction with colleagues, a relaxed research environment, and contact with live organisms.[32]

A Biological Resort: Community, Relaxation, Nature

By 1890, with the Fish Commission under serious attack, Whitman recognized that the potential of the MBL was much greater than its originators expected. His own commitment to the institution had been reinforced by his appointment a year earlier to lead the new biology program at Clark University in nearby Worcester. Whitman's attempt to outline his views on the nature and functions of the MBL smoothly linked his personal concerns with broader disciplinary aims. He complained that "isolation in work" was becoming "more and more unendurable," and in a speech that inaugurated the MBL's published series of *Biological Lectures* he explained that the laboratory could play a crucial role as a biological community.[33]

Whitman's analysis (entitled "Specialization and Organization, Companion Principles of All Progress—The Most Important Need of American Biology") began with the assumption that "a society is an organism" and "an organism is a society."[34] He outlined the process of evolution from the free but lowly "nomadic" protozoa to cells that "herded together" to form colonies, and thence from the "Hydra community of cells" to the complex division of labor in the higher organisms. According to Whitman, the evolution of science followed the same sequence. "Cosmogonists" who "engaged single-handed with all the mysteries of the universe" had given way to the present, seemingly rapid, fragmentation of biological science into specialties. While sanguine that no science was in danger of "flying into disconnected atoms," he wished to counteract the "danger of narrowness that may lurk in the tendency to specialize." Whitman sought "that kind of organic association which permits each unit to work for itself while making it the servant of all the rest." Although journals (and presumably groups such as the ASN) were essential means of communication, "the unity of action in so extended a body cannot be complete." What was necessary was a way to bring together the leading American naturalists and place them in "intimate helpful relations." A national marine biological station, as the best means to accomplish this goal, was "unquestionably *the* great desideratum of American biology." The result would be to accelerate "that '*moving equilibrium*' of our specialized forces which constitutes progressive scientific life"; it was a form of progress that, for Whitman as for Herbert Spencer, extended beyond mere adaptation toward perfection.[35]

The MBL in its early years embodied the characteristics Whitman called for in this important essay. He had particularly emphasized that the effectiveness of a "social or an organic body" resulted from having its "points of union multiply," and Woods Hole in its early years was notable for the density of its community network.[36] Professors and graduate students worked in unfinished rooms on one floor of the MBL laboratory, where they could share interesting results quickly and discuss general issues on a continuing basis. The weekly biological lectures provided a sequence of specific foci for discussion.[37] The younger male scientists became close friends as they swam naked off the old wharf of the guano works; both men and women joined in periodic collecting ventures, sailing excursions, and picnics. The more established professors soon began to rent houses and bring their families; the presence of wives and children added a further dimension to each summer's social interaction. Perhaps the most significant aspect of community life was the creation of a "dining club," soon nicknamed the "Mess" by male workers, where people could meet and talk informally. The dining club was particularly important because it was also open to the men from Harvard and Hopkins who continued to work at the Fish Commission; it thus provided a means to minimize the institutional gap between the two laboratories.[38]

At the same time that MBL scientists were generating a network of personal interactions, they worked in a more formal fashion to develop their village into Whitman's "strong common center" for American biology. The number of investigators grew rapidly from 1891 to 1896, as representatives of the liberal arts colleges and other institutions joined the nucleus of participants affiliated with the large urban biology programs at Clark, Columbia, and the University of Chicago (see Figure 4.1). Educational programs also expanded to encompass both elementary and graduate study and to include not only marine zoology and embryology but also botany and physiology. Whitman united these groups through appeals to "biology" and to the rhetoric of the cooperative movement, a strong force in America in the depression years of the mid 1890s. In 1894, researchers and their families and friends formed the Biological Association to lobby and raise funds for the laboratory. Whitman justified these actions on the grounds that "growth and development must be regulated by the animating principle from within, not by arbitrary dicta of extra-ovate concoction."[39]

Whitman's expansionism soon resulted in conflict between the Biological Association and the "extra-ovate" Bostonians who had created the MBL and who continued to control its finances. The Woman's Education Association and its professorial allies wanted a small-scale operation that could be visited for short periods of work and quiet study; the ambience of Whitman's summer science center was much more vulgarly professional than they had envisioned. A series of disagreements about its speed of development culminated in 1897 in a few weeks of intense political maneuvering. The balance of power shifted when a number of the laboratory's trustees living in Boston were forced to leave Woods Hole in the middle of a meeting in order to catch the last train back to the city on a Sunday evening; after they had departed, a rump assembly dominated by the Biological Association's summer residents called for a special meeting to change the bylaws. They reorganized the laboratory into a cooperative association in which legal power was vested in the MBL Corporation, composed of the active members of the community. The Bostonians—whether women, amateurs, or weekenders—were pushed to the margins.[40]

While Whitman's major interest was to create a biological community at Woods Hole, he and other participants were also concerned to insure that the MBL would be a place where scientists could relax and engage in free scientific play. Whitman had earlier opposed the plans of E. P. Allis, Jr., to organize his private "Lake Laboratory" in Milwaukee on the authoritarian principles of "scientific management"; at some risk to his position, Whitman persuaded Allis that every employee in that laboratory should be given the time and opportunity to do some research on his own.[41] At Woods Hole, this freedom for "rest and investigation" was insured, because the MBL was separated, both geographically and organizationally, from the routines of university teaching and the formalities of academic administration. Although the MBL sponsored elementary classes, it did not link them to any formal degree program nor did it provide more than minimal pay to its teachers. Formal hierarchy at the laboratory was kept to a minimum; when the Carnegie Institution of Washington proposed to absorb the MBL in 1902, Whitman's major argument was that becoming a "department" with a salaried staff would create distinctions among researchers that would destroy individual independence.

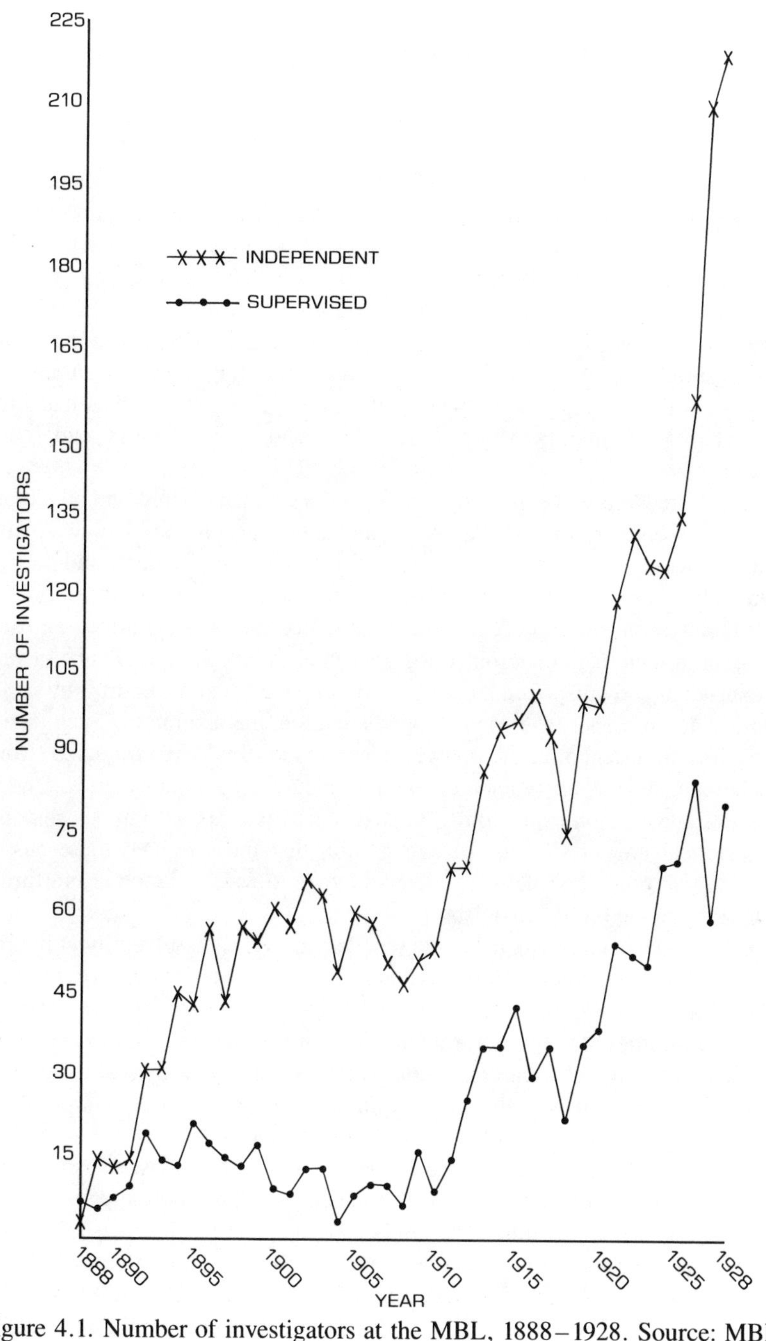

Figure 4.1. Number of investigators at the MBL, 1888–1928. Source: MBL *Annual Report*.

Biologists, as participants in a voluntary endeavor with minimal duties, were to be able to spend their summers as they wished, pursuing individual research wherever it led.[42]

Woods Hole biologists could relax not only because they were free from the demands of work but also because they were supported in their activities and protected from outsiders. The laboratory's facilities enabled scientists to study living organisms without having to "rough it"; biologists would no longer be restricted by age or temperament to the naturalist's closet. Within the substantial Woods Hole summer colony, scientists could engage in their delicate and sometimes peculiar activities without anxiety. The lower-class "natives" of Woods Hole might make disparaging comments among themselves about those "little fellows" who came from the city every summer to engage in the unmanly activity of "bug hunting," but such jibes had little impact in a community where it was clear who held economic and social power.[43]

Such freedom was especially important for women. Falmouth had a reputation as a proper resort for unaccompanied ladies; in contrast to more isolated areas, it was a place where female scientists could walk, collect, and participate in biology without danger to their reputations. They were subject only to community control. E. G. Conklin, for instance, jovially recounted an early incident in which the men joined together to humiliate Julia L. Platt, "a very independent and outspoken person . . . of ample girth and weight," by dunking her in the harbor; she soon stopped attending the laboratory. Such tactics were less important after the first years; as a resort, the MBL embodied strong social inducements for women to take up "natural" community roles. Underlying Conklin's notorious comment that the MBL was a "famous institute of practical eugenics" was the recognition that most women there passed smoothly through the roles of student and lover to wife and mother, nurturing the next generation of biologists.[44]

Woods Hole scientists talked a great deal among themselves about the importance of social contact and individual freedom. Yet the overt purpose of the MBL was not social but natural; biologists justified their summers in terms of the opportunities they found for interaction with fauna. As a biological resort, Woods Hole offered a carefully constructed setting in which to experience living organisms. It was also an environment within which certain problems and concepts gained particular communal importance.

The beginner, arriving in Woods Hole by train or boat, saw a small village with remnants of its former primitive industries, along with a few estates in the recently reforested hills. The rambling headquarters of the Fish Commission dominated the harbor, and an early stop for new arrivals would be its public aquarium, "stocked with the most interesting and attractive animals of the region." Until 1913 the MBL was located in plain frame buildings nearby; each student would be assigned a table and would have access to supplies, organisms stored in aquaria, and salt and fresh water. The place of work thus contrasted with both the substantial construction of universities (especially the increasingly closed and artificial environments of the anatomy laboratories and hospitals of the leading medical schools) and the discomfort and uncertainty of sites for fieldwork. The MBL was the marine faunal equivalent of a hothouse, a structure designed specifically to recreate environments that would sustain certain organisms during the period of study.[45]

For most students, the Woods Hole organisms were exotic in comparison with previous experiences centered around insects and vertebrates. Such animals as sea urchins, horseshoe crabs, and limpets possessed unusual anatomical and physiological qualities, and thus rapidly broadened a student's conception of the nature of life. Because marine organisms were both more durable and less personable than higher animals, they were easier to manipulate while alive; they were, figuratively and sometimes literally, transparent to the sufficiently careful observer.

The MBL program made this new world particularly easy to study. Students became acquainted with the fauna through the elementary course in invertebrate zoology. They did not need to worry about locating material, because it was preselected and generally precollected. They could draw on years of accumulated experience in managing and culturing organisms. Through a process of gradual assimilation, a student could develop from novice to "embryo investigator" to mature researcher.[46]

The result of this combination of rustic yet comfortable facilities, compliant organisms, and a fund of local traditions was that biologists could potentially accomplish more each year at Woods Hole than at nearly any other research location. Life there had neither the danger nor the sublimity of the wilderness, but it opened up a new intensity of experience with nature. In the retrospective assessment of George Howard Parker of Harvard University, who was affiliated with the Fish Commission and the MBL for nearly fifty years, the American marine laboratories' chief importance was in providing the opportunity for intimate acquaintance with "real animals"; such genuine, if narrow, experience was the greatest desire of urban biologists at the end of the nineteenth century.[47]

What about science itself? Were the problems and concepts of the MBL scientists shaped by their resort environment? Such a connection seems likely, given the degree to which social and biological concepts merged in the thinking of both Whitman and Frank Lillie, his successor as MBL director.[48] On the organizational level, Woods Hole leaders were, like other resort entrepreneurs, centrally concerned with development—how, in a place where power was fragmented and interests divergent, change could be guided toward an "ideal" that was only dimly envisioned. Those interests reinforced and shaped—if they did not initiate—thinking about biology. The MBL community's overriding scientific interest in the 1890s was to understand how the different parts of small, isolated cell communities (called marine embryos) functioned in cooperative fashion. More specifically, they sought to determine how such cell communities maintained their organic unity as they matured, that is, as the community became more complex and the activities of individual cells more specialized. The consensus was that while external, "extra-ovate" forces were important modifiers of development, the cell community (like the biological community) was fundamentally autonomous and self-directed. To exaggerate only slightly, scientists at Woods Hole sought to provide an account of life at a resort.[49]

Woods Hole and American Biology

Scientists associated with the Woods Hole summer colony produced research of major significance. The cell lineage studies of the 1890s formed one of the first instances of an independent and conceptually fruitful tradition of American laboratory science. Interest in experimental embryology led by the end of the century to Jacques Loeb's spectacular invention of artificial parthenogenesis, and, in the years prior to World War I, to Frank Lillie's creation of the foundations for modern understanding of sexual reproduction. Cytologists such as E. B. Wilson and Nettie Stevens linked chromosomal mechanics to the phenomena of Mendelian heredity; T. H. Morgan's interactions with these colleagues formed the background for his work on *Drosophila* genetics, which earned him a Nobel Prize. Biologists associated with Woods Hole were the first group of American laboratory scientists to become international research leaders, both experimentally and theoretically; this new status became evident after 1900 as the flow of American biology students to Germany ended, and Europeans began to come to the United States for postgraduate study.[50]

The one significant limitation of Woods Hole research was its academic nature. The MBL was originally set off from the Fish Commission Laboratory, as well as from the recently established agricultural experiment stations, by its emphasis on pure science. This separation increased during the 1890s as the Fish Commission focused more exclusively on applied work and as the universities expanded. After the turn of the century, however, the academic nature of biology was largely a function of the structure provided by Woods Hole. Medical scientists and government zoologists were increasingly involved with such problems as tenement ventilation, mental hygiene, nutrition, eugenics, and insect-borne diseases. Biologists, who left the cities each summer for the seacoast, were detached from these urban concerns. At Woods Hole the organisms, the desire to relax, and the community all reinforced an academic search into the mysteries of unspoiled nature.[51]

Woods Hole's importance extended beyond the research performed there.

Its greatest impact was as the center of the unified science of biology. Yet its role in defining that discipline was unique. Chemists, psychologists, and other scientists were isolated most of their lives in their university departments, government bureaus, or corporate laboratories. They interacted with colleagues through reading, correspondence, and the yearly meetings of their professional societies, where formal papers alternated with the momentarily intense interactions of conventioneers. Such rhythms pushed younger scientists in most disciplines to define their work carefully so that it would stand out, and induced each field's leaders, in reviewing membership lists and preparing programs, to make explicit decisions about definitions and boundaries. Admission to societies and programmatic statements became major issues for these disciplined academics.[52]

Biology, centered around the annual gathering at Woods Hole, was different. Apart from their focus on pure research, biologists were extremely catholic in their enthusiasms. They discussed a wide variety of subjects, all of which they considered within the scope of their science. These included heredity, evolution, cell cleavage, neuroanatomy, and behavior; in most cases, moreover, biologists ranged freely, working with organisms from protozoa to primates. As Jane Maienschein has emphasized, biologists also gladly accepted multiple causation in each of these areas; their only strong antagonism was toward explanations that seemed to reduce a wide range of phenomena to a single unifying factor. As a result of their eclecticism, American biologists were not interested in providing a comprehensive definition of the whole of their science. The only article to deal with this subject in the 1890s was a popular essay by the anatomist F. P. Mall. When H. S. Jennings tried to establish an explicit theoretical basis for biology in his *Behavior of the Lower Organisms* (1906) there was a nearly total lack of response.[53]

Biology was difficult, if not impossible, to define. Fortunately, however, definition was not necessary. Individuals who had established positions within the community centered around Woods Hole, and who had demonstrated that they had good minds and did interesting research, were, barring strong grounds to the contrary, biologists. Biology was the totality of subjects that Woods Hole scientists discussed. Whitman's inaugural lecture asserted the evolutionary conviction that organic social organization would eventually result in intellectual unification. Until that progressive development was completed, biology would be defined in terms of its community.[54]

Woods Hole importance was thus ultimately more social than intellectual. It formed the center for a cohesive academic elite, through which biology rapidly became established as the core of the life sciences in America. Between 1892 and 1910 an average of sixty-three investigators worked each summer at the MBL, and sometimes up to thirty more at the Fish Commission.[55] In addition, an average of 134 students attended the MBL's introductory course. While investigators came from an average of more than twenty different institutions, the active local leadership—embodied in the MBL's Board of Trustees—was in the hands of professors from the University of Chicago and Columbia, the two leading metropolitan universities; by about 1905 Harvard, Hopkins, Princeton, and Pennsylvania were also playing major roles.

A number of measures indicate the extent to which the Woods Hole elite dominated the life sciences. The fifty highest-ranking zoologists in the first edition of *American Men of Science* (tabulated in 1903) included twenty-three who served as MBL trustees; if individuals closely associated with the Fish Commission and those who worked for a substantial number of years at the MBL (but left before being elected trustee) are added, that figure rises to thirty-three. The American Society of Naturalists, the nation's closest approximation to a biological society, had twenty-four presidents between 1891 and 1915; seventy-one percent of these were trustees of the MBL. All except one president of the American Morphological Society (1890–1902) were MBL trustees, as were six of the first seven presidents of the amalgamated American Society of Zoologists (1914–1920). Leaders of the MBL edited the *Biological Bulletin, Journal of Morphology, Journal of Experimental Zoology, Journal of General Physiology,* and other major publications; the Columbia psychologist James McKeen Cattell, who edited *Science, Popular Science Monthly,* and *American Naturalist,* was a long-time resident of the village.[56]

The wealthy businessman Crane understood the significance of describing the MBL as a "biologist's club." It was a place where a select group of people became friends, talked about their long-range plans, hammered out difficult disputes, and established a consensus to present forcibly to the outside world; even those who were not there each year could trust their old swimming companions to make decisions maturely and with due consideration of all the views that mattered. Because biology was centered around a club, formal organization was unnecessary. As Tobey Appel notes, the ASN became an anomalous professional society by the turn of the century, and the "American Biological Society" proposed in 1906 received no significant support. The leaders of biology worked out their field's problems informally during the summers.[57]

The Problem of Development

The "golden age" of American biology lasted from the early 1890s to the mid-1910s. Biologists' lives, centered around Woods Hole, were interesting, pleasant, and fulfilling. Yet biologists were frustrated in their attempts to resolve the problem of development—both with regard to marine invertebrates and, more immediately, with regard to the discipline of biology. The number of practitioners in the life sciences grew, and new specialties developed rapidly. Some people (who rarely defined themselves as biologists) were able to restructure farm and urban research settings in such a way that they could work productively and deliberately to generate basic knowledge. Biologists' dominance became tenuous, and with World War I they realized that they were no longer adapted to their surroundings. Attempts to maintain the MBL's unique position were unsuccessful, and by the 1920s biology was becoming formalized—both institutionally and intellectually—as a more typical disciplinary unit.

Whitman noted repeatedly during the first decade of the MBL that the institution needed to expand continuously in order to maintain its unique position in the life sciences. By the beginning of the 1900s, however, as both the laboratories and local housing reached capacity, residents found themselves unable to reconcile growth with their interest in maintaining the comfortable resort ambience of Woods Hole. Scientists, who occupied nearly all the available seasonal rooms in the village, were gradually buying up the small number of houses and displacing local residents to less convenient areas. Village property was expensive, and all the nearby open land was closely held. Expansion would require money, and thus outside interests were introduced to replace the Bostonians who had been expelled in 1897; any major change was bound to upset a significant part of the community.[58]

In 1901 and 1902 two plans to attract funds were developed. The first involved unofficial absorption of the laboratory by the University of Chicago, and the second, promoted vigorously by the Columbia professors E. B. Wilson and H. F. Osborn, was a formal takeover by the new Carnegie Institution of Washington. Each plan collapsed in the face of opposition from an important minority of scientists; no one was sufficiently committed to expansion to override opposition, because doing so would destroy the neighborliness that had become so important over the last decade. As a result, the laboratory's level of funding and activity stayed constant, and its direction remained indefinite, until the end of the decade.[59]

The static situation of the Woods Hole biologists became increasingly anomalous in the face of the continuing growth of the life sciences. Despite its declining significance, membership in the ASN doubled between 1894 and 1918. Specialties such as physiology, neurology, and psychology grew even more substantially, and younger individuals began to occupy such new fields as biochemistry, biometry, and, above all, genetics. By the 1910s at least two contexts—agriculture and medicine—supported new ways to interact with nature. These areas also carried with them the important "masculine" appeal to utility.

Botanists and breeders were able, after 1905, to draw on funds from state universities, federal agriculture programs, and the Carnegie Institution to produce basic research in an agricultural context. Ecology, as developed by Frederic Clements with the aid of the automobile, transcended the sharp botanical dichotomy between field and herbarium and displayed the broad cultural relevance of botanical knowledge within a defeminized context. Agricultural breeders shed their image of amateur rural vulgarity as they allied themselves with such urban academics as Columbia's *Drosophila* group to form a new focus for research that was independent of the balance of interests "traditional" in biology.[60]

Medical scientists realized similar opportunities during the 1910s. The promise of a system of laboratory-based medical research was gradually realized as the Rockefeller family began to devote a major portion of its philanthropy to this area. Bacteriologists, anatomists, and biochemists—at Harvard, Johns Hopkins, and especially the new Rockefeller Institute for Medical Research, located in Manhattan—were able to build major research programs within the new air-conditioned urban laboratories. This highly refined and delicate work took on heroic proportions within the rhetoric of the "war" against disease; that manly image was widespread in the popular press long before Paul De Kruif utilized it in the 1920s as his motif in the best-selling *Microbe Hunters*.[61]

Biologists' confidence in the unity and significance of their science was shaken with the outbreak of World War I. Physicists, chemists, and even psychologists were able to push themselves forward in the war effort through programs of applied research. Biologists seemed superfluous, largely because utilitarian aspects of the life sciences were already well represented in various federal and state agencies. Raymond Pearl complained that government officials consulted with everyone except biologists. Conklin, the initial head of the National Research Council (NRC) Committee on Zoology, apologized that his group represented "the residue of zoological science after [the] more practical subjects have been taken away." His group could point to few achievements beyond formulating short courses for the Students' Army Training Corps, which apparently were seldom implemented. In early 1918 the committee was collapsed into a much broader division that included agriculture, botany, forestry, and fisheries. Furthermore, because involvement with military "preparedness" carried with it the full weight of traditional masculine values, biologists' lack of participation led them once again to appear effeminate. Conklin acknowledged this as early as 1915 when he took a strong stand in a symposium on preparedness against the view that "academic biology should be classed with embroidery."[62]

While the war was the event that brought home the difficult cultural position of biology, Woods Hole scientists had been aware for some time that their enterprise was losing ground. In 1908 the trustees regretted that a number of "lines of work" were no longer centered in Woods Hole and expressed hope that such research would be reintroduced and thus reintegrated into biology.[63] On becoming director that year, Frank Lillie led a major effort to provide facilities for increased numbers of participants, and thus to reestablish the laboratory's preeminence. After three years of committee meetings, a brick laboratory, described in the local newspaper as "the pretentious new building" of the MBL, was built with money from C. R. Crane. Following the war,

Lillie led a much larger expansion effort, obtaining over $1.5 million from the Rockefeller Foundation, Carnegie Corporation, Crane, and John D. Rockefeller, Jr., to build new laboratories and to endow their operation.[64]

Throughout this expansion Lillie was anxious to maintain Woods Hole's identity as a family resort for biologists. He therefore involved the MBL in real estate development. In 1916 the laboratory purchased a twenty-one-acre tract and subdivided it into building lots; these were sold at low prices to the more established biologists and were kept under community control with a restrictive covenant giving the MBL veto over future transfers. As owners of homes that could only be sold within the scientific community, biologists now had a substantial personal stake in the maintenance of the MBL. A much larger tract was purchased for the same purpose in 1925; only when the pattern of home ownership was set did Lillie respond to pressure from the General Education Board to build an apartment house for younger and more transient scientists.[65]

Many biologists recalled pleasant summers spent in Woods Hole during the 1920s and after. Yet Lillie's efforts to maintain community through controlled development were only partially successful. Some senior scientists felt that after the expansion program Woods Hole was a less attractive place for scientific work.[66] Moreover, given its size and Lillie's dominant role, the laboratory remained a cooperative endeavor only in a formal sense. The scientific activities and social lives of most biologists became much less closely connected with the governance of the MBL. Crane implied more about the laboratory's organization than he meant when he referred to it in 1926 not as a club, but as "a form of Soviet": the trustees perpetuated themselves through repeated renomination and uncontested elections. A further result of the laboratory's real estate policy was to create an all-too-permanent community; people who came for research stayed to live in an exclusive resort. With a rapid increase in property values during the 1920s younger scientists could not easily purchase summer homes. Predominance in the community was no longer directly linked to one's standing in biology.[67]

The expansion of the MBL was, in any case, too little too late. During the war years, the center of efforts in biology shifted from Woods Hole to more formal venues, including the NRC, the Rockefeller Foundation, and the leading universities. In 1919 biologists reorganized the NRC Division of Agriculture, Botany, Forestry, Zoology, and Fisheries to form a new Division of Biology and Agriculture; because agriculture was already fully funded, the division's first priorities were "biological publication," "research fellowships for biologists," and the MBL. Biologists, especially Lillie, controlled such Rockefeller-funded organizations as the Committee for Research in Problems of Sex. Of greater significance, the research universities began to create biology departments and divisions for the first time since the 1890s. The California Institute of Technology established a division of biology in 1925, and Stanford followed suit in 1929. Raymond Pearl became head of a new Institute of Biological Research within the Johns Hopkins Medical School in 1925. At Harvard, where the life sciences had been unusually fragmented, a division of biology and a new biological laboratory were created between 1925 and 1929; the division became a unified department in 1934.[68]

The most fundamental change, however, was that biologists in the late 1910s and 1920s no longer relied on an intuitive, communal understanding of the meaning of their science: there were a number of important attempts to state explicitly *what* biology was. Men such as Lillie, W. M. Wheeler, W. E. Ritter, and L. J. Henderson joined H. S. Jennings's effort to set out the nature and boundaries of a unified, autonomous science of biology. They fixed on the concept of holism and, more explicitly, on the importance of understanding biological organization. What for Whitman had been a habitual, almost unconscious, mode of expression became, in the late 1920s, an explicit body of theory; the holists' central argument was that biology could be defined as the science of natural communities.[69]

These formal elements, while important, paled in comparison with the much richer, if less definite, experience of the early part of the century. That experience, more than the later formulations, accounts for the strength of the biological ideal within the American scientific world. In spite of the rapid development of numerous specialties throughout the century, scientists and administrators have believed in the unity of the science of biology and sought to maintain it. Just as the ideal of community has guided social policy through much of this century, the ideal of biology has guided science policy. The two ideals have long been intertwined; in the American context, they were both shaped substantially by life at Woods Hole.[70]

Acknowledgments

In the preparation of this paper I have benefited greatly from the comments of John Burnham, Mary Ann James, Sharon Kingsland, Gary McDonogh, and Michael Moffatt, as well as from discussion with the other participants in the Friday Harbor conference. I wish particularly to thank Donna Haraway, whose unpublished work introduced me to the history of the MBL. The MBL Archives and the Woods Hole Historical Collection provided important information.

Notes

1. Philip J. Pauly, "The Appearance of Academic Biology in Late Nineteenth-Century America," *Journal of the History of Biology,* 1984, *17:* 369–397.

2. Ibid.; Appel, in this volume.

3. The classic work on the concept of community is Ferdinand Tönnies, *Gemeinschaft und Gesellschaft* (Leipzig: Rues.'s Verlag, 1887). In the American context, see Charles H. Cooley, *Social Organization* (New York: Scribners, 1909); John Dewey, *The Public and its Problems* (New York: Henry Holt, 1927); and, more recently, Conrad M. Arensberg and Solon T. Kimball, *Culture and Community* (New York: Harcourt, Brace, and World, 1965); and Colin Bell and Howard Newby, *Community Studies* (New York: Praeger, 1972). An introduction to the history of the concept can be found in Thomas Bender, *Community and Social Change in America* (New Brunswick, N.J.: Rutgers University Press, 1978).

4. C. R. Crane to J. D. Rockefeller, Jr., 22 December 1923, quoted in Frank R. Lillie, *The Woods Hole Marine Biological Laboratory* (Chicago: University of Chicago Press, 1944), p. 73. For important insights on sociable groupings in science, see Susan F. Cannon, *Science in Culture* (New York: Science History Publications, 1978), pp. 161–162. Howard Mumford Jones, *The Age of Energy: Varieties of American Experience 1865–1915* (New York: Viking Press, 1971), pp. xii, 306, 333–336, introduced the idea of studying the history of the life sciences in America in terms of the changing environments in which they were practiced. His argument, however—that the crucial transformation was from nature to the urban laboratory, from landscape to microscope—was only a first approximation.

5. Lillie, *MBL;* Donna J. Haraway, "The Marine Biological Laboratory of Woods Hole: An Ideology of Biological Expansion," unpublished paper, 1975; Jane Maienschein "Agassiz, Hyatt, Whitman, and the Birth of the Marine Biological Laboratory," *Biological Bulletin,* 1985, *168* (suppl.): 26–34; Jane Maienschein, ed., *Defining Biology: Lectures from the 1890s* (Cambridge, Mass.: Harvard University Press, 1986), pp. 3–20; Keith R. Benson, "Laboratories on the New England Shore: The 'Somewhat Different Direction' of American Marine Biology," *New England Quarterly,* 1988 (in press). I became aware of Angela Metropulos O'Rand, "Knowledge, Form and Scientific Community: Early Experimental Biology and the MBL," in Gernot Böhme and Nico Stehr, eds., *The Knowledge Society* (Dordrecht: D. Reidel, 1986), pp. 183–202, too late for consideration here.

6. On conditions in American cities in the nineteenth century, see Jacob Riis's classic *How the Other Half Lives* (New York: Charles Scribner's Sons, 1890); a graphic but unscholarly survey is Otto Bettmann, *The Good Old Days: They Were Terrible!* (New York: Random House, 1974).

7. Henry Adams, *The Education of Henry Adams: An Autobiography* (Boston: Houghton Mifflin, 1918), pp. 7–16. While most histories of American resorts are superficial or antiquarian, two excellent studies of mass resorts, Charles E. Funnell, *By the Beautiful Sea: The Rise and High Times of That Great American Resort, Atlantic City* (New York: Knopf, 1975), and John F. Kasson, *Amusing the Million: Coney Island at the Turn of the Century* (New York: Hill & Wang, 1978), indicate what can be done.

8. Barbara Novak, *Nature and Culture: American Landscape and Painting, 1825–1875* (New York: Oxford University Press, 1980), provides an introduction to American concepts of the natural landscape; see also the Architectural League of New York and the Gallery Association of New York State, *Resorts of the Catskills* (New York: St. Martin's Press, 1979).

9. Peter J. Schmitt, *Back to Nature: The Arcadian Myth in Urban America* (New York: Oxford University Press, 1969), pp. 3–32.

10. Cleveland Amory, *The Last Resorts* (New York: Harper, 1951). Art colonies such as Cornish, New Hampshire, have been the subject of some study: see Hugh Mason Wade, *A Brief History of Cornish 1763–1974* (Hanover, N.H.: University Press of New England, 1976), pp. 43–94.

11. Jason Epstein and Elizabeth Barlow, *East Hampton: History and Guide* (New York: Random House, 1985); John Higham, *Send These to Me: Jews and Other Immigrants in Urban America* (New York: Atheneum, 1975), pp. 138–173; Kasson, *Amusing the Million,* pp. 29–34.

12. "The New Biological Laboratory of the Johns Hopkins University," *Science,* 1884, *3:* 350–354; E. A. Andrews, "The Old Laboratory," in "Hopkins Biology News-Letter," June 1948, pp. 1–4, mimeograph, Records of the Department of Biology, Ferdinand Hamburger Jr. Archives, Johns Hopkins University, Baltimore; more generally see Larry Owens, "Pure and Sound Government: Laboratories, Gymnasia, and Playing Fields in Nineteenth-Century America," *Isis,* 1985, *76:* 182–194.

13. Benson, "Laboratories."

14. The summer expeditions continued until 1897, when a professor and a postdoctoral fellow died of yellow fever; purely academic trips were not considered worth such risks. See W. K. Brooks, "Notes from the Biological Laboratory," *Johns Hopkins University Circular,* 1897, *17:* 1–2; H. L. Clark to D. C. Gilman, 14 September 1897, Daniel Coit Gilman Papers, Eisenhower Library Department of Special Collections, Johns Hopkins University, Baltimore (hereafter, Gilman Papers).

15. See, e.g., Joe L. Dubbert, "Progressivism and the Masculinity Crisis," in Elizabeth Pleck and Joseph Pleck, eds., *The American Man* (Englewood Cliffs, N.J.: Prentice-Hall, 1980), pp. 303–320 (I am grateful to Michael Budd for this reference).

16. Jane Maienschein, "Birth of the MBL," p. 27; "Brooks, William Keith," *Dictionary of Scientific Biography,* vol. 2 (New York: Charles Scribner's Sons, 1970), pp. 501–502; Dorothy Ross, *G. Stanley Hall: The Psychologist as Prophet* (Chicago: University of Chicago Press, 1972), p. 137; William A. Setchell, "William Gilson Farlow," *National Academy of Sciences Biographical Memoirs,* 1926, *21* (4): 7–8. See also Thurman Wilkins, *Clarence King: A Biography* (New York: Macmillan, 1958), pp. 106–107, on Harvard's Sereno Watson.

17. Theodore Roosevelt, *The Works of Theodore Roosevelt,* vol. 22: *An Autobiography* (New York: Scribner's, 1925), p. 31.

18. Margaret W. Rossiter, *Women Scientists in America: Struggles and Strategies to 1940* (Baltimore: Johns Hopkins University Press, 1982), pp. 57–60; Emanuel D. Rudolph, "Women in Nineteenth-Century American Botany: A Generally Unrecognized Constituency," *American Journal of Botany,* 1982, *69:* 1346–1355; J. F. A. Adams, "Is Botany a Suitable Study for Young Men?" *Sci.,* 1887, *9:* 116–117. Prominent in his positive reply to this question was the argument that the adolescent male botanist had a chance to "penetrate the woods to their secret depths, scramble through the swamps, and climb the hills"; when he "leaves the world behind, and seeks, amid the solitudes of Nature, to penetrate her wondrous mysteries, he feels the quickening of a higher life." On such sexual imagery in nineteenth century boys' advice books, see G. J. Barker-Benfield, *The Horrors of the Half-Known Life* (New York: Harper and Row, 1975), esp. pp. 135–228.

19. Eugene Cittadino, "Ecology and the Professionalization of Botany in America, 1890–1905," *Studies in History of Biology,* 1980, *4:* 171–198; Pauly, "Appearance of Biology," p. 392.

20. Appel, in this volume.

21. For introduction to the history of marine zoology, see "The Naples Zoological Station and the Marine Biological Laboratory: One Hundred Years of Biology," special supplement to vol. 168 of the *Biological Bulletin* (1985).

22. Alexander Agassiz, "The Abandonment of Penikese," *Popular Science Monthly*, 1892, *42:* 123; Benson, "Laboratories"; Lillie, *MBL*, pp. 15–35.

23. R. W. Dexter, "The Annisquam Laboratory of Alpheus Hyatt," *Scientific Monthly*, 1952, *74:* 112–116; Benson, "Laboratories;" Alpheus Hyatt, "Report of the Curator," *Proceedings of the Boston Society of Natural History*, 1882, *22:* 9–11; 1887, *23:* 361–362, 371; Woman's Education Association, *Annual Report*, 1886, 8–11.

24. The spelling of the village name varied in the nineteenth century as social predominance shifted from locals to summer people. Joseph Story Fay, a Boston businessman and the largest property owner in the area, believed that Falmouth was the location of the pre-Columbian Norse settlements in North America, and argued that the Vikings had taught the local Indians to use the term "holl," Norse for "hill"; this was presumably altered by the English colonists to "hole." From the 1870s to the 1890s Fay persuaded the Post Office to adopt the pseudohistoric "Wood's Holl"; the MBL leaders, who were grateful to Fay for allowing them to purchase land and to rent buildings at low rates, also used this spelling in the early years. After complaints that "Holl" was "the meaningless corruption of Wood's Hole by finical summer visitors," the federal Board of Geographic Names decreed that the spelling should be "Woods Hole." See Joseph Story Fay, "The Track of the Norseman," *Magazine of American History*, 1882, *8:* 431–434; Frederik A. Fernald, "Hole or Holl?" *Pop. Sci. Mthly.*, 1892, *42:* 123; Lillie, *MBL*, p. 5.

25. Wealthy Bostonians began to buy waterfront property in Falmouth in the 1850s, and this trend accelerated with the completion of a railroad line to Woods Hole in 1872 and the establishment of the "Dude Train," a semiprivate summer express from Boston in 1886. In the 1890s the property of the guano works became the exclusive residential neighborhood of Penzance. On the development of Woods Hole as a resort, see Millard C. Faught, *Falmouth Massachusetts: Problems of a Resort Community* (New York: Columbia University Press, 1945), pp. 5–37.

26. The Old Colony Railroad, which had lost its major customer with the closing of the guano works, was the largest contributor. See Dean Conrad Allard, Jr., *Spencer Fullerton Baird and the U.S. Fish Commission* (New York: Arno Press, 1978), pp. 322–341; U.S. Department of the Interior, Bureau of Commercial Fisheries, *The Story of the Bureau of Commercial Fisheries Biological Laboratory, Woods Hole, Massachusetts*, by Paul C. Galtsoff, Circular no. 145 (Washington, D.C.: Government Printing Office, 1962).

27. The original seven-member board of trustees included the young MIT biologist W. T. Sedgwick, the Harvard botanist Farlow, and the struggling Harvard Medical School embryologist C. S. Minot, in addition to two representatives of the Woman's Education Association. Sedgwick's wife was a member of the WEA's executive committee. Hyatt, "Report of the Curator," 1887, p. 362; Benson, "Laboratories"; Maienschein, "Birth of the MBL"; Lillie, *MBL*, pp. 34–36, 204–206. For an earlier venture in summer edification on a somewhat broader scale, see Theodore Morrison, *Chautauqua: A Center for Education, Religion, and the Arts in America* (Chicago: University of Chicago Press, 1974), pp. 31–52.

28. Lillie, *MBL*, pp. 34–36; see also Alexander Agassiz to Marshall McDonald,

24 April 1888; McDonald to Agassiz, 25 April 1888; McDonald to C. S. Minot, "May 1888"; Alpheus Hyatt to McDonald, 8 June 1888, 26 June 1888; McDonald to Hyatt, 22 June 1888; all copies in bound typescript, "Important Correspondence Regarding Woods Hole Laboratory, Relations with the Marine Biological Laboratory, Outline of Policy, Privileges Extended to Other Laboratories, etc.—1888–1891," series 117 (Fish Commission Stations, 1875–1929), U.S. Fish Commission Records, Record Group 22, National Archives, Washington, D.C. (hereafter, Fish Commission Records).

29. Lillie, *MBL,* p. 26; W. K. Brooks to D. C. Gilman, 1 August [1888], Gilman Papers; "Another Investigation: The Senate Committee After the Fish Commission," *Baltimore News-American,* 1 June 1890; "Prof. Brooks Defends It," *Baltimore News-American,* 30 June 1890. In 1892 Agassiz was asking about rumors that the government planned to sell its laboratory to the MBL; see Agassiz to Marshall McDonald, 16 December 1892, series 23 (Letters Received by Marshall McDonald, 1879–1895), and McDonald to Agassiz, n.d. [December 1892], series 24 (Copies of Confidential Letters Sent by Commissioner McDonald, 1888–1895), Fish Commission Records. Professional representation on the MBL board increased substantially after the 1889 season; see Trustees Minutes, 14 November 1889, Marine Biological Laboratory Archives, Woods Hole, Mass.

30. See the following by C. O. Whitman: "Death," *The Radical,* 1868, *4:* 372; "Free Inquiry," *Radical,* 1869, *5:* 394–396; "The Idea of Immortality," *Radical,* 1871, *9:* 53–63; "Progress Has No Goal," *Radical,* 1871, *9:* 201–204. *The Radical* was a publication of the Universalist Free Religious Association, an early Universalist group. On Whitman, see F. R. Lillie, "Charles Otis Whitman," *Journal of Morphology,* 1911, *22:* xv–lxxiii; E. S. Morse, "Charles Otis Whitman," *Natl. Acad. Sci. Biog. Mems.,* 1912, *7:* 269–289; C. B. Davenport, "The Personality, Heredity and Work of Charles Otis Whitman, 1843–1910," *American Naturalist,* 1917, *51:* 5–30; Maienschein, *Defining Biology,* pp. 10–17.

31. Spencer F. Baird to C. O. Whitman, 29 November 1882, 9 February 1883 (letters 6327, 8741), series 15 (Letters Sent, 1871–1906), and Whitman to Baird, 6 February 1883 (letter 14616), series 8 (Letters Received), Fish Commission Records.

32. On Whitman's relations with his employers, see the series of letters from Whitman to Alexander Agassiz in the Alexander Agassiz Papers, Library of the Harvard Museum of Comparative Zoology, Harvard University, Cambridge, Mass. (especially 8 December 1885, 14 July 1886, 27 July 1888); E. P. Allis in E. J. Dornfeld, "The Allis Lake Laboratory, 1886–1893," *Marquette Medical Review,* 1956, *21:* 115–144. Allis recalled (p. 127) that when a local collector was not available, he ordered Whitman to wade into a millrace to search for mudpuppies. Whitman accepted the MBL directorship, despite the difficulty of leaving his position in Milwaukee, as part of his continuing campaign to find a secure academic berth. He was chosen only after both Brooks and Samuel F. Clarke of Williams College declined to serve. See Maienschein, "Birth of the MBL"; E. G. Conklin in "Marine Biological Laboratory Dedication Exercises, July 3, 1925," *Sci.,* 1925, *62:* 278. On Whitman's impulses regarding animals, see Burkhardt, in this volume.

33. C. O. Whitman, "Preface," and "Specialization and Organization, Companion Principles of All Progress—The Most Important Need of American Biology," in *Biological Lectures Delivered at the Marine Biological Laboratory of Wood's Holl,* 1891, *1890:* 1–26.

34. Whitman, "Specialization and Organization," pp. 1–2. August Comte and Herbert Spencer, whom Whitman cited in his *Radical* essays, were probably the basic sources for his organicism, especially in linking the evolution of individuation with that of mutual dependence. His teacher Leuckart reinforced this view. Leuckart had devoted his life to understanding parasitism, that most intimate relationship between organisms, in which the social collapsed completely into the biological. Leuckart's major public address, delivered at the University of Leipzig while Whitman was his student, described universities as organisms, and it argued that the chief value of Darwin's *Origin of Species* was in showing that the interactions between species extended beyond matters of mere mutual self-interest to form "a general, and fundamentally a family, affair," thus ending the alienation of humanity from nature. See Rudolf Leuckart, "Ueber die Einheitsbestrebungen in der Zoologie" (1877), reprinted in Klaus Wunderlich, *Rudolf Leuckart* (Jena: Gustav Fischer Verlag, 1978).

35. Whitman, "Specialization and Organization," esp. pp. 21–25.

36. Ibid., p. 14.

37. Maienschein, *Defining Biology,* reprints a number of the important lectures and discusses their context.

38. For early discussions of life at the MBL, see J. S. Kingsley, "The Marine Biological Laboratory," *Pop. Sci. Mnthly.*, 1892, *42:* 605–615; also the early issues of the *Marine Biological Laboratory Report.* The relations between the MBL and the Fish Commission were not without tension. It was a great coup when E. B. Wilson and Whitman detached the promising Hopkins graduate student E. G. Conklin from the Fish Commission Laboratory and affiliated him with the MBL. Characteristically, the groups ritualized their competition by arranging boat races and by playing baseball against each other. On community life, see articles by E. G. Conklin, "Early Days at Woods Hole," *American Scientist,* 1968, *56:* 112–120; "M. B. L. Stories," *Am. Sci.*, 1968, *56:* 121–128. The name of the dining club/mess was one element in the conflict over the MBL's gender identity; see especially "Report of the Trustees," *MBL Report, 1894–1895,* pp. 10–11.

39. On these themes, see Whitman's "Report of the Director," *MBL Report, 1894–1895,* esp. pp. 23–26, 46–49. They were promoted to wider audiences in his "A Marine Biological Observatory," *Pop. Sci. Mnthly.*, 1893, *42:* 459–471, and "A Marine Observatory the Prime Need of American Biologists," *Atlantic Monthly,* June 1893, pp. 235–242.

40. Lillie, *MBL,* pp. 40–46, 169; S. H. Scudder, et al., "A Statement Concerning the Marine Biological Laboratory at Woods Hole, Massachusetts," *Sci.*, 1897, *6:* 529–534; S. F. Clarke, et al., *A Reply to the Statement of the Former Trustees of the Marine Biological Laboratory* (Boston: Alfred Mudge & Sons, 1897). The now-redundant Biological Association immediately disbanded.

41. See Allis's reminiscences in Dornfeld, "Allis Lake Laboratory," pp. 120–121.

42. "Report of the Director," in *MBL Report, 1889,* p. 30; C. O. Whitman, "The Impending Crisis in the History of the Marine Biological Laboratory," *Sci.*, 1902, *16:* 529–533.

43. Lillie, *MBL,* p. 176.

44. Conklin, "Early Days," p. 120; Lillie, *MBL,* p. 172.

45. Kingsley, "MBL."

46. Ibid., p. 611; Kenneth Manning, *Black Apollo of Science: The Life of Er-*

nest Everett Just (New York: Oxford University Press, 1983), pp. 67–114, discusses work practices at the MBL in the 1910s.

47. G. H. Parker to George R. Agassiz, 24 May 1927, UAV.217.10, Harvard University Archives, Cambridge, Mass.

48. Lillie's organicism, while more muted than that of Whitman, is evident in his "Address at the Dedication of the New Buildings of the Marine Biological Laboratory," *Sci.*, 1914, *40:* 229–230; see also "The Mechanistic View of Vital Phenomena," lecture to University of Chicago Philosophy Club, 4 December 1919, F. R. Lillie Papers, Marine Biological Laboratory Archives, Woods Hole, Mass.

49. Whitman, "Specialization and Organization"; and "Report of the Director," *MBL Report, 1895,* pp. 17–27. Also Maienschein, *Defining Biology,* pp. 21–26.

50. Jeffrey Werdinger, "Embryology at Woods Hole: The Emergence of a New American Biology" (Ph.D. dissertation, Indiana University, 1980); Lillie, *MBL,* pp. 115–156; John M. Farley, *Gametes and Spores: Ideas about Sexual Reproduction, 1750–1914* (Baltimore: Johns Hopkins University Press, 1982), pp. 235–251; Scott F. Gilbert, "The Embryological Origins of the Gene Theory," *J. Hist. Biol.*, 1978, *11:* 307–351; Philip J. Pauly, *Controlling Life: Jacques Loeb and the Engineering Ideal in Biology* (New York: Oxford University Press, 1987), p. 109.

51. See John C. Burnham, "Psychiatry, Psychology, and the Progressive Movement," *American Quarterly,* 1960, *12:* 457–465; John Ettling, *The Germ of Laziness* (Cambridge, Mass.: Harvard University Press, 1981); Naomi Aronson, "Fuel for the Human Machine: The Industrialization of Eating in America" (Ph.D. dissertation, Brandeis University, 1979).

52. See, e.g., John M. O'Donnell, *The Origins of Behaviorism: American Psychology, 1870–1920* (New York: New York University Press, 1985), pp. 141–145.

53. Maienschein, *Defining Biology,* pp. 21–26; Mall, "What Is Biology?" *Chautauquan,* 1894, *18:* 411–414; Jennings, *Behavior of the Lower Organisms* (New York: Columbia University Press, 1906), pp. 338–350. Although Europeans such as Hans Driesch found Jennings's work profound, his major American reviewer, G. H. Parker, treated the book solely on the empirical level; see G. H. P.[arker], "The Behavior of the Lower Organisms," *Sci.*, 1907, *26:* 548–549.

54. Whitman, "Specialization and Organization," pp. 22–23.

55. See Figure 4.1. The Fish Commission Laboratory only rarely provided data, probably because directors sought to obscure from congressional investigators the fact that private individuals were using the laboratory. U.S. Commission of Fish and Fisheries, *Report of the Commissioner for 1888,* p. 513, reported eighteen workers; H. C. Bumpus, "The Work of the Biological Laboratory of the U.S. Fish Commission at Woods Hole," *Sci.*, 1898, *8:* 96, referred to twenty-four; F. B. Sumner, "The Summer's Work at the Woods Hole Laboratory of the Bureau of Fisheries," *Sci.*, 1904, *19:* 242, listed thirty. For the summers of 1896 and 1897 the laboratory was closed to investigators.

56. Information compiled from Lillie, *MBL,* pp. 252–254; "Leading Men of Science in the United States in 1903," *American Men of Science,* 5th ed. (New York: Science Press, 1933), pp. 1269–1278; *Records of the American Society of Naturalists,* 1930; "Proceedings of the American Society of Zoologists," *Anatomical Record,* 1929, *41:* 123–125.

57. On clubs see E. Digby Baltzell, *Philadelphia Gentlemen* (Chicago: Quad-

rangle Books, 1971), pp. 335–363; G. William Domhoff, *The Bohemian Grove and Other Retreats: A Study in Ruling-Class Cohesiveness* (New York: Harper and Row, 1974).

58. Faught, *Falmouth Massachusetts,* p. 26.

59. Lillie, *MBL,* pp. 47–61; Jane Maienschein, "Early Struggles at the Marine Biological Laboratory over Mission and Money," *Biol. Bull.,* 1985, *168* (suppl.): 192–196. Whitman remained director in name until 1908, but he ended active involvement in 1903. Associate director Frank Lillie functioned as chief administrator; his brother-in-law, C. R. Crane, became president of the trustees and agreed to make up operating deficits. The lack of expansion meant that graduate students were squeezed out in the struggle for space; the number of "investigators under instruction" declined significantly during the first decade of the new century (see Figure 4.1).

60. Ronald C. Tobey, *Saving the Prairies* (Berkeley: University of California Press, 1981), pp. 110–154; Edward H. Beardsley, *Harry L. Russell and Agricultural Science in Wisconsin* (Madison: University of Wisconsin Press, 1969); Kimmelman and Paul, in this volume; Hagen, in this volume. The sudden interest of biologists such as Parker and Conklin in eugenics around 1914 was in part an attempt to combat the declining status of their subject by demonstrating its relevance. See G. H. Parker, *Biology and Social Problems* (Boston: Houghton Mifflin, 1914); E. G. Conklin, *Heredity and Environment in the Development of Man* (Princeton: Princeton University Press, 1915).

61. Howard S. Berliner, *A System of Scientific Medicine* (New York: Tavistock Publications, 1985); F. R. Sabin, *Franklin Paine Mall: The Story of a Mind* (Baltimore: Johns Hopkins University Press, 1934); Alejandra C. Laszlo, "Physiology of the Future: Institutional Styles at Columbia and Harvard," in Gerald L. Geison, ed., *Physiology in the American Context, 1850–1940* (Bethesda, Md.: American Physiological Society, 1987), pp. 67–96.

62. [Raymond Pearl], "Tentative Plan for Symposium at the Afternoon Session of the American Society of Naturalists . . . December 29, 1916," Jacques Loeb Papers, Raymond Pearl file, Manuscript Division, Library of Congress, Washington, D.C. (hereafter, Loeb Papers); [E. G. Conklin], "First Report of the Committee on Zoology," *Proceedings of the National Academy of Sciences,* 1917, *3:* 725–731; National Academy of Sciences, *Annual Report,* 1918, pp. 95–97, 108–109. Members of the original committee were Conklin, G. H. Parker, T. H. Morgan, F. R. Lillie, S. A. Forbes (Illinois), C. A. Kofoid (California), J. Reighard (Michigan), and H. U. Smith (Commissioner of Fisheries). On "embroidery," see E. G. Conklin, "The Value of Zoology to Humanity: The Cultural Value of Zoology," *Sci.,* 1915, *41:* 334. In spite of his title, he noted, typically, that he was speaking "for all the biological sciences and not for zoology alone" (p. 333).

63. "Report of the Trustees," *MBL Report, 1908,* in *Biol. Bull.,* 1909, *17:* 11: see also F. R. Lillie, "Cooperation in Biological Research," *Sci.,* 1908, *27:* 269–372.

64. Lillie, *MBL,* pp. 62–75; "Dedication Exercises of New Laboratory Building," Falmouth *Enterprise,* 11 July 1914.

65. Lillie, *MBL,* pp. 62–67; Faught, *Falmouth Massachusetts,* p. 27; F. R. Lillie to H. J. Thorkelson, 20 October 1925, 20 November 1925, and Thorkelson to Lillie, 23 November 1925, Records of the General Education Board, box 357, F.3680, Rockefeller Archive Center, Pocantico Hills, N.Y. Exceptions to this policy were stu-

dent quarters; a woman's dormitory was constructed in 1915, and some cottages were converted to student use in the 1920s, probably as local rooming houses became private homes.

66. Manning, *Black Apollo,* pp. 67–114; Jacques Loeb to Simon Flexner, draft on Flexner to Loeb, 15 January 1923, Box 4, Loeb Papers.

67. C. R. Crane, in "MBL Dedication Exercises," p. 271; Faught, *Falmouth Massachusetts,* pp. 27, 176–177; problems during the interwar years were noted in Luther J. Carter, "Woods Hole: Summer Mecca for Marine Biology," *Sci.*, 1967, *157:* 1288–1292.

68. National Academy of Sciences, *Annual Report,* 1918, p. 108, 1919, pp. 100–101; Diana E. Long, "Physiological Identity of American Sex Researchers between the two World Wars," in Geison, *Physiology in the American Context,* pp. 263–278; Robert E. Kohler, *From Medical Chemistry to Biochemistry* (Cambridge: Cambridge University Press, 1982), pp. 307–321.

69. L.J. Henderson, *The Order of Nature* (Cambridge, Mass.: Harvard University Press, 1917); F. R. Lillie, "The Mechanistic View of Vital Phenomena"; W. E. Ritter, *The Unity of the Organism,* 2 vols. (Boston: R. G. Badger, 1919); W. M. Wheeler, *Emergent Evolution and the Development of Societies* (New York: Norton, 1928). An intriguing instance of unconscious self-reflexiveness is the holist Lewis Thomas's use of Wheeler's concept of the superorganism in *Lives of a Cell* (New York: Viking Press, 1974), pp. 58–63, to describe the MBL. For a minority viewpoint during these discussions, see E. B. Wilson, *The Physical Basis of Life* (New Haven: Yale University Press, 1923).

70. Most major American theorists of community between 1890 and 1940, including George Herbert Mead, John Dewey, Robert MacIver, Robert Park, and Robert Redfield, worked at either the University of Chicago or Columbia, and were aware of both the writings and the lives of their biological colleagues. See, e.g., Martin Bulmer, *The Chicago School of Sociology* (Chicago: University of Chicago Press, 1984), pp. 129–150.

Jane Maienschein

5 Whitman at Chicago: Establishing a Chicago Style of Biology?

In the 1860s or early 1870s, an American interested in a professional career in biology would likely have been drawn to Harvard University, to study botany with Asa Gray or zoology with Louis Agassiz, or possibly to Yale. After 1876, such an American would have found the new Johns Hopkins University particularly attractive, with its much-publicized emphasis on the medically related biological and physical sciences. Other alternatives existed by this time, including a visit to European laboratories, but male students interested in a program in life sciences would nonetheless probably have found Johns Hopkins the most exciting, while after 1884 women might well have migrated to Bryn Mawr College. After 1890, our motivated student would have found yet other new opportunities, at research-oriented Clark University (after 1889) or at Columbia University's College of Pure Sciences (after 1891), for example. Each of these institutions offered programs of study in the biological sciences, although not all were explicitly labeled as "biology." Each produced a collection of outstanding students. The University of Chicago then entered the competition in a major way.

The University of Chicago was legally established and construction began in 1892, as Chicago planned its great Columbian Exhibition for the next year. Indeed, the University virtually backed onto the fairgrounds, so that a ride on the Ferris wheel provided a fine view of the developing university campus. Debates had surrounded the construction of both the Fair and the University, demonstrating that what some saw as progressive, others regarded as retrogressive. For example, architect Louis Sullivan lamented that with the Fair, "architecture died in the land of the free and the home of the brave—in a land declaring its democracy, inventiveness, unique daring, enterprise, and progress. Thus ever works the pallid academic mind, denying the real, exalting the fictitious and false. The damage wrought by the World's Fair will last for half a century from its date, if not longer."[1] According to this view, the university followed with its deplorable "Collegiate Gothic" buildings of "City Gray." Yet others found the Fair, the University buildings, and art in Chicago exciting. Sculptor Augustus Saint-Gaudens, for example, saw things much more positively, asking about the Exhibition preparations, "Do you realize that this is the greatest gathering of artists since the fifteenth century?"[2] Out of this excitement and disagreement came what was called a "Chicago style" of architecture.

More recently it has become unfashionable in some circles to speak of styles in architecture, and perhaps rightfully so. Yet there is some level of generalization that helps clarify patterns of historical development, unifying study of individuals and of institutions, while also considering the sort of work done. For science, the sociologist/historians of science at the Tremont Research Institute have identified what they label as the "style of work."[3] This sense of style concerns what scientists ask, what problems they consider worth solving, what techniques they employ, what approaches they adopt, what organisms they choose. In short, what work do they do and how do they do it? If there is a definable style of work for a particular set of researchers, then they should share many of the ways of working with others in their group but not as many with workers outside. Yet the style need not be as localized as a research institution or research school might be. Closer in some ways to a research tradition, a style is a subset of such units, influenced also by local setting, individuals, and organization and by non-rational factors. If a Chicago style of biology exists in this sense, we should be able to identify research work that, at least with a high probability, has its origins in Chicago rather than somewhere else.

I am not fully convinced that this sense of "style" is the best possible unit of historical study. Yet the phenomenon did occur that Charles Otis Whitman and others with similar convictions about biological work created a community at Chicago that produced students who pursued work of just the same sort and in the same way as their advisors. Explaining that phenomenon calls for studying at some different level of analysis in science, more than simply a consideration of either the individuals or the institution involved, and perhaps the style of work is the appropriate place to look. It cannot be coincidental that researchers such as Whitman, Frank Lillie, William Morton Wheeler, Charles Manning Child, Ernest Everett Just, and others of similar scientific approach all gathered in Chicago. Chicago was extremely influential in biology and had considerable glamour and prestige. It produced results that people associated with Chicago. Perhaps that particular character did lie in the work done, by individuals with a particular vision of biology, within a peculiarly promising institutional setting.

This paper provides a preliminary exploration of a Chicago style of biology by considering the origins of the University of Chicago as an institution, biology there, and especially the first chairman of biology, Charles Otis Whitman, who had a driving sense of what biology should be and an autocratic approach to putting his vision into effect. To a remarkable extent, zoology in particular followed Whitman's direction. This paper therefore concentrates on Whitman's role as director and exemplar for Chicago biology.

The story told here suggests that there was a characteristic Chicago style of biology initiated by Whitman, which extended beyond Whitman. The full argument and evidence for that broader claim remains beyond the scope of this particular paper and tantalizingly suggests a much larger program of study which would extend to all subfields of biology at Chicago. In addition, any claim that a Chicago style of biology was unique would have to compare work there with work elsewhere. In this paper I want to establish what happened at Chicago and to offer preliminary suggestions for interpreting its significance.

In the Beginning

An early effort to establish a Baptist University of Chicago failed for financial reasons.[4] After foreclosure on the building loans brought an end to the initial effort, a small group of prominent Chicago Baptists determined in 1886 to try once more, this time in a new location and with a sound financial footing. They acquired land and sought outside funding, hoping to establish a "western Yale." Loyal Baptist supporter Thomas Wakefield Goodspeed, who spearheaded the project, sought the support of that wealthy Baptist John D. Rockefeller.[5]

In effect, Rockefeller made the new University of Chicago possible. After months of negotiation and careful deliberation, Rockefeller was persuaded to give $600,000 by the positive reports from his assistant Frederick T. Gates and by conversations with sympathetic project supporter William Rainey Harper. He thereby founded a Baptist college, rather than a more ambitious research university, with the condition that his contribution be matched by $400,000 from local Chicago supporters. By September 1890, those supporters had pledged more than the requisite amount, and the new University of Chicago was incorporated on 10 September. Shortly thereafter, Goodspeed and others sought to lure Harper away from his position as biblical scholar at Yale University to accept the presidency of the new western university. But Harper worried both that the job would force him to abandon the biblical study he loved and that the $1,000,000 from Rockefeller and the Chicago supporters would not prove sufficient to build a first-class institution. Evidently responding to this concern Rockefeller provided a second million dollars, of which part was to support a seminary and $800,000 was designated for graduate support for the University. Harper accepted the presidency in February 1891, despite being pressured to remain at Yale.[6]

Harper then settled down to the difficult task of securing the best faculty, a task exacerbated by recruitment efforts at Columbia University and Stanford University at the same time.[7] With the goal of securing a strong arts and science faculty of established scholars and surrendering, at least for the moment, hopes for a technical school as well, Harper began to make appointments.

In the biological sciences, he first recruited Clarence L. Herrick, a biologist at the University of Cincinnati who had been at Denison University where Harper had earlier spent some time. This hiring ultimately proved problematic.[8] Recall that Harper was a Biblical scholar, not a scientist. Recall also that he was originally hiring for a Baptist college rather than for a major research university. In his first appointment in science, he looked to a midwesterner. Herrick, born in Minnesota and educated there and in Germany, did not then have a Ph.D. (although he did complete one later). After a period at Denison, he went to the University of Cincinnati in 1889 as chair in biology. With his special interest in psychobiology, or that borderland between physiology and psychology and even philosophy, Herrick could offer Chicago a modern and popular area of study. He would also bring his *Journal of Experimental Neurology*. For a Baptist school supported by religious interests, the appointment of someone who studied the biology of mind made sense; in the bargain, Herrick was a good Baptist. After some negotiation about his precise role and about financial details, Harper formally offered Herrick a position, evidently as Professor of Comparative Psychology, in June 1891.[9]

In his initial proposals for the department at Chicago, Herrick had stressed the importance of undergraduate teaching in particular. Herrick clearly took his appointment as evidence that Harper was at least favorably disposed toward his plans, and there is even some evidence that Harper may have initially considered Herrick for the chairmanship for the biology department.[10] It seems that Herrick expected to have charge of at least the anatomy and physiology sections.[11] On the basis of what he felt was a strong commitment to his ideas as well as a firm offer from Harper, Herrick left his position at Cincinnati and set off for Europe for a year. Yet Harper had actually refrained from making any concrete commitments beyond the offer of a faculty appointment. Harper had even informed Herrick in May, and hence before his job offer, that he had also entered into negotiation with zoologist Charles Otis Whitman of Clark University.

In fact, Harper had corresponded with Whitman about the development of biology at Chicago and had received rather different suggestions from those Herrick offered. Whitman stressed the importance of graduate education and of both faculty and student research. In December 1891, Whitman wrote to Harper that he was "ready to consider the offer" that Harper had made, presumably for Whitman to chair the biological sciences at Chicago, if Harper could promise him at least $50,000 income for biology each year and would give him control. He thought that Chicago offered "the opportunity to start an organization in one of the most advantageous regions of the entire country."

He then advertised his qualifications for the job, presumably in part to convince Harper to work harder to obtain what Whitman requested. As Whitman pointed out, he would bring with him the leading American zoological journal and control of the only national marine laboratory, both considerable attractions. Whitman also hoped to add an inland lake laboratory sponsored by the University of Chicago to the collection. As to organization, he would set up zoology, botany, paleontology, and physiology as the four divisions of the biological sciences, with anthropology to follow soon after. Whitman felt certain that others from Clark would join the move if Harper invited them.[12]

When it became clear that Chicago could become a full-scale university rather than a more modest Baptist college, thanks to a generous donation after 1891 from the estate of William B. Ogden to build scientific laboratories, Harper began to consider seriously Whitman's suggestions for first-rate biological research.[13] Harper turned increasingly to Whitman and to Clark University for inspiration and for quality material with which to build his faculty. In fact, the serious problems at Clark in 1891–92 probably gave Harper his first major successes in recruiting in the sciences.

Clark University, like the Johns Hopkins and the University of Chicago, had been established with grand hopes for providing the best in education based on a strong scientific and research-oriented foundation. Yet Clark's benefactor, Jonas Gilman Clark, turned out to be somewhat less generous in his financial support of the new institution than some, at least, had expected and less generous than was actually needed. Thus President Granville Stanley Hall, with a Ph.D. and teaching experience in experimental psychology at Johns Hopkins, had high ambitions and great enthusiasm but also, inevitably, problems. In this unique institution designed specifically to offer quality graduate education, Hall had gathered an impressive group of faculty and students in the very first year, 1889.

Financial and ideological differences surfaced quickly, however, so that Whitman seriously considered an offer to go to Stanford in 1891. He decided to stay after Hall promised to improve the situation at Clark.[14] But by the third year, Whitman and most of the rest of the faculty recognized deeper trouble. After several unsatisfactory meetings in the fall of 1891, in January 1892 a majority of the full Clark teaching staff signed a vote of no-confidence in their president and formally resigned their positions in protest. They felt that, because various promises had been broken, they could no longer trust their chief administrator.[15] Hall worked desperately to keep them, and they did withdraw their resignations at least temporarily. Yet, as one historian remarked, Chicago offered powerful attractions, for "I think we may be rather sure that even if they had been on the best possible terms with Dr. Hall and the Board few of them could have refused the opportunity to go into new laboratories, in beautiful buildings fitted with every possible convenience, with much more to spend for equipment, books, laboratory assistants, etc.; with the background of a big city containing rare libraries, medical schools and other facilities that Clark could never duplicate, and with no Founder dropping in every day or so."[16] At least such accouterments are what Harper promised.

Yet Whitman had no desire to jump from one problem at Clark to another at Chicago. He was an excellent and ambitious administrator and sought to clarify many of the numerous and superficially tedious details that he recognized as important. Although he had resolved that the Clark situation was intolerable, he also worried about the situation at Chicago. Correspondence between Whitman's friend, anatomist Franklin Paine Mall, and Harper reveals the critical issues. As early as 27 January 1892, only a week after the no-confidence meeting at Clark, Mall reported to Harper that

> I have also constantly had my fears that the biological scheme might not develop. The amount he [Whitman] suggests is not great if all the departments are included; the physiological department alone at Columbia has a salary list of 15,000 and at Berlin over 50,000. Yet with these things clearly in view, I have constantly urged Prof. Whitman, and his enthusiasm has most of the time been the highest. When you wrote to him last he felt a little downhearted but the idea of making a biological department with various branches (but not full departments) represented seemed first to me and then to him a way out of the difficulty. Now I feel more hopeful and he tells me that he has written a hopeful letter to you.
>
> On account of much freedom here, in spite of our trouble (confidential), we cling to our ideals. You know of Whitman's organizing abilities. I may add that his students idolize him. —I yet believe that if the ideals which biologists prize so much are again plainly laid before him that he will consider the place most favorably.[17]

In March, Whitman expressed noticeably greater enthusiasm in a letter to Harper. He suggested that, although Hall had asked him to withhold his final decision, he remained quite interested in Chicago. In fact, he would like to take a colleague or two with him if he went. Clearly he knew by then of the Ogden gift and the resulting improved prospects for a biology building. With assurances that a new and modern biological laboratory would be forthcoming, he explained that he could surely decide in favor of Chicago. Yet he remained cautious.[18]

By 7 April, 1892, the situation had become more heated and letters were flying. Mall expressed fears, perhaps calculated to push Harper toward committing his support to Whitman, that Whitman was giving up on Chicago. Chicago offered nothing more than a duplication of what Clark had already given, he pointed out. "Now," he worried, "I believe that nothing short of a laboratory or its absolute assurance within the near future will induce him to accept." Hall wanted to hold onto his faculty members, and they found it difficult to leave. Also, Whitman was a skilled negotiator determined to get the resources he thought necessary to pursue first-rate biological research. As a result of Whitman's requests, as well as Mall's, and made possible by the increased availability of funds for the sciences from the Ogden gift, Harper did promise a laboratory.

Whitman expressed his enthusiasm for Chicago once more in a letter to Harper on 7 April, while once again urging that "the laboratory is simply indispensable." Finally, on 9 April, Mall wrote to Harper that Whitman had, after several hours of arguing and despite all efforts to keep him at Clark, consented to accept Chicago's offer.[19] Harper arrived on the Clark campus eight days later and conducted his famous and brilliantly successful "raid."[20] Meeting at Whitman's house with a majority of Clark's teaching staff, from fellows to instructors to full faculty members, Harper made them an offer that many eventually accepted. Approximately two-thirds of the entire faculty and seventy percent of the students left Clark in 1892, about half to Chicago.[21] Hall reported later that Harper even went so far as to try to persuade him to join the "act of wreckage," but that Hall naturally declined.[22] Of the sixteen biologists at Clark in 1891–92, Harper reportedly arranged to take all but four to Chicago, although a few later accepted offers elsewhere. Leaders among these new recruits from Clark included Whitman, of course, as head of biology and zoology, Mall in anatomy, Henry H. Donaldson in neurology, and Charles A. Strong in psychology.

Herrick no longer would hold primacy even in his own area of psychobiology, a hegemony that he thought Harper had guaranteed him from the beginning. On learning of the new appointments already in effect, about which he had not been consulted, Herrick was evidently furious and resigned his professorship before ever really entering it. He complained of Harper's lack of good faith. Although it may be that Harper exercised imperfect tact in the situation, Storr, Blake, and the archival records all show that, contrary to Herrick's belief, Harper did *not* mislead Herrick.[23] Harper had perhaps alienated Herrick and left him embittered as well as unemployed, but Whitman's version of biology at Chicago had quite reasonably prevailed in the new environment of improved resources and research objectives.[24] Harper had, in fact, obtained a real bargain with the Clark staff, which far surpassed anything that Herrick had to offer.

Whitman at Chicago

When Whitman moved to Chicago, he took with him George Baur, Charles Lawrence Bristol, Henry Herbert Donaldson, Edwin O. Jordan, Frank Rattray Lillie, Franklin Paine Mall, Albert Davis Mead, Charles Augustus Strong (in psychology), Shosaburo Watase, and William Morton Wheeler. Physiologist Jacques Loeb joined the group soon after.

In addition to the faculty, Whitman also took his own ideas about what biology should be like and how it should be organized. He had begun to set those ideas forth publicly in 1887 with an article considering "Biological Instruction in the Universities."[25] At that time, Whitman was director of the Allis Lake Laboratory near Milwaukee, Wisconsin, following a year as professor of zoology at the Imperial University of Tokyo and two years as assistant in zoology at Harvard. The variety of those jobs, with their several leadership and subordinate roles, against the background of Whitman's graduate work in Germany with Rudolf Leuckart, gave Whitman comparative perspectives from which to reflect on what biology should be like.[26]

At the annual meeting of the American Society of Naturalists in 1886, he had also presented his views on biological instruction. There, Whitman had been stimulated to respond to suggestions that botanist William Gilson Farlow had made the previous year. Farlow had maintained that a university student must be treated, in effect, as a schoolboy, subject to lectures and rote learning "since his capacity for observing and investigating natural objects has been blunted by a one-sided course of instruction at school."[27] Although Whitman agreed with Farlow that observation and investigation were important, he responded that he had greater confidence in the abilities of able students to conduct individual research. Thus, for Whitman, prospective biologists should not be treated as schoolboys but should be put to work doing research.

Although he offered nothing radically new or controversial, Whitman insisted firmly that Americans should follow the successful German model wherein students engaged in active research and began to specialize at an early stage. Biology could not advance with mere lectures and without direct participation. Nor could biology advance if each biologist attempted to cover the entire field. Such a generalist approach reflected an archaic Linnaean attempt to encompass the entire "Systema Naturae" at once. Whitman acknowledged that, regrettably, most Americans calling themselves biologists operated on such hindsight rather than foresight. He believed that

> argument will never dislodge them; they can be reached only through the leavening influence of high examples. A single biological department organized on a basis broad enough to represent every important branch at its best, and provided with the means necessary to the freest exercise of its higher functions, would furnish just the example we stand in need of. It is clear enough where we ought to look for such examples, but it is not so clear where or when we shall find them. We have often heard of the 'coming university,' but still it comes not. Men and money are all that is required to create such a department, and the country has both. We wait only for the rare conjunction of wisdom, will, and means for the realization of the long-postponed expectation.[28]

Let Americans build a system of specialized researchers, with biology including the areas of botany, zoology, physiology, anatomy, and pathology, and with a range of researchers of different ranks within each area.

In 1890, Whitman had his chance to effect his proposals when Clark University promised to provide the necessary "men and money," with Whitman as head of the new biology department there. In his role at Clark and in his capacity as head of the Marine Biological Laboratory (MBL) at Woods Hole, Massachusetts (since 1888), Whitman continued to preach his missionary message of specialization in the biological sciences. But he now explicitly called for organization and cooperation among the specializing researchers.[29] That call he put into effect at the MBL and at Clark. Yet, as we have seen, the Clark experiment did not succeed in graduate biology, at least, and the MBL remained essentially a summer station that could not serve as an example for all of American biological education. The great example would have to be built at Chicago if Whitman was to play the part he sought.

In considering plans for Chicago, Whitman endorsed his earlier view that biology ought to be divided into separate institutes or departments, following the German model. Each of these would then cooperate with the others as a part of a coordinated biological sciences program. Thus, Whitman saw biology as an integrated organic unity with specialized parts and not just as the arithmetic sum of different subdisciplines.[30] In a letter of December 1891, Whitman expressed his enthusiasm for the "new era in the Biology of this country" and his conviction that zoology, botany, paleontology, and physiology should be the four separate departments with which Chicago would begin, with anthropology to follow shortly. Yet the same letter endorsed the selection of Franklin Paine Mall as head of anatomy, so presumably he intended anatomy to be included as one of the specialties as well.[31]

In another letter written after he had accepted Chicago's offer, Whitman expressed his vision for organization more visually as follows:

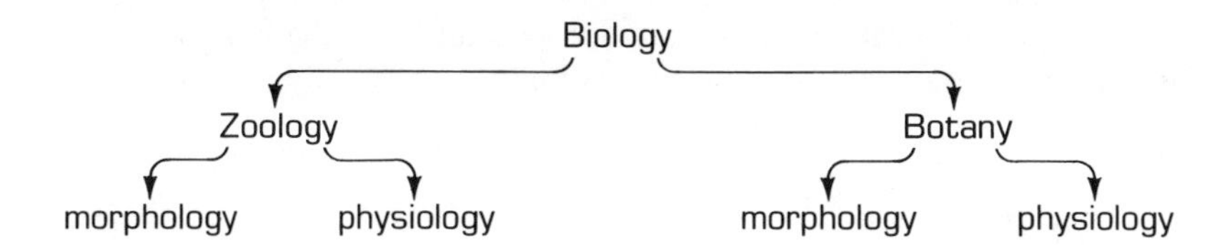

The zoological morphology side was divided further into zoology, anatomy, histology, neurology, paleontology, and pathology, he contended, with anthropology, cellular biology, and experimental biology to follow. Zoological physiology was divided into human, general, and chemical physiology, along with hygiene and psychology. Botany he did not discuss in as much detail, because he felt that one botanical institute could cover all the important work. Clearly, then, Whitman saw a broad set of biological sciences, with separate organizational and research units under the inclusive rubric of biology.

Actually, Whitman seems to have experienced some ambivalence or lack of clarity in these early efforts to define biology. Whitman seems to have had trouble deciding between a taxonomic organization, as suggested by his diagram, or the more functionally oriented classification of departments listed earlier. Clearly, he wanted to stress the value of separate, independent units of organization that worked together to constitute biology. As to exactly how that was to happen, Whitman initially remained indecisive. His sense of biology was to be inclusive rather than exclusive and open rather than tightly limited. But, again, the specialty units of the biological whole should retain their autonomy and should have definition, as cells do within an organic body.

Following the German ideal of research as he saw it, Whitman believed that at the institutional level these units should exist as administratively separate departments.[32] Harper did not. Harper evidently decided that, at least for the first year, biology would remain as one department, with Whitman as head and with division to come later. Finally, as Mall reported to Harper, Whitman agreed to one initial department with various branches, because several independent departments would have been too expensive.[33] In the second year, biology did divide into separate, though coordinated, departments of zoology, anatomy, physiology, neurology, and paleontology. At that time, Whitman became head of the biological division and of the department of zoology, positions that he retained until his death in 1910.[34] After receiving full departmental status, the several departments began to branch off in various ways so that each became more or less autonomous, despite Whitman's initial calls for strong coordination. To understand the full story of biology at Chicago after the first year, then, we would have to look at the evolution of each of the separate units.[35] Instead of pursuing that study here, I propose in this preliminary effort to concentrate on Whitman and the zoology program, with consideration of those departments most closely allied with zoology. The driving questions are: did Whitman manage to provide an example of the new biology as he had hoped to do; and what was the work like?

Whitman in Charge

As the sober and pious Yankee that one biographer saw, Whitman exhibited a composite of stubbornness and commitment—or what some would call pigheadedness—to what he regarded as justified goals.[36] Having been promised by Harper that biology would not have to struggle along with the inadequate conditions he had endured elsewhere, Whitman continued to work for better laboratory conditions, which he felt had failed to materialize. Having set his sights on a financially solid MBL, supported by Chicago, and on an inland biological laboratory and farm, he lobbied hard for support. Having been led to believe that he would continue to have a secure, high-quality group of faculty and students, Whitman also fought to retain his best people and to obtain better conditions and resources for those researchers. In all these arenas, he experienced frustrations and setbacks. Never a patient man and not one to accept compromises easily, he suffered as a result of his battles both at Chicago and at the MBL.[37]

Concerning buildings, Whitman was willing, though reluctant, to accept some crowding in the very first year with the full expectation of having that situation remedied in the second. Yet the second year found the then multiple departments moving from one set of crowded quarters into another crowded section of the Kent Chemical Laboratory.[38] Whitman wrote again and again to Harper that if the university could provide for chemistry, then why not for biology? In fact, Harper had promised $150,000 for biology buildings and the board of trustees had stated their intention to appropriate the designated funds. But intention proved stronger than action for a few years. In 1894 Harper acknowledged the need for adequate housing for biology, with its special requirement of "the most carefully adjusted accommodations," as the greatest need of the university.[39] Yet who would build this building?

Helen Culver did. This remarkable woman had inherited the considerable estate of Charles J. Hull and had contributed to Chicago's well-being in other ways before she determined to help the University. On 14 December 1895 she wrote to President Harper and declared her intention to make a gift "devoted to the increase and spread of knowledge within the field of the biological sciences."

> I mean to provide: (1) That the gift shall develop the work now represented in the several biological departments of the University of Chicago by the expansion of their present resources; (2) That it shall be applied in part to an inland experimental station and to a marine biological laboratory; (3) That a portion of the instruction supported by this gift shall take the form of University Extension Lectures on the West Side of Chicago. These lectures shall communicate in form as free from technicalities as possible the results of biological research.[40]

Actually, it seems that Harper had to talk her into giving the money to biology rather than to the arts as she had originally intended, but they did agree finally.[41]

Whitman was especially excited at the prospects for his inland laboratory and for the MBL as well as for Chicago. But the various interested parties eventually decided to spend all the money for biological buildings at Chicago instead. The West Side lectures seem to have evaporated as well. Rather than erecting one building, they decided to construct a quadrangle of four showcase buildings. The zoology, physiology, anatomy and botany departments each acquired its separate building, united by walkways and cloisters, which allowed some autonomy as well. At Helen Culver's request, the university designated the unit as Hull Court, a cluster of well-designed buildings also equipped with modern apparatus through this benefactor's further donations. As Whitman recorded at the grand cornerstone laying,

> The Culver gift came to us all as a grand surprise. Our earliest days in the University were spent in the garrets and kitchens of a tenement house. We were then tenderly transferred to the unused corners of Kent Chemical Laboratory where . . . we struggled for three years for bare existence. . . . Just as our hopes had cooled to near the freezing point came . . . the story, told in all the brevity and gravity that befit great deeds: 'A gift of a million to Biology.'[42]

The buildings opened in 1897, and biology was placed on a solid ground for the first time at Chicago.[43]

With buildings eventually in place, Whitman still worried about the MBL and his plans for an inland experimental station. In 1902 he almost succeeded in putting together a plan to provide financial security for the MBL, but the plan involved giving some control—or at least apparent control—to a group of people from Chicago, an idea that the MBL trustees vehemently opposed.[44] In fact, some of those trustees complained that Whitman and Chicago were trying to take over the MBL. This charge hurt Whitman deeply because he had worked so hard, and at considerable personal cost, to maintain the MBL as a truly national facility not connected with or controlled by any one university or financial group. He believed that the offer from a group of Chicago supporters did not in any way threaten the MBL's continued independence and felt betrayed when others found his intentions suspect.[45] In turn, Whitman opposed an alternative plan for the Carnegie Institution to support the MBL, which would have entailed a shift away from the strong teaching tradition to a research focus, as well as some loss of control. After the difficult negotiations of 1902, Whitman in effect gave up the directorship of the MBL to his assistant Frank Lillie, even though he did not officially retire from the position until the laboratory's twenty-first birthday, in 1908.[46]

While the MBL grew increasingly successful despite its financial controversies and its lack of support from the University of Chicago, Whitman's long-anticipated inland biological station never materialized. Probably ever since his time at the Allis Lake Laboratory, Whitman had envisioned an inland experimental research station, which, he added later, could also provide animals and plants for the laboratory research in Chicago, a function that he later assigned to his prospective biological farm. Even in his earliest correspondence with Harper outlining the directions Chicago ought to follow in biology, Whitman stressed the marvelous opportunity offered by having a lake biological observatory, as he then envisioned it. As he pointed out, "Our location combines so many natural advantages in the way of lakes and rivers that we can easily lead the world in this work. A *Lake Biological Observatory* such as I have suggested in our 'Programme,' in combination with *A Marine Observatory* will make us masters of the situation, and place the sciences of life, Physiology, Medicine, and all the rest on a footing that will simply surpass anything hitherto known in the world." We have the MBL already, he pointed out, so

> shall we fail to take the finishing step—that of planning a *Lake Observatory* for experimental research? I am sure your plans are too large to let this opportunity slip. This is something that to my mind will add much more to the enduring fame of this university than the establishment of an astronomical observatory . . . [because] the former stands for something that *other universities are not likely to duplicate,* and for something that the biological sciences the world over will pay homage to.[47]

Despite repeated pleas such as this, Whitman never persuaded Harper or anyone else to undertake the establishment of his inland laboratory. Helen Culver had intended to support the project, and presumably would have done so had not the economy caused the buildings for the Chicago campus to absorb so much of her gift. Only later, under Whitman's successor Frank Lillie and with Lillie's money, did Chicago begin to approach the facility that Whitman had envisioned, though by then in a different form more along the lines of a biological farm.[48]

In addition to his problems with buildings, money, and attempts to develop new facilities such as the inland laboratory, Whitman also had the usual troubles with and about faculty and students. Whitman seems to have inspired great loyalty from many of his faculty members, especially in the early years when he took a more active role in departmental administration and when the group remained relatively small and congenial. But he did not always have such success with Harper, and there were numerous skirmishes over the number of faculty positions, the number of fellowships for graduate students, and salary support for faculty. In 1894, for example, Lillie graduated and left for Michigan and then Vassar College, moves that Whitman very much regretted but could not prevent. Fortunately for Whitman, he did bring Lillie back to Chicago as an assistant professor in 1900.

Yet that success of 1900 followed on a particularly trying year, in which Whitman, probably not coincidently, had begun to withdraw quite considerably from departmental affairs. In 1899, Chicago lost both William Morton Wheeler and Shosaburo Watase. Harper had complained that Watase, in particular, was an "expensive luxury." His small class enrollments condemned him, in Harper's practical eyes. Yet to Whitman, Watase was "the broadest and soundest student of cellular biology in America," even surpassing Edmund Beecher Wilson. In addition, Watase was rare in his combination of physiological and morphological work, and of plant and animal studies. "All advanced students in cellular biology, whether in Botany, Zoology, or Anatomy, ought to go to Watase," Whitman insisted.[49] Because Watase gave "honor to the University," Whitman felt that Harper must keep him. Yet Harper remained sufficiently noncommittal that when Watase received a solid offer to return to Japan to teach, he went unhesitatingly. Harper similarly failed to understand why keeping Wheeler was a high priority, and Wheeler left for a distinguished career at the University of Texas and then at Harvard.

Also in 1899, Whitman needed to replace the paleontologist George Baur who had originally come with Whitman from Clark and who had recently died. Whitman had lost three positions by 1899, partly through what he probably regarded as Harper's lack of proper support. To add further insult, Harper proposed to give Whitman only two positions to make up for the three. Presumably Harper thought that Watase, with few students, did not have to be replaced. In addition, the continued arguments over whether paleontology properly belonged in biology or geology had left Baur with few students. Whitman was outraged. In a lengthy letter lambasting Harper's descent to "the level of the Mississippi Valley," he deplored the loss of men and the unfulfilled promises for equipment.[50] Chicago had once been the best, he wrote, but no longer. In the face of what he saw as a hopeless lack of support, he nonetheless still proposed to hire the very best scholars available, namely Frank Lillie and Thomas Montgomery. In fact, he did not succeed in appointing Montgomery, who was then comfortable at Pennsylvania, but hired Charles Davenport instead. The disaster was perhaps not so horrific as Whitman presented it, but replacing Wheeler and Watase with Lillie and Davenport did change the emphasis of the department somewhat, especially in moving away from cytological and behavioral studies.

In all these episodes of hiring and of lobbying for resources generally, Whitman played a strong directive role and held his convictions as inviolate. He hated to compromise when he felt he was right. Although he undoubtedly managed at times to irritate Harper and other administrators, it must have been clear that Whitman was doing an exceptional job. The department of zoology produced the largest number of Ph.D.'s, and succeeded in finding them jobs, often very good positions not limited to the zoological subset of biology. Zoology had an impressive research and publication rate, as revealed in the university's *Decennial Reports*. With Whitman's editorship of the *Journal of Morphology* and the *Biological Lectures,* as well as his role as founder of the *Biological Bulletin,* in addition to Whitman's and then Lillie's directorship of the MBL, Chicago visibly excelled throughout the United States.

At Chicago, Whitman dominated the zoology department with his autocratic style of leadership. Whitman was never "one of the boys."[51] He made the major decisions, and he suffered or rejoiced largely alone over the resulting failures or successes. Dedicated above all to superior research, Whitman never worried much about formal lectures or other typical aspects of university life.[52] He ignored many of the formalities and other trappings of academia as well. Most of his colleagues seem to have respected his approach and to have admired his high ideals. Yet if conditions were then anything like today, we can be sure that his aloof and dictatorial approach did not please everyone, especially administrators. Perhaps Whitman failed to get as much as he wanted from Harper at times because he did not really work with Harper. Yet perhaps he did well at other times because he so clearly believed in what he was doing and so clearly obtained impressive results.

Whatever the attitudes toward Whitman's leadership in the early years, as Chicago became better established and as other departments began competing more successfully for funds, Whitman withdrew increasingly from administration and concentrated on his own research with pigeons. Moreover, he established his pigeon colonies at his home rather than on campus (for various reasons), and he increasingly absented himself from the university. As Lillie took over the actual running of the department not long after his arrival in 1900, he moved toward a more participatory form of control that worked well for the different groups with their evolving goals. Much of the work remained the same under Whitman and Lillie but with changes of emphasis and detail.

Whitman as Teacher

Mall cited Whitman's excellence as a teacher as yet another reason that Harper should make every effort to obtain Whitman for Chicago. Yet Whitman, like Mall himself, clearly excelled in some aspects of teaching and not in others.[53] Whitman disliked undergraduate teaching and generally avoided it. Even graduate instruction he accomplished more through visits to the students at their laboratory desks than in a formal classroom setting, thereby following the model of his own German mentors such as Leuckart.

The earliest evidence of Whitman's approach to teaching comes from his visit to the Imperial University of Tokyo. There, in a very short time, Whitman attracted great respect from his four students, each of whom continued his zoological research after Whitman left. As those students recalled later, Whitman emphasized the importance of careful, detailed work, with technical accuracy the basis for success. As one student recorded, "His way of supervising our work was very kind and earnest. Twice a day, once in forenoon and once in afternoon, he asked about the progress of the work and then gave us necessary criticism and suggestions. To look after the work of the students twice a day, is the way common in German universities. Weismann in Freiburg used to do the same."[54] In Tokyo, Whitman helped students to learn English and German; he introduced them to fundamental problems of the day and to current techniques; he taught them how to carry out research and shape the results into publishable form. Only occasionally did he lecture to his students, and then primarily as an introduction to recent books or important ideas. His training system worked well for his Tokyo group.

His approach also worked well in the early years of the MBL. There Whitman collected a group of promising young researchers and arranged for them to teach the courses, which he insisted should form a vital and essential part of the MBL program. From the very first year, when the MBL had a total of only seventeen people in attendance, Whitman relied on others to do the direct teaching.[55] He organized the laboratory and its set of courses, then taught by example and by looking over the shoulders of researchers involved in work he found interesting. Cornelia Clapp, one of the students of that first year at the MBL and a much admired researcher there for the rest of her life, reported that she especially loved the MBL that first year. It was quiet and she appreciated the low-profile support that Whitman gave the students.[56] In fact, Clapp admired Whitman and his work sufficiently that she decided to attend the University of Chicago for her Ph.D., the first such research degree readily available in biology to women.[57] It became part of the yearly routine for Whitman to pack up his necessary materials and his students and head for what British zoologist E. Ray Lankester had labeled the "spasmodic descent upon the seashore."[58] Thus, work at the MBL and at Chicago blended together during the early 1890s, and the combination helped to cement the feeling of community and cooperation in research that Whitman regarded as fundamental.[59]

At the MBL after the first year or two, Whitman began to build a research program that new students were expected to join. As Edmund Beecher Wilson, Edwin Grant Conklin, and Whitman began to recognize the potential significance of their overlapping researches into the details of early cell lineages of developing organisms, Whitman encouraged other students to participate with studies of different organisms.[60] Cell-lineage work, inaugurated by Whitman's work on leeches, involved the tedious, meticulous tracing of exactly what each cell does as it undergoes cell division after cell division: What is each cell's lineage? Individuals working on different organisms then compared their results and discussed alternative theoretical interpretations of the differences and similarities in developmental patterns and processes. This cell-lineage and related work on early development became the hallmark of MBL and Chicago research during the first decade.

When Frank Lillie decided in 1891 to go to Clark University to study with Whitman, Whitman urged him to begin that summer by attending the Woods Hole session. There Lillie learned that he was expected to join in and pursue cell-lineage work. The only real question concerned which organism he would choose. He decided to work on the freshwater clam, *Unio,* thinking that it would provide useful comparative information. Indeed it did. But Lillie had to lug his heavy buckets, boots, and other apparatus by train back and forth to the nearby town of Falmouth to obtain the freshwater species.[61] Notwithstanding the extra effort, Lillie clearly felt privileged to be a part of such an active and friendly working group of researchers as he encountered in Woods Hole. Whitman's approach to graduate supervision succeeded in attracting such enthusiastic, eager, hard-working, and loyal students.

At Clark University, despite its orientation toward graduate work, Whitman had had only a few graduate students during his two years on the faculty. As Hall noted at Clark, Whitman "had never taught and found the requirement of our minimum of two hours weekly somewhat irksome."[62] Presumably, in the laboratory he continued his over-the-shoulder approach to teaching.

Whitman acknowledged in his first Decennial report to Harper that the zoology department at Chicago, as at Clark, had always emphasized research with considerable success. Also as at Clark, Whitman paid little attention to undergraduate education at Chicago and cared little about enrollments and lecture offerings. By 1897, he rarely appeared in his office because he was working at home with his pigeons.[63] His approach to graduate students continued along his earlier lines, only perhaps with not quite such uniform success in the larger environment of Chicago. As one student, Horatio Hackett Newman, reported:

> Dr. Whitman's treatment of his graduate students was somewhat harsh. His plan was to let each student work out his own salvation. In brief, he used the "sink or swim" method. Sometimes his result was good, sometimes bad. The present writer's experience was similar to that of several others. When I was first appointed a Fellow, Dr. Whitman by correspondence immediately assigned me a research program: "The Origin of Metamerism," using the annelid *Podarke,* as material. I was told to go to Woods Hole, to study the development of this species and to preserve all stages for microscopic examination when I came to Chicago in the autumn. No reading was suggested and no directions were given as to how to go about the work. I did as I was told, collected stages of development all summer. When these were examined in Chicago it was found that no metamerism had occurred, but that larvae had retrogressed in later stages. Evidently some special food was required for later development. Here a little guidance would have obviated the failure. Of course, I was too young and ignorant in biology to have been put on my own, and this initial failure nearly made me give up trying to be a biologist. Fortunately for them, many other graduate students of Dr. Whitman came through the ordeal without damage to their self-confidence.[64]

Actually, so did Newman, or at least he recovered his self-confidence sufficiently to complete his degree at Chicago in 1905 and to join the faculty thereafter (in 1911).

In fact, the zoology department had by far the largest number of students and largest number of graduates of the biological departments at Chicago throughout Whitman's time there, as he enjoyed pointing out to Harper. The list of Ph.D. degrees awarded in zoology, under Whitman's leadership, demonstrates his success in attracting students both to Chicago and into his own areas of research (see Appendix). The particular set of subjects covered was unique to Chicago.[65] A few of the Ph.D. students may have worked with advisors other than Whitman, yet Whitman exerted an important impact on most, introducing them to a research community and to his set of research problems and methods that then guided their own work. My cursory and unsystematic look at theses from Chicago and from Johns Hopkins, Harvard, Princeton, and Yale at the same time reveals a decidedly stronger emphasis at Chicago on early development, cytology, relations of embryology to evolution, behavior, and study of organisms as a whole, all of which characterized Whitman's work.

Whitman clearly worked hard to establish the sort of setting he found appropriate for a major research university. He worked for buildings, equipment, a marine laboratory and funds for students to do their research there, and to obtain and retain the best faculty. The Chicago department produced an impressive number of Ph.D.'s, a remarkable number of whom became professional biologists and continued their research after graduate school.[66] As an administrator and graduate teacher, Whitman was exceptionally successful. Yet none of this successfully addresses the question whether Whitman generated an identifiable Chicago style of work in biology, or at least in zoology. For an answer, we must look at Whitman's particular research and at the impact it had on colleagues and students at Chicago. Were there problems, techniques, assumptions, frameworks, or general approaches that characterize Whitman's work as well as the work of the Chicago zoological community? Did Whitman inaugurate a Chicago style of work?

Whitman as Researcher

In his earliest study of living organisms, as a boy, Whitman was attracted to birds. At age twelve, in 1854, he had a pet blue jay, and young Charles taught himself taxidermy in order to stuff and mount his pet when it died. He developed a strong fascination with natural history generally.[67] After failing the physical examination in the draft for the Civil War, presumably because of poor eyesight, Whitman entered an academic career, studying at Bowdoin College and teaching school to earn his way. It was only after attending Louis Agassiz's Anderson School of Natural History at Penikese Island in 1873 and 1874 that Whitman decided to study biology. During those summer sessions he kept largely to himself but explored and questioned.[68] On deciding to become a professional biologist, he went to Germany in 1875 to study natural history. He chose to work in Rudolf Leuckart's laboratory in Leipzig, where he learned the most advanced techniques for sectioning, staining, and preparing materials for microscopic study.[69]

For his Ph.D. dissertation, Whitman decided to study the embryology of several species of the leech *Clepsine* rather than the ascidians he had examined at Penikese.[70] He obviously spent considerable time in the library, reviewing earlier work on the leech as well as work on the developmental problems he found interesting. He probably concentrated on embryology, as so many others did at the time, because of the excitement about early development, stimulated in part by discussions of evolutionary questions and in part by technical advances that occurred between the 1850s and 1870s. In his dissertation, Whitman made it clear that he felt he had entered that vital world of morphological research and all stages of early development that Leuckart and others inhabited.[71] From the very formation of the egg cell to full development of the differentiated neurular stage, Whitman used the most advanced techniques and meticulously detailed his observations and their significance in light of other studies on the same and related species. Comparisons with vertebrates in particular revealed the degree of similarity and probable genealogical relationships of these leeches to other life forms.

Of special interest in Whitman's dissertation, given his later emphasis, is his discussion of cleavage. Recent histological work with improved techniques had demonstrated the developmental significance of cleavage for a number of researchers such as Alexander Kowalevsky, Edouard van Beneden, Wilhelm His, Carl Rabl, Leuckart, and others. Presumably, researchers in Leuckart's laboratory were predisposed to believe that, as Whitman wrote,

> in the fecundated egg slumbers potentially the future embryo. While we cannot say that the embryo is predelineated, we can say that it is predetermined. The "Histogenetic sundering" of embryonic elements begins with the cleavage, and every step in the process bears a definite and invariable relation to antecedent and subsequent steps. . . . It is, therefore, not surprising to find certain important histological differentiations and fundamental structural relations anticipated in the early phases of cleavage, and foreshadowed even before cleavage begins.

Such a position specifically denies that the embryo lies strictly preformed in the egg and must simply grow. Rather, the egg parts are more like building blocks that must be put into their proper places to have value for future differentiation. Such features as bilateral symmetry appear early, while other features follow later because "the egg is, in a certain sense, a quarry out of which, without waste, a complicated structure is to be built up; but more than this, in so far as it is the architect of its own destiny. The raw material is first split into two, four, or more huge masses, and some or all of these into secondary masses, and some or all of these into tertiary masses, &c., and out of these more or less unlike fragments the embryonal building-stones are cut, and transported to their places of destination."[72] That cleavage processes and patterns have a significant role in effecting differentiation Whitman did not doubt, but that fact played only a small part in his research of 1878.

In 1879–80, Whitman went to the Imperial University of Tokyo where he pursued his developmental study of leeches. Then a visit to the Naples Zoological Station during his return trip to the United States in 1882 and an assistantship in zoology at Harvard University's Museum of Comparative Zoology under Alexander Agassiz from 1882 to 1886 took him to other, related studies. Whitman published a series of articles on microscopical methods, leading eventually to a very useful book outlining research techniques as well as to several substantive papers from his work in Japan, Naples, and the United States.[73] Only in 1886 did he return directly to his studies of early developmental stages and to the significance of cleavage. At that time, he was director of the Allis Lake Laboratory near Milwaukee, Wisconsin, and had convinced Edward Phelps Allis to support an American biological journal.[74]

When that new *Journal of Morphology* appeared in 1887, Whitman included two of his own studies, which considered respectively the origin and fate of the germ layers in the leech *Clepsine,* and oökinesis (or cell development). Both papers stressed the importance of cytoplasm as well as the nucleus in development, insisting that any hypothesis that stressed the role of the nucleus to the exclusion of the cytoplasm simply could not explain the facts.[75] Both nuclear and cytoplasmic forces work together, Whitman insisted; we must therefore recognize that a variety of forces, influenced by heredity as well as by present conditions, direct all developmental stages and that we must move beyond the study of patterns to look also at processes if we are to understand development.[76] At the MBL, Clark University, and Chicago, Whitman interpreted this concern with both patterns and processes at the same time as a call for both morphological and physiological work. Individual researchers might concentrate on one side or the other, just as individuals would choose different organisms, but these specialists must then work cooperatively in order to carry out proper biological work.[77]

What Whitman offered in his work before arriving at Chicago was a very solid demonstration of the most advanced microscopical techniques and methods, a deep acquaintance with the English, German, and French literature concerning development, and an awareness that many important questions remained open with merely suggestive lines of attack or preliminary hypotheses.[78] Whitman was fully aware that researchers disagreed on many points and on interpretation. Careful technique, thorough familiarity with other work, cautious sorting out of possibilities, and working toward facts and solid interpretations should characterize biological work. With this set of approaches, he framed a style of biological work that was adopted by a growing community.

At the MBL, Edmund Beecher Wilson had begun to conduct his own studies of early development of the annelid *Nereis*. Like Whitman, Wilson became convinced that the earliest stages of egg formation and especially cleavage had significance for later differentiation.[79] Whitman and Wilson discussed their respective work, and Whitman encouraged Wilson to contribute his results to the new *Journal of Morphology.* As Wilson lectured to the embryology course each summer, he carried his conviction about early development to new biologists as well. In 1891, when Wilson learned that Edwin Grant Conklin, then at the U.S. Fish Commission, was also studying early development, namely of the gastropod *Crepidula,* Wilson went to talk with Conklin and invited him to the MBL to speak with Whitman. Whitman encouraged Conklin to contribute to the *Journal,* then invited him to join the MBL staff for the next summer. This move took Conklin away from the emphasis on later, germ layer stages of development, which his Ph.D. dissertation advisor William Keith Brooks emphasized, and toward Whitman's concerns. Thus, at the MBL Whitman began to develop a community of researchers with common goals, carrying out their individual studies on different organisms and then comparing results. Such comparisons would yield useful information for establishing genealogical or evolutionary relationships as well, Whitman, Wilson, and Conklin believed, and they all regarded improved understanding of evolution as vitally important. As the 1890s progressed, Whitman also began to focus his attention and that of his scientific community on major theoretical problems directly concerning development and heredity.

One such problem concerned the status of the cell theory. As Whitman recognized, "Each cell leads a double life; an independent one, pertaining to its own development alone; and another incidental, in so far as it has become an integral part of a plant."[80] This view had recently reemerged with force in researches claiming that parts of an organism, such as an isolated blastomere, could separate themselves from the organism and still develop independently. Yet development also requires organization. The cells do not operate completely independently, but must be integrated as part of a whole organism. The relations among the cells are as important as the cells themselves, so that "every elementary part possesses a power of its own, an independent life, by means of which it would be enabled to develop independently, if the relations which it bore to external parts were but similar to those in which it stands in the organism."[81]

As a result, organization is key. Evidence from half and quarter embryo experiments, which suggested that cell division divided predelineated areas of the egg, did nothing to undercut Whitman's absolute conviction that organization of the whole organism is necessary for normal development to occur. Whitman endorsed Thomas Henry Huxley's view that "they [the cells] are no more the producers of the vital phenomena than the shells scattered along the sea-beach are the instruments by which the gravitative force of the moon acts upon the ocean. Like these, the cells mark only where the vital tides have been, and how they have acted."[82] Given his emphasis on organization of the whole, on what Whitman called an organismal viewpoint, the way to understanding development necessarily lay with addressing processes, such as cleavage, of the developing whole. Patterns such as occur in the production of metamerism served as obvious paths of inquiry. Such concentration on the organization of the whole, approached in a variety of ways, characterizes the work of many Chicago biologists.[83]

With the *Biological Lectures* presented to the MBL community and intended to address shared concerns, Whitman encouraged discussion of preformation and epigenesis, of the relative roles of the nucleus and cytoplasm, and of the validity of "organic physics" to parallel organic chemistry. Most of the lectures fit neatly together around the general themes of understanding the patterns and processes of development, particularly early development. Most of the Ph.D. students from Chicago worked on problems within that general framework as well, with many going on to advocate an organismal approach to biology thereafter. As Newman noted, Whitman usually suggested the students' beginning research topics and the approaches to them and continued to exert a powerful influence thereafter.

The influence also becomes apparent as some of the students' dissertation projects reflect Whitman's rather abrupt shift of emphasis after 1897. In 1897 the MBL suffered a severe financial and ideological crisis in which Whitman insisted unyieldingly that the laboratory must become a truly national rather than merely local organization and that it must have sufficient operating funds. Evidently, it was in 1897 also that Whitman returned to full-time research on birds, after beginning to develop a pigeon colony in 1895. After considering early development studies and the evolutionary relationships that these studies revealed, Whitman may have decided, as others did, that more productive results in developmental studies lay elsewhere. All the comparative studies had established much about development and the significance of cleavage but less about larger questions, especially those concerning evolution. Embryologists had begun to move toward other, move manipulative, experimental approaches, whereas Whitman chose to move in other directions.

Even while stressing development in his early work, Whitman had also examined those other aspects of the natural history tradition perpetuated by Agassiz: heredity, behavior, life history, and the anatomical details of organisms. The behavior of leeches, for example, had raised for Whitman the question of whether they acted instinctively or had some other form of intelligence. By 1897, he had begun to focus on that work and to turn especially to pigeons for further evidence about behavior and evolution. Pigeons made sense at Chicago as marine organisms did not, because he now wished to address problems requiring live rather than prepared organisms. At first, Whitman transported his pigeons between Chicago and Woods Hole by train, carrying the birds with his other baggage so he could properly care for them. When the railroad officials finally forbade this practice and required the birds to travel by separate express, many more died—too many. "Indeed," Lillie recalled, "the transfer became an intolerable burden, and he relinquished his charge of affairs at Woods Hole rather than curtail his own research."[84] As Burkhardt (in this volume) shows, studying behavior of living organisms, and especially of birds, entailed considerable cost.

Although Whitman actually published very little of his work on pigeons, he did amass tremendous amounts of research results, which his student Oscar Riddle shepherded into press after Whitman's death. Riddle was just one of the several students who turned with Whitman to pigeons and to problems of heredity and evolution. Yet Riddle was undoubtedly the closest and most loyal follower, who spent most of his career working with Whitman's pigeon colony at Cold Spring Harbor where they were, in effect, banished after Whitman's death.[85]

The Chicago zoology department never fully embraced Whitman's particular brand of evolutionary or behavioral study, or his pigeons, in the way that they had the work on development and heredity, even though the university had begun with a strong interest in neurobiology and psychobiology. Students such as Riddle and Wallace Craig, who moved into behavioral studies, did not fit into zoology or even into biology more generally, as Whitman himself did not after 1897 or so. Perhaps this division resulted because Whitman had so effectively established the study of development and heredity as appropriate that the zoology department had trouble accepting this shift to another set of problems. Perhaps, as Burkhardt suggests, the department had trouble accepting Whitman's failure to publish and his reclusive retreat to his backyard bird cages. The reasons for Whitman's shift, as well as the subsequent resistance by the department that he had himself built so painstakingly and with such investment of personal resources, deserve further investigation, which will carry the story into the twentieth century.

Conclusion: The Chicago Style

In my view, a biological style is characterized by a shared set of problems regarded as appropriate, techniques regarded as useful, and approaches regarded as productive. Those sharing a style participate in similar sorts of day-to-day activities because they have similar attitudes and approaches. Chicago did develop a style of work, I believe, a style based on commitment to the study of organization of whole organisms (and populations) and to cooperative and comparative study.

The list of researches and researchers in zoology, and associated fields, also reveals a continued strong commitment to problems of development and heredity, as do the lists of courses and seminar topics. Similar questions were asked, namely, what morphological patterns occur, and what physiological processes shape the development of the whole individual? Specifically, how do the organism and its parts act as a whole? Using a combination of traditional and up-to-date histological and microscopical techniques, as well as study of the organisms in their natural environments with summer trips to the MBL, these researchers pursued a wide range of separate but related questions. Their repeated sojourns to Woods Hole undoubtedly helped to create the sense of shared mission that was lacking at other schools, for no other research institution sent so many students, younger investigators, and instructors to the MBL each year as Chicago did (or to any other research laboratory, it seems).[86]

Equally important in allowing Chicago to develop a persistent style of work was Whitman's neglect of undergraduate instruction, which permitted a highly specialized program of study. Other schools had to be more practical because, in addition to offering undergraduate study, they might also serve medical schools or agricultural interests. Chicago, as Pauly has shown, flirted with but never embraced medicine.[87] Also, those with older programs often were dominated by traditional natural historians, concerned with broad evolutionary questions and not specifically with early development. Some newer programs stressed experimental work, often on very specialized subjects, so that the leading researchers moved toward more radically manipulative experimental programs.[88] At Chicago, Whitman generally avoided hiring people with such concerns and simply did not listen if he was advised to pursue interests he did not find important. He wanted people who could work cooperatively together, who had shared convictions about problems, methods, and approaches but not necessarily any body of shared doctrine. By the time Lillie, who was much more open in approach and cooperative with administrators, took over, the pattern was successful and well established. Chicago's research style remained largely intact with such researchers as Lillie, Charles Manning Child, and Charles Benedict Davenport.[89]

What Lillie changed in the department at Chicago came primarily with his move to a more participatory approach to government and his concern with undergraduate as well as graduate teaching. He did not significantly change the style of research, although details and emphasis did shift, following similar lines of research with similar problems and similar approaches but in a new context of additional techniques and additional ideas. Lillie did add ecology to the department's offerings, but otherwise the departmental research and graduate training continued to concentrate on the study of development and heredity, with related work on evolution. Researchers such as Lewis Victor Heilbrunn, Albert William Bellamy, Benjamin Harrison Willier, William John Crozier, Warder Clyde Allee, Sewall Wright, Paul A. Weiss, Graham Phillips DuShane, Libby Hyman, Dorothy Price, Lincoln Domm, and Mary Juhn all fit in nicely at Chicago. The group at Chicago remained a cohesive community, at least for a while.[90]

In 1926, Lillie became sufficiently frustrated by the lack of adequate space for the growing group of researchers that he and his wife, Francis Crane Lillie, gave money to Chicago to construct the Whitman Laboratory. Designed as a research facility, separate from the teaching offices and classrooms, this laboratory attracted able researchers and undoubtedly helped to maintain Chicago's edge over other programs and its sense of "glamour."

Equipped with animal facilities, chicken yards, and other such advantages that Whitman and Lillie had long sought under the label of "inland laboratory" or "biological farm," the building became a workplace for the group studying various problems of sex inheritance and differentiation, as well as for some of the evolutionary and genetics work such as that done by Sewall Wright. Whitman would have approved of the laboratory and of the work that emerged from it.

A style of work thus characterized Chicago's contributions in zoology at a time when researchers there and elsewhere sought to define what biology should be like. The group shared problems and approaches and an attitude rather than a commitment to particular theories or conclusions. Elements of the style existed outside of Chicago as well, but I am aware of nowhere else where they came together in precisely the same coherent and cooperative way. Whitman surely deserves credit (or blame) for establishing this style at Chicago, which persisted beyond Whitman, into the twentieth century, although the extent remains to be explored more fully. Thus the style is not exclusively Whitman's but is, more properly, a Chicago style of biological work.

What this tells us about biology, and particularly biology in America, is that the particular combination of a strong-willed visionary individual with a formative resource-rich institution could produce a distinct tradition of biological work. Whether we label this characteristic work as a style or tradition does not matter. As long as we continue to examine what we mean by the label selected and to extend the exploration to research beyond Whitman, beyond Chicago, and even beyond America, the study of biological work and its context and its participants will help to illuminate the nature of biology.

Appendix: List of Chicago Ph.D. Degrees in Zoology. From Zoology Department Records, Chicago.

1894 Herbert Parlin Johnson, "A Contribution to the Morphology and Biology of the Stentors."
Frank Rattray Lillie, "The Embryology of the Unionidae."

1895 Albert Chauncey Eycleshymer, "Early Development of Amblystoma with Observations on Some Other Vertebrates."
William Albert Locy, "Contribution to the Structure and Development of the Vertebrate Head."

1896 Howard Stedman Brode, "A Contribution to the Morphology of Dero vaga."
Cornelia Clapp, "The Lateral Line System of Batrachus Tau."
Agnes Mary Claypole (Mrs. Robert O. Moody), "The Embryology and Oögenesis of Anurida marktima Guen."
Albert Davis Mead, "The Early Development of Marine Annelids."

1897 Charles Lawrence Bristol, "The Metamerism of Nephalis."
Samuel J. Holmes, "The Early Development of Planorbis trivolvis."
John P. Munson, "The Ovarian Egg of Limulus: A Contribution to the Problem of the Centrosome and Yolk-Nucleus."

1899 Emily Ray Gregory, "Observations on the Development of the Excretory System in Turtles."
Aaron Louis Treadwell, "The Cytogeny of Podarke."

1900 Michael Frederick Guyer, "The Spermatogenesis of Normal and Hybrid Pigeons."

1901 Elliott Rowland Downing, "The Spermatogenesis in Hydra."
Wilhelmina Entemann (Mrs. W. E. Key), "Coloration of Polistes (the common Paper Wasp)."
Ralph Stayner Lillie, "Excretory Organs of Arenicola cristata."
Virgil Everett McCaskill, "The Metamerism of Hirudo Medicinalis."
John McClelland Prather, "The Skeleton of Salaux Microdon."

1902 Eugene Howard Harper, "History of the Fertilization and Early Development of the Pigeon's Egg."

1903 Bennet Mills Allen, "The Development of the Ovary and the Testis in the Mammals."
William J. Moenkhaus, "The Development of the Hybrids between Fundulus heteroclitus and Menidia notata with Especial Reference to the Behavior of the Maternal and Paternal Chromatin."

1904 Charles Dwight Marsh, "The Plankton of Lake Winnebago and Green Lake."
John William Scott, "Studies in the Experimental Embryology of Some Marine Annelids."
Charles Zeleny, "Studies in Regulation and Regeneration."

1905 Lynds Jones, "The Development of the First Down and Its Relation to the Definitive Feather."
Horatio Hackett Newman, "The Morphogeny of the Chelonian Carapace."

1906 James Francis Abbott, "The Morphology of Coeloplana."

1907 Frank Eugene Lutz, "The Variations and Correlations of the Taxonomic Characters of Gryllus."
Oscar Riddle, "The Genesis of Fault-Bars in Feathers and the Cause of Alterations of Light and Dark Fundamental Bars."
Victor Ernest Shelford, "The Life-Histories and Larval Habits of the Tiger Beetles."
Charles Henry Turner, "The Homing of Ants: An Experimental Study of Ant Behavior."

1908 Charles Christopher Adams, "The Geographic Variations and Relations of Io."
Mary Blount, "The Early Development of the Pigeon's Egg from Fertilization to the Organization of the Periblast."

Wallace Craig, "Expression of Emotions in the Pigeon."
John Thomas Patterson, "Gastrulation in the Pigeon's Egg."
Katashi Takahashi, "Histogenesis of the Lateral Line System in Necturus."
George Washington Tannreuther, "History of the Germ Cells and Early Embryology of Certain Aphids."

1909 Marian Lydia Shorey, "The Effect of the Destruction of Peripheral Areas on the Differentiation of the Neuroblasts."
H. L. Wieman, "A Study in the Germ Cells of Leptinotarsa signaticollis."

1910 George William Bartelmez, "The Bilaterality of the Pigeon's Egg: A Study in Egg Organization."

Acknowledgments

I wish to thank Adele Clarke, Richard Creath, Anne Gully, Gregg Mitman, and the participants in the Friday Harbor conference for their reading of preliminary drafts and for their excellent suggestions. The Archives of the Joseph Regenstein Library at the University of Chicago generously granted permission to quote from their holdings. Research for this project has in the past been supported by grants from the National Science Foundation and Arizona State University.

Notes

1. J. M. Richards, *Modern Architecture* (Baltimore: Penguin Books, 1940, revised 1962), p. 71.

2. John F. Dryfhout, *The Work of Augustus Saint-Gaudens* (Hanover, N.H.: University Press of New England, 1982), p. 30.

3. Elihu Gerson, "Styles of Scientific Work and the Population Realignment in Biology, 1880–1930," presented at the Conference on History and Philosophy of Biology, Granville, Ohio, 1983; idem, "Scientific Work and Social Worlds," *Knowledge: Creation, Diffusion, Utilization,* 1983, *4:* 357–377.

4. Richard J. Storr, *Harper's University. The Beginnings* (Chicago: University of Chicago Press, 1966), pp. 3–6.

5. Storr, *Harper's University,* p. 6.

6. W. Carson Ryan, *Studies in Early Graduate Education. The Johns Hopkins University, Clark University, the University of Chicago* (New York: Carnegie Foundation, 1939), pp. 105–106, 108; Storr, *Harper's University,* p. 47.

7. Stanford was begun officially in 1887 but only really began recruiting faculty in 1891. As David Starr Jordan wrote to Harper, "I find that we are likely to cross each other's path more than once in the selection of our faculty," (28 July 1891); Storr, *Harper's University,* p. 69. Ronald Rainger points out, through personal correspondence and based on his work with the Columbia Archives, that Henry Fairfield Osborn also recognized the competition for good faculty at the same time.

8. Lincoln C. Blake, "The Concept and Development of Science at the University of Chicago, 1890–1905" (Ph.D. dissertation, University of Chicago, 1966), chap. 3, discusses a number of documents available in the University of Chicago Archives.

9. Harper to Herrick, 15 June 1892, Whitman Collection, Department of Special Collections, The Joseph Regenstein Library, University of Chicago (hereafter Whitman Collection, Chicago). For a discussion of Herrick's hiring and subsequent problems at Chicago, see also the biography by his brother: Charles Judson Herrick, "Clarence Luther Herrick. Pioneer Naturalist, Teacher, and Psychobiologist," *Transactions of the American Philosophical Society,* n.s., 1955, *47:* 1–85.

10. Herrick to Harper, especially 9 May 1892, Whitman Collection, Chicago; also letters in William Rainey Harper Papers, Chicago (hereafter Harper Papers).

11. Blake, "The Concept," p. 77.

12. Whitman to Harper, 19 December 1891, Whitman Collection, Chicago.

13. Storr, *Harper's University,* pp. 78–79.

14. Jordan to Whitman, May or June 1891 [Whitman rarely dated his letters], David Starr Jordan Papers, Stanford University Archives, Stanford, Calif. I thank Keith Benson for bringing this document to my attention. Orrin Leslie Elliot, *Stanford University. The First Twenty-five Years* (Stanford, Calif.: Stanford University Press, 1937), p. 62, confirms that Whitman was officially offered but declined a position.

15. Ryan, *Studies,* pp. 56–57, 59.

16. Ibid., pp. 62–63. Ryan cites a contemporary photograph of Frank Lillie's as evidence of the number who went with Whitman to Chicago.

17. Mall to Harper, 27 January 1892, Harper Papers, Chicago. Ironically, after all Mall's efforts to help Harper attract Whitman, Mall remained at Chicago only one year himself before moving to the Johns Hopkins Medical School as chairman of Anatomy when it opened in 1893. Mall corresponded with Harper's brother Robert, a friend of Mall's in Leipzig, rather than directly with William Rainey, as discussed by Dorothy Ross, *G. Stanley Hall: The Psychologist as Prophet* (Chicago: University of Chicago Press, 1972), p. 226, and decided early during his Clark stay that he wanted to leave and that Chicago was attractive.

18. Whitman to Harper, 21 March 1892, Harper Papers, Chicago.

19. Mall to Harper and Whitman to Harper, January through May 1892, Harper Papers, Chicago.

20. Ross, *G. Stanley Hall,* p. 226; letters, Harper Papers, Chicago; G. Stanley Hall, *Life and Confessions of a Psychologist* (New York: D. Appleton, 1924), pp. 294–297, reveal Hall's disgust and disappointment with Harper's move, which he compared to the action "of a housekeeper who would steal in at the back door to engage servants at a higher price," p. 296.

21. Ross, *G. Stanley Hall,* p. 227.

22. Ryan, *Studies,* p. 62.

23. Storr, *Harper's University,* p. 83; Blake, "The Concept," chap. 3. The charge against Harper was that he had opportunistically seized on the Clark group and had not kept promises to Herrick.

24. Herrick to Harper, 14 November 1898, Whitman Collection, Chicago. Herrick went to New Mexico and in 1898 wanted Chicago to set up a biological station for him on Lake Chapala in central Mexico, essentially to compensate for Harper's past in-

justices. He claimed that the station would follow along the lines of the Naples Zoological Station and requested a salary for himself and a small stipend for an assistant. Harper replied briefly and negatively.

25. Whitman, "Biological Instruction in the Universities," *American Naturalist,* 1887, *21:* 507–519.

26. Edward S. Morse, "Charles Otis Whitman," *National Academy of Sciences Biographical Memoirs,* 1912, *7:* 269–288; Frank R. Lillie, "Charles Otis Whitman," *Journal of Morphology,* 1911, *22:* xv–lxxvii; Jane Maienschein, introduction to *Defining Biology. Lectures from the 1890s* (Cambridge, Mass.: Harvard University Press, 1986).

27. Whitman, "Biological Instruction," p. 507.

28. Ibid., pp. 516–517.

29. Whitman, "Specialization and Organization," *Biological Lectures delivered at the Marine Biological Laboratory of Wood's Holl,* 1891, *1890:* 1–26; "The Naturalist's Occupation," *Biol. Lect.,* 1891, *1890:* 27–52.

30. Philip Pauly, "The Appearance of Academic Biology in Late Nineteenth-Century America," *Journal of the History of Biology,* 1984, *17:* 369–397, on pp. 382–387; Donna Jeane Haraway, "The Marine Biological Laboratory of Woods Hole: An Ideology of Biological Expansion," unpublished manuscript.

31. Whitman to Harper, 19 December 1891, Harper Papers, Chicago.

32. Whitman to Harper, n.d., probably early 1892, Harper Papers, Chicago.

33. Mall to Harper, 27 January 1892, Harper Papers, Chicago.

34. Blake, "The Concept," p. 135.

35. Gerald Geison has begun to examine the Chicago program in physiology, for example; see "International Relations and Domestic Elites in American Physiology, 1900–1940," in Gerald L. Geison, ed., *Physiology in the American Context, 1850–1940* (Bethesda, Md.: American Physiological Society, 1987), pp. 115–154.

36. Morse, "Whitman," p. 269.

37. Helen Frost to Lillie about Whitman's deep disappointment with the MBL's lack of support, 4 October 1911, Lillie Papers, Chicago: Whitman to Conklin and Wilson, letters throughout the 1890s, Whitman Collection, MBL Archives, Woods Hole, Mass.

38. Blake, "The Concept," p. 135.

39. Thomas Wakefield Goodspeed, "Helen Culver," *The University of Chicago Biographical Sketches,* vol. 2 (Chicago: University of Chicago Press, 1925), pp. 77–99, on p. 94.

40. Goodspeed, *The University,* p. 95; Helen Culver to Harper, 14 December 1895, Whitman Collection, Chicago.

41. Howard S. Miller, *Dollars for Research* (Seattle: University of Washington Press, 1970).

42. Goodspeed, *The University,* p. 96.

43. The new complex generated some of the usual controversy surrounding biological work at the time. Even Helen Culver received at least one letter criticizing her gift to the university. As antivivisectionist Mary Totten wrote, "We know, from the testimony of the operators themselves, that the greatest and most inexcusable cruelties that now have place on this earth take place in such laboratories under the name of 'scientific research'—and that, in such an institution as is proposed to be established

by your gift, these cruelties will be on a scale more appalling than ever." Such traditional attempts to sidetrack biology had little practical effect on Helen Culver or on the university.

44. Frank R. Lillie, *The Woods Hole Marine Biological Laboratory* (Chicago: University of Chicago Press, 1944), esp. p. 60; Jane Maienschein, "Early Struggles at the Marine Biological Laboratory over Mission and Money," *Biological Bulletin,* 1985, *168 suppl:* 192–196.

45. Whitman to Conklin, 2 March 1902, Whitman Collection, MBL Archives.

46. The MBL Archives and *Trustees Minutes* makes clear that Whitman had de facto retired after 1902. See "The Resignation of Prof. Whitman as Director of the Marine Biological Laboratory at Woods Hole, Mass.," *Anatomical Record,* 1908, *2:* 380–382.

47. Whitman to Harper, n.d., probably late 1891 or early 1892, Harper Papers, Chicago.

48. Whitman, et al., to the President and Board of Trustees, 3 December 1906, Presidential (Judson) Papers, Chicago; Adele E. Clarke, "A Biological Case Study: The Reproductive Research Enterprise at the University of Chicago," unpublished manuscript, 1984; idem, "Research Materials and Reproductive Physiology in the United States, 1910–1940," in Geison, *Physiology in the American Context,* pp. 323–350. Both articles discuss the Chicago efforts to establish a biological farm.

49. Whitman to Harper, 3 May 1899, Harper Papers, Chicago. Also, as Philip Pauly has pointed out, in personal correspondence, one should never underestimate the importance of prejudice against Watase as a non-aristocratic Japanese scientist.

50. Whitman to Harper, 12 September 1899, Harper Papers, Chicago.

51. As evident from letters between Lillie and others after Whitman's death, as Lillie was constructing a biography.

52. Blake, "The Concept," pp. 124–125; H. H. Newman, "History of the Department of Zoology in the University of Chicago," *BIOS,* 1948, *19:* 215–239.

53. Donald Fleming, *William H. Welch* (Boston: Little, Brown and Company, 1954), discusses Mall's teaching.

54. Chiyomatsu Ishikawa, "Professor Charles Otis Whitman," translated by Shigro Yamanouchi from *Magazine of Zoology,* 1911, *23:* 14, Whitman Papers, Chicago.

55. The MBL *Annual Reports* list the teachers for each session, as well as investigators and students.

56. Cornelia Clapp, "Some Recollections of the First Summer at Woods Hole, 1888," *Collecting Net,* 1927, *2 (4):* 3, 10.

57. Ann H. Morgan, et al., "Cornelia Maria Clapp," *Mount Holyoke Alumnae Quarterly,* 1935, *19:* 1–9.

58. E. Ray Lankester, "An American Sea-side Laboratory," *Nature,* 1880, *25:* 497–499.

59. Whitman often wrote about the need for specialization and cooperation working together in the same community. See, for example, "Specialization and Organization."

60. On the various cell-lineage studies, see Jane Maienschein, "Cell Lineage, Ancestral Reminiscence, and the Biogenetic Law," *J. Hist. Biol.,* 1978, *11:* 129–158. Wilson, Conklin, Lillie, and others often mentioned the way in which Whitman encouraged them to work together, comparing the results of their cell-lineage studies.

61. Lillie, "Autobiography," 1926, unpublished manuscript, Lillie Collection, MBL Archives.

62. Hall, *Life and Confessions,* p. 289.

63. Ishikawa, "Whitman," wrote that he was told when he visited Chicago that Whitman rarely ever came to the University to work but that he stayed with his pigeons at home; Blake, "The Concept," p. 124, commented that Whitman "justified his withdrawal from classroom contact with his students by insisting that American institutions were too patronizing and paternalistic." Also see Newman, "A History," pp. 221–222.

64. Newman, "A History," p. 221.

65. The full argument for this claim demands a much larger study, of course, but I feel confident of its accuracy. See list of Ph.D. degrees awarded, Zoology Department Records, Chicago, in the Appendix.

66. The list of Chicago Ph.D.'s in zoology shows that thirty-three remained professionally in zoology, including most of the men; Zoology Department Papers, Chicago.

67. Morse, "Whitman," p. 271.

68. Clapp to Lillie, 19 April 1911, Lillie Papers, Chicago; Lillie, "Whitman," p. xviii; Morse, "Whitman," p. 274.

69. Morse, "Whitman," p. 274; On Leuckart, see Klaus Wunderlich, *Rudolf Leuckart* (Jena: Gustav Fischer Verlag, 1978); a list of his students appears on pp. 41–51.

70. Charles Otis Whitman, "The Embryology of Clepsine," *Quarterly Journal of Microscopical Science,* 1878, *18:* 215–315; Morse, "Whitman," p. 274.

71. Lynn Nyhardt, "The Career of Experimental Embryology in Germany, 1890–1925" (Ph.D. dissertation, University of Pennsylvania, 1986), discusses the German morphological tradition, as does E. S. Russell, *Form and Function* (London: John Murray, 1916), and *The Interpretation of Development and Heredity,* 1930 (Freeport, N.Y.: Books for Libraries Press, 1972).

72. Whitman, "The Embryology," pp. 263–264.

73. Charles Otis Whitman, *Methods of Research in Microscopical Anatomy and Embryology* (Boston: S. E. Cassino, 1885); Lillie, "Whitman," pp. lxxiv–lxxvii, contains the list of other works.

74. For discussion of that laboratory, see Ernst J. Dornfeld, "The Allis Lake Laboratory," *Marquette Medical Review,* 1956, *21:* 115–144.

75. Charles Otis Whitman, "Oökinesis," *J. Morph.*, 1887, *1:* 228–252.

76. Whitman often argued for the importance of studying heredity and development, physiology and morphology. Especially see MBL *Annual Reports.* Also discussed in Jane Maienschein, "Physiology, Biology, and the Advent of Physiological Morphology," in Geison, *Physiology in the American Context,* pp. 177–193.

77. For discussion of Whitman's emphasis on cooperation, see, for example, Whitman, "Specialization and Organization;" and Maienschein, *Defining Biology,* esp. pp. 17–21.

78. Keith Benson, "Naples Stazione Zoologica and Its Impact on the Emergence of American Marine Biology: Entwicklungsmechanik and Cell Lineage Studies," *J. Hist. Biol.* (forthcoming), discusses the importance of technical developments and Whitman's role in introducing the latest techniques from Europe to the United States.

79. Alice Levine Baxter, "Edmund Beecher Wilson and the Problem of Development" (Ph.D. dissertation, Yale University, 1974), and later articles.

80. Charles Otis Whitman, "The Inadequacy of the Cell Theory of Development," *Biol. Lect.*, 1894, *1893:* 105–124, on p. 105.

81. Whitman, "Inadequacy," p. 106 (emphasis in the original).

82. Ibid., p. 124, quoting Huxley.

83. See, for example, Frank R. Lillie, *Problems of Fertilization* (Chicago: University of Chicago Press, 1919); Charles Manning Child, *Patterns and Problems of Development* (Chicago: University of Chicago Press, 1941); Jacques Loeb, *The Organism as a Whole* (New York: G. P. Putnam's Sons, 1916).

84. Lillie, "Whitman," p. lxiii; Ishikawa, "Whitman," p. 18.

85. George W. Corner, "Oscar Riddle," *Natl. Acad. Sci. Biog. Mems.*, 1974, *45:* 427–465; Jane Maienschein, "Riddle, Oscar" *Dictionary of Scientific Biography*, supp. *2* (N.Y.: Charles Scribners Sons, forthcoming); Oscar Riddle, ed., *Posthumous Works of Charles Otis Whitman,* vol. 1, "Orthogenetic Evolution in Pigeons"; vol. 2, "Inheritance, Fertility, and the Dominance of Sex and Color in Hybrids of Wild Species of Pigeons"; and Harvey A. Carr, vol. 3, "The Behavior of Pigeons" (Washington, D.C.: Carnegie Institution of Washington, 1919).

86. The MBL *Annual Reports* record where the MBL participants came from.

87. Pauly "The Appearance," pp. 382–387.

88. The series of histories of biology departments in *BIOS* details many of the early programs and how they changed.

89. Victor Chandler Twitty, *Of Scientists and Salamanders* (San Francisco: W. H. Freeman and Company, 1966). The flavor of the department, as well as the sense of importance of the work there, trickled out to prospective graduate students. As Victor Twitty recorded, when he began to apply to graduate schools some appealed "because of the glamour of their very names." Chicago and Harvard fit into that category. Twitty had just finished reading a book by the "celebrated (and controversial)" Child and found it exciting. Chicago offered Twitty a tuition scholarship "but left unresolved the matters of food and shelter," while his Harvard application was "returned without comment" because he had forgotten to sign it. When he received a better offer from Yale, he decided to go there rather than to Chicago. In retrospect, he realized that this was a fortunate choice, but at the time Chicago clearly offered more glamorous and exciting prospects than Yale. Various letters in the Harrison Collection, Yale University Manuscripts and Archives, as well as in the Lillie Collection, Chicago, 1919, show that Yale's chair of zoology, Ross Granville Harrison, sent one daughter to Chicago for undergraduate work and considered sending another to graduate school, partly because of the intellectual atmosphere there.

90. Newman, "History," pp. 226–228. Gregg Mitman, personal correspondence, points out that this cohesiveness should not be surprising, because Chicago experienced considerable "inbreeding." A high percentage of those hired by Chicago were Chicago products, and some department members were actually related. Mitman also suggests that as time went on and as some sorts of study (embryology, genetics, sex research) received greater support than others (ecology), some researchers began to feel alienated from the dominant Chicago style of work.

Part Three

Working at the Boundaries of Biology

Richard W. Burkhardt, Jr.

6 Charles Otis Whitman, Wallace Craig, and the Biological Study of Animal Behavior in the United States, 1898–1925

The modern scientific discipline of ethology, which identifies its subject as the biology of animal behavior, took shape in Europe in the 1930s and 1940s under the leadership of the Austrian naturalist Konrad Lorenz and the Dutch naturalist Nikolaas Tinbergen. When in 1953 the American psychologist Daniel Lehrman assessed the theoretical foundations of the scientific system Lorenz and his European colleagues had developed, Lehrman noted that the disciplinary settings of animal behavior study in Europe and America differed from each other. In Europe most of the students of animal behavior were "zoologists, physiologists, zoo curators or naturalists," whereas in America they were psychologists. However, of those investigators whom Lorenz himself identified as the "three great pioneers of ethology" of the first part of the century, only one, Lorenz's own mentor Oscar Heinroth, was European. The other two, Charles Otis Whitman and Wallace Craig, were both American, and both were trained as zoologists.[1]

Whitman and Craig were not the only Americans interested in the study of animal behavior in the early twentieth century. There was an efflorescence of behavior studies in the United States from the end of the nineteenth century up to World War I. Indeed the amount, the quality, and the diversity of animal behavior studies conducted in the United States in this period might lead one to think that by the second or third decade of the century, behavior studies could have become, like genetics, an area in which America enjoyed a genuine preeminence in the world scientific community. Only in Germany were behavior studies being pursued on such a wide variety of fronts. In addition to Whitman and his behaviorally oriented students—a group that included Craig, William Morton Wheeler, Samuel J. Holmes, and Oscar Riddle—the United States could claim both of the principals of the famous Loeb-Jennings debate; Charles H. Turner, the black biologist whose experimental investigations of the color sense and form sense in bees predated Karl von Frisch's work on these topics; Francis H. Herrick and his pioneering studies on the domestic life of birds; the animal ecologists C. C. Adams and Victor Shelford; the animal psychologists Margaret F. Washburn, Robert M. Yerkes, and John Broadus Watson; and, beginning in 1911, the *Journal of Animal Behavior,* the first scientific journal devoted exclusively to the study of animal behavior.

It cannot be assumed that the various American investigators of animal behavior at the beginning of the twentieth century shared a common view of either the way in which animal behavior should be studied or how the study of animal behavior related to the broader disciplines of biology or psychology. Whitman's approach to behavior was not the same as Loeb's. Watson's approach to behavior was not the same as Yerkes's or Washburn's. For the most part, however, the more prominent of these investigators appear to have believed that their particular approaches to behavior constituted the means by which the broader disciplines of which they were a part could be restructured and developed and that such a restructuring was not only to the advantage of those disciplines but also in the interests of the individual investigators themselves. They did not seek to establish the study of animal behavior as an independent discipline. They sought instead to make behavioral studies an integral part of biology or psychology and to reform these broader enterprises in the process. The different approaches they took in this regard reflected not only their respective disciplinary affiliations and institutional situations but also their personal aspirations.

Whitman's own position with regard to the development of new specialties could not have been more clear. He consistently maintained that specialized studies were only to be pursued as parts of the organized whole that was biology. In 1897, in his presidential address to the Society of American Naturalists that described what the functions of a biological station should be, he urged that "we need to get more deeply saturated with the meaning of the word 'biological', and to keep renewing our faith in it as a governing conception. Our centrifugal specialties have no justification except in the *ensemble.*" He promoted this same view of biology in 1902 when he spelled out the need for a special experiment station—a "biological farm" where "the study of life-histories, habits, instincts and intelligence" would be conducted along with "the experimental investigation of heredity, variation, and evolution." In the same year Whitman's former student, William Morton Wheeler, proposed that the term "ethology" be used for that part of zoology "which must deal very largely with instincts, and intelligence as well as with 'habits' and 'habitus' of animals." Wheeler declared that scientists were "on the eve of a renascence in zoology," a new era in which this broadly defined ethology would occupy a central position with respect to the other zoological disciplines.[2]

Whitman, it should be underscored, appreciated that biology could not be defined solely in terms of its parts or even the whole of its subject matter. Biology as he came to perceive it involved a particular approach or governing conception, and it needed special institutions if it was to flourish. He identified his approach as "experimental natural history" in contradistinction to what he saw as the more narrow, physiological approach of Loeb in which, Whitman complained, "instinct reduces itself in the last analysis to heliotropism, stereotropism, and the like" and "the whole course of evolution drops out of sight altogether." The work of Loeb, G. H. Parker, and others on tropisms, other reactions to stimuli, sensory discrimination, and sensory physiology were well suited to the new zoological laboratories that were being developed in university departments around the country. Whitman sought a biological farm because, as he put it, "the laboratory is too narrow, and the world too wide for the continuous study of living organisms, under conditions that can be definitely known and controlled."[3]

Whitman's biological farm was never established, and the hopes of Whitman and Wheeler to make behavior studies an integral part of biology were not realized. Animal behavior studies failed to gain a secure place in American biology. The biologists who studied behavior did not develop a common view of the aims, methods, or even subject matter of behavior studies. Nor did they establish the kinds of institutional bases that might have served to provide their activities with a greater unity, clearer focus, or more secure future. The psychologists in the meantime fared scarcely better. They too were seriously divided with regard to the aims, methods, and subject matter of their own discipline. By the end of the second decade of the century, biologists and psychologists alike, with but few exceptions, seem to have concluded that animal behavior studies were not the most attractive area in which to invest their energies and resources.[4]

This paper will focus primarily on the work and careers of the biologist Charles Otis Whitman and his student, Wallace Craig. Their work and careers constitute only part of the story of animal behavior studies in the United States in the early twentieth century. Their cases illustrate the problems of pursuing behavioral studies of a naturalistic, experimental, and evolutionary nature at a time when the institutional support for such studies did not exist and when other models of research were more successful in laying claim to the various resources that the discipline of biology, the American scientific community, and the broader American society had to offer.

Charles Otis Whitman

In 1898, when he delivered his now classic lecture at Woods Hole on animal behavior, Charles Otis Whitman was arguably the most influential biologist in America. He had directed the Marine Biological Laboratory at Woods Hole since its founding in 1888. He had been the head of the Department of Biology (and later the Department of Zoology) at the University of Chicago since Chicago opened its doors in 1892. Before assuming the headship at Chicago, he had served as the first chair of the department of zoology at Clark University. He had founded the *Journal of Morphology* in 1887, and had been the key figure in establishing in 1890 the American Morphological Society (American Society of Zoologists after 1902; see Appel in this volume). The range of his interests and expertise was virtually unexcelled among his zoologist or biologist colleagues.[5]

Ethologists have typically reported that their fascination with animal life began when they were children, long before they were old enough to consider what it meant to be a scientist. Whitman's personal history fits this model. As a boy, he was an ardent collector of birds, distinguishing himself by his skill in mounting the specimens he shot. He also kept a host of live animals as pets, including amphibians, birds, and mammals. Among the birds were pigeons. He is said to have "sat and watched them by the hour, intensely interested in their feeding, their young, and in everything that they did." It does not follow, however, that when Whitman later studied pigeon behavior as a professional zoologist he was, as Lorenz blithely described him, "happily free from even a working hypothesis," or that he arrived at his conclusions by "just observing the pigeons . . . [he] loved." Whitman's thoughts on animal behavior were intimately tied to his broader concerns as a biologist.[6]

The doctoral dissertation Whitman wrote under Rudolf Leuckart at Leipzig in 1878 was an embryological study, focusing on the development of the leech *Clepsine*. Over the course of the next two decades, Whitman's studies of leeches expanded to encompass their systematics, phylogeny, ecology, and behavior. *Clepsine* eventually figured prominently in his animal behavior lecture of 1898. So too did *Necturus,* the fresh-water salamander Whitman began studying when he was director of the Allis Lake Laboratory in Milwaukee (1886–89), and pigeons, the birds that had fascinated him as a boy and that were to be the almost constant focus of his research for the last fifteen years of his life. His idea of proper biological research was to take a single species or a group of closely related species and study it in all aspects of its existence. In 1899 he defined biology as: "The life-histories of animals, from the primordial germ-cell to the end of the life-cycle; their daily, periodical, and seasonal routines; their habits, instincts, intelligence, and peculiarities of behavior under varying conditions; their geographical distribution, genetic relations and oecological interrelations; their physiological activities, individually and collectively; their variations, adaptations, breeding and crossing." His focus on the living animal in all its aspects is epitomized in the claim that his students greeted each other not with the question, "What is your special field?" but instead the question, "What is your beast?"[7]

Whitman's famous paper "Animal Behavior" was delivered as two lectures in the biological lecture series at the Marine Biological Laboratory in Woods Hole in the summer of 1898. The paper was a masterpiece and now stands as a classic in the history of ethology. Of this paper Frank R. Lillie wrote: "No other of [Whitman's] papers illustrates better the qualities of his genius: the selection of a fundamental problem; painstaking study; publication only after years of observation and reflection; skill in laying bare the simple basis of an apparently complex group of phenomena; a grasp of the subject in all its bearings; and the use of the comparative or phyletic method of attack."[8] In this fifty-four-page tour de force, Whitman set forth his views on the proper methods of studying behavior, the nature of instinct and the importance of studying instinct from a phylogenetic standpoint, the means by which behavior has evolved, and the relations—both ontogenetic and phylogenetic—between instinct and intelligence. In the context of understanding instinct, he also continued to sound his old theme of organization, which had been central both to his earlier studies of morphogenesis and to his views of how the discipline of biology ought to develop.

Whitman maintained that if one were to understand the behavior of any animal species, one had to have a thorough knowledge of the animal's entire behavioral repertoire. The deceptive quiet of *Clepsine,* he pointed out, was such that an observer unfamiliar with *Clepsine's* habits would almost certainly draw the wrong conclusions about its sensitivity to stimuli. Similarly, he explained, it had taken him considerable experience with *Necturus* adults and two whole seasons rearing *Necturus* young before he appreciated "the extreme timidity of these animals," a timidity that was "so deep-seated and persistent that one can form only a poor idea of it without considerable actual contact with it."[9]

Whitman also insisted that in order to understand animal behavior one had to study it under natural conditions, one had to "observe and experiment under conditions that ensure *free behavior.*" He would return to this point four years later in his "biological farm" proposal, stressing that being able to study animals under natural conditions in fact required special institutional arrangements, not those of the zoological or botanical parks familiar to, and frequented by, the public. Only under special circumstances could the investigator "have the unbroken quiet required in delicate observation, or expect natural behavior from the forms occupying his attention."[10]

Whitman was particularly concerned with instinctive behavior and how instinct and intelligence were related. For him, the "first criterion of instinct" was that it could "be performed by the animal without learning by experience, instruction, or imitation." The fact that an animal's acts were adapted to purposeful ends, he explained, was not in itself proof of intelligence on the animal's part. In describing the food-taking behavior of *Necturus,* he wrote: "Its movements in approaching and snatching a piece of meat, as if it were a living object are, then, those characteristic of the species, not because they are measured and adapted to a definite end by intelligent experience, but because they are organically determined; in other words, depend essentially upon a specific organization."[11]

Whitman did not suppose that in sneaking up on its prey, *Necturus* had any appreciation of the importance of stealth. It was quite blind to the significance of its action. If *Necturus* had to depend on its intelligence, Whitman said, it was "difficult to see how it could escape immediate extinction." The instincts of the creature were what assured its welfare. Likewise Whitman found pigeons incubating their eggs to be oblivious to the biological function of their actions. He maintained:

> It is quite certain that pigeons are totally blind to the meanings which we discover in incubation. . . . They sit because they feel like it, begin when they feel impelled to do so, and stop when the feeling is satisfied. Their time is generally correct, but they measure it as blindly as a child measures its hours of sleep. A bird that sits after failing to lay an egg, or after its eggs have been removed, is not acting from "expectation," but because she finds it agreeable to do so and disagreeable not to do so. The same holds true of the feeding instinct. The young are not fed from any desire to do them any good, but solely for the relief of the parent.[12]

To Whitman, it was clear that instinct and organization were "two aspects of one and the same thing." Both were to be understood not only in terms of their development in the individual but also in terms of their evolutionary past. As he put it, "Instinct and structure are to be studied from the common standpoint of phyletic descent."[13]

Instincts were like organs, Whitman maintained, in the way they sometimes remained unchanged through a whole group of organisms. They therefore could be used, like organs, to reconstruct phylogenies. Most contemporary evolutionists, Whitman believed, did not understand how far back the phylogenies of instincts had to be traced. They did not appreciate how deeply rooted instincts were. Writers like G. J. Romanes were too ready to view instincts as "disconnected phenomena of independent origin" and not to see their "connected genealogical history." A major source of this view, Whitman thought, was the idea that instincts arose through the inheritance of acquired characters, that "an instinct could become gradually stamped into organization by long-continued uniform reactions to environmental influences." Even when biologists thought in terms of natural selection, Whitman said, the central question had been:

> How can intelligence and natural selection, or natural selection alone, initiate action and convert it successively into habit, automatism, and congenital instinct? In other words, the genealogical history of the structure basis being completely ignored, how can the instinct be mechanically

> rubbed into the ready-made organism? Involution instead of evolution; mechanization instead of organization; improvisation rather than organic growth; specific *versus* phyletic origin. . . . How long this blunder-miracle had to be repeated before it happened all the time does not matter. Purely imaginary things can happen on demand.[14]

Whitman's own approach was exemplified in his analysis of the way *Necturus* pauses before grasping its prey. This particular behavior pattern, he believed, represented "an instinct the history of which may be coextensive with the evolution of the animal." In his view, "Very early in the vertebrate phylum, possibly at its dawn, the chief characters of the instinct, as we now find it, were probably fixed in structural elements differing from those in *Necturus* only in superficial details." Similarly, though agreeing with Darwin that the pointing of dogs was not initially the result of training, Whitman was not satisfied with Darwin's explanation that the original tendency to point in dogs arose as an accidental variation, the cause of which was unknown. Whitman thought it was more appropriate to consider such variations as "manifestations of instinct roots of more or less remote origin."[15]

Whitman clearly believed that the study of animal behavior could illuminate not only phylogenetic questions but also the issue of the mechanisms by which evolution takes place. In addition to embracing an orthogenetic view of evolution that stressed that the modifications in instincts that appeared were a function of the organic constitution of the organism, he was keen to attack the Lamarckian idea of the inheritance of acquired characters. Darwin had argued in the *Origin of Species* that the instincts of neuter castes of insects could not be explained by the inheritance of acquired characters, but Darwin had still believed that certain other instincts were best explained in those terms. Whitman took the stronger, neo-Darwinian position of August Weismann, claiming: "Repetition may become habit and produce marked effects on the nervous mechanism or other organs; but the individual structure so affected is not continued from generation to generation, so that the effects are cancelled with each term of life, and there is no conceivable way by which they could be stamped upon the germs and so carried on cumulatively."[16] The importance to Whitman of denying the inheritance of acquired characters was that it meant that instinct could not be interpreted as "lapsed intelligence." Whitman's position was that "instinct preceeds intelligence both in ontogeny and phylogeny, and it has furnished all the structural foundations employed by intelligence." Instinct was thus "the actual germ of mind."[17]

To study how instinct graded into intelligence, it was necessary, Whitman believed, to study animals with complex instincts, the automatic character of which was uncontested. He regarded pigeons as perfect for this purpose, because pigeons not only possessed such instincts but also could be studied comparatively. He tested how the wild passenger pigeon, the tamer ring-neck pigeon, and the domesticated dovecot pigeon reacted to having their eggs placed just outside their nests. He found that it was only the most domesticated of these three species, the dovecot pigeon, that would reclaim as many as two eggs placed outside the nest. He explained this more fully as follows: "Under conditions of domestication the action of natural selection has been relaxed, with the result that the rigor or instinctive coordinations which bars alternative action is more or less reduced. Not only is the door to choice thus unlocked, but more varied opportunities and provocations arise, and thus the internal mechanisms and the external conditions and stimuli work both in the same direction to favor greater freedom of action."[18]

Whitman did not regard what went on under domestication, however, as the antithesis of what went on in nature. He supposed to the contrary that "domestication merely bunches nature's opportunities and thus concentrates results in forms accessible to observation." His concluding remarks on the subject, supported by statements from Conway Lloyd Morgan, William James, and Herbert Spencer, bear repeating here: "Superiority in instinct endowments and concurring advantages of environment would tend to liberate the possessors from the severities of natural selection; and thus nature, like domestication, would furnish conditions inviting to greater freedom of action, and with the same result, namely, that the instincts would become more plastic and tractable. Plasticity of instinct is not intelligence, but it is the open door through which the great educator, experience, comes in and works every wonder of intelligence."[19]

Such were the major conceptual thrusts of Whitman's animal behavior paper of 1898. Throughout, the paper was rich in observations and insights bearing on fundamental questions regarding instinct and the evolutionary process. It was based on careful, long-term observations of naturally occurring behavior, supplemented by experiments. Although oriented toward theoretical matters, the paper also included some excellent descriptions. On the face of it, it would seem to have been the sort of paper that could have served as a powerful model for further biologically oriented researches on animal behavior.

Whitman's model, however, was not without its drawbacks. The enterprise of reconstructing phylogenies had less appeal at the beginning of the twentieth century than it had had in the previous quarter century, and Whitman's organismic and orthogenetic viewpoint lacked the attractive simplicity of contemporary views of heredity and evolution that focused on unit characters. Doubts about whether the animal mind was accessible to scientific investigation may also have cast a shadow on Whitman's interest in the evolution of animal intelligence. On top of these problems, the practice that Whitman's model of research required was formidable. The time and expense involved in long-term detailed studies of the complete behavioral repertoires of different animal species among the higher animals were such as to dissuade all but the most committed investigators (or the most ardent animal lovers) from proceeding very far in such work. Casting a critical eye on the recent forays of psychologists into behavior studies, Whitman observed in 1899 that it would not be a simple matter for comparative psychology to be established as a science in its own right, because

> any attempt to soar to "the nature and development of animal intelligence," except through the aid of long schooling in the study of animal life, is doomed to be an Icarian flight. . . . The qualification absolutely indispensable to reliable diagnosis of an animal's conduct is an intimate acquaintance with the creature's normal life, its habits and instincts. Little can be expected in this most important field of comparative psychology until investigators realize that such qualification is not furnished by parlor psychology. It means nothing less than years of close study,—the long-continued patient observation, experiment, and reflexion, best exemplified in Darwin's work."[20]

Whitman believed that while Darwin's *ideas* had come to be generally appreciated, the "real secret of [Darwin's] success" was yet to be generally recognized:

> He was no hustler on the jump for notoriety, no rapid-fire writer; but a cool, patient, indefatigable investigator, counting not the years devoted to preliminary work, but weighing rather the facts collected by his tireless industry, and testing his thoughts and inferences over and over again, until well-assured that they would stand. Such a method was altogether too laborious and searching to be imitated by students ambitious to reach the heights of comparative psychology through a few hours of parlor diversion with caged animals, or by a few experiments on domestic animals. We are too apt to measure the road and count the steps beforehand. Darwin allowed the subject itself to settle all such matters, while he forgot time in complete absorption with his theme.[21]

Significantly, Whitman's description of Darwin applied less to Darwin than it did to Whitman himself. Whitman had recast Darwin in his own image. It was Whitman who forgot time as he became completely absorbed in his pigeon studies. Darwin's production of scientific books and papers in the last fifteen years of his life was in fact prodigious. Whitman's, in contrast, was meagre. In his researches on pigeons, Whitman discovered, among other things, the phenomenon that later investigators would call imprinting. Indeed his success in breeding different species of pigeons with one another came from his exploitation of this phenomenon. But he did not exploit the phenomenon, as later ethologists would, by using it to attract the attention of other professionals or the public to the interest—or the charm—of studying animal behavior in general or the interrelations of instinct and intelligence in particular. Although he alluded to the phenomenon in his animal behavior paper, he never published an article on the subject.[22]

Whitman began his systematic study of pigeons around 1895. F. R. Lillie reports:

> He gradually collected a large number of species of pigeons from all parts of the world, and in the latter part of his life the collection comprised some 550 individuals representing about thirty species. His house was surrounded by pigeon cotes, and he always had some birds under observation indoors, so that the cooing of doves was for years a dominant sound in his house. He took care of the birds for the most part himself, though he usually had the assistance of one or two maids. He thus actually lived with his birds constantly, and very rarely was absent from them even for a single day. He made observations and kept notes on all aspects of the life and behavior of each species, as well as of such hybrids as he was able to produce. He always had one Japanese artist at work continuously drawing pigeons, and for several years two.[23]

When his associates remarked on his insight in understanding his birds, Whitman is said to have responded: "Live with the birds day and night year in and year out."[24]

Whitman's goal in his pigeon studies was to reconstruct the evolutionary history of pigeons by careful study of their variation, heredity, and development; he looked at their color patterns, their behavior, the fertility of their hybrids, and so forth. For many summers, he took his pigeons with him from Chicago to Woods Hole and back again. This, however, proved to be both expensive and worrisome, as he lost some birds in the process. The difficulties involved in transporting the birds to and from Woods Hole were the primary reason he eventually resigned the directorship of the Marine Biological Laboratory.[25]

The biological farm Whitman called for was never established. In the last years of his life, Whitman devoted all his energies and resources to developing a facility of his own at his home in Chicago. As his wife later explained, "with the co-operation of his family [he] had let nothing stand in his way." Whitman and his wife cashed in their life insurance to help pay for the expanded plant. "Outside friends," including Mrs. Whitman's wealthy brother, L. L. Nunn, contributed some $20,000 in support of the work. Whitman thus finally secured all the facilities he thought were needed for his pigeon work—barns, cages, fountains, and the like. Because he spent very little time at the university, his graduate students had to seek him out at home. He planned to write his monograph on the behavior of pigeons after he completed his analysis of their heredity and evolution. In the end, however, he finished neither. On 1 December 1910, a cold wave struck Chicago. Whitman spent the whole afternoon out of doors putting his beloved birds in their winter quarters. The next day he was found in a coma. He developed pneumonia, and on 6 December he died.[26]

The University of Chicago was not prepared to support the continuation of Whitman's work or the publication of his manuscript notes. In September 1911, Whitman's former student, Oscar Riddle, contacted Robert S. Woodward, the president of the Carnegie Institution of Washington, inquiring whether the institution might be willing to fund the work necessary to keep Whitman's researches from being lost to science. Woodward had little enthusiasm for the project at first. As he later told C. B. Davenport, he had "entertained the hypothesis that [Whitman] was one of those who potter and who from over-refinement or other reasons do not bring their works to the point of publication." Davenport acknowledged that Whitman had not set the best of examples late in his career: "For 13 years he worked almost without publishing, in a fine disdain of the modern craze for rushing into print, until many of those who knew him best felt some doubt whether he had anything to say; though all had to admit two things, that he had a breadth of view and a thoughtfulness that put him first among biologists and, secondly, that he was everlastingly at his work with the most single-eyed devotion."[27]

Riddle, Davenport, and Albert P. Mathews urged the importance of Whitman's work on Woodward. Mathews, who described Whitman's studies as the most important work done on evolution by an American, explained the existing financial situation to Woodward as follows: "Dr. Whitman left no estate. All his income had gone for years into this work. He left no life insurance. Mrs. Whitman has kept the birds alive and together at great personal sacrifice as she is poor and the University pays her only $1500 a year pension. We cannot permit this splendid work to be lost, but I don't know where to turn for help if you are unable to aid. It is desirable also that the birds be kept alive and together for constant reference during the preparation of the work for publication."[28] In February 1912 the Carnegie Institution of Washington agreed to support Riddle in preparing Whitman's manuscripts for publication and continuing studies on Whitman's birds. A sum of $2,000 annually was to be provided for the care and maintenance of the pigeons. Riddle was to receive a salary of $2,400. Articles of agreement were drawn up with Mrs. Whitman, stating that the Carnegie Institution would pay $2,000 yearly for the maintenance of the pigeons. Mrs. Whitman was to allow Riddle free access to the pigeons and, when the work was done, to turn over the pigeons to the Carnegie Institution of Washington for transportation to the Station for Experimental Biology at Cold Spring Harbor.

After a variety of difficulties in attempting to conduct his researches and secure the cooperation of Mrs. Whitman, Riddle transferred the operation to Cold Spring Harbor in the fall of 1913. It was not until 1919, however, that Whitman's posthumous works appeared as three large volumes in the publications of the Carnegie Institution of Washington. The first volume was on "Orthogenetic Evolution in Pigeons." The second was on "Inheritance, Fertility and the Dominance of Sex and Color in Hybrids of Wild Species of Pigeons." The third, the one that had to be pieced together from the notes that were in the most fragmentary state, was on "The Behavior of Pigeons."[29]

Wallace Craig

The case of Wallace Craig is as instructive as that of Whitman. Craig's influence on the development of modern ethology was significant, but it was effected more through his interaction with Lorenz than through any direct and substantial impact on the course of behavioral studies in America. Craig made contact with Lorenz through Margaret Morse Nice in the mid 1930s. Craig's views were reflected in the theoretical conclusions to Lorenz's pathbreaking paper of 1935, "Companions as Factors in the Bird's Environment," and were of even greater importance for Lorenz's programmatic paper of 1937 entitled "The Establishment of the Instinct Concept." Of the second paper Lorenz later wrote: "It is hardly an exaggeration to say that this paper is about half written by Craig himself because it is really the distillate of the extensive correspondence we had in the years 1935–37."[30]

Particularly significant for the development of Lorenz's thinking was Craig's observation that instinctive behavior patterns did not consist simply of chain reflexes but also involved elements of appetite and aversion. Craig defined an appetite as "a state of agitation which continues so long as a certain stimulus, which may be called the appeted stimulus, is absent. When the appeted stimulus is at length received it stimulates a consummatory reaction, after which the appetitive behavior ceases and is succeeded by a state of relative rest." He defined an aversion as "a state of agitation which continues so long as a certain stimulus, referred to as the disturbing stimulus, is present; but which ceases, being replaced by a state of relative rest, when the stimulus has ceased to act on the sense-organs." As Craig explained in his paper of 1918 on this subject, "in . . . most supposedly innate chain reflexes, the reactions of the beginning or middle part of the series are not innate, or not completely innate, but must be learned by trial." What was "always innate" was "the end action of the series, the consummatory action." In other words, a bird had to "*learn* to obtain the adequate stimulus for a complete consummatory reaction, and thus to satisfy its own appetites."[31]

Lorenz found Craig's distinction between appetitive behavior and consummatory acts particularly instructive. He used it to sharpen the definition of instinct (by restricting instinct to innate, nonmodifiable actions and not counting appetitive and aversive behavior as instinctive). Craig's work together with that of Erich von Holst also helped Lorenz to see that the active organism was not simply responding to external stimuli through a chain of reflexes but was itself producing centrally coordinated impulses that formed the basis of its fixed motor patterns. Lorenz was particularly pleased with this perspective because it helped him account for the apparent spontaneity of behavior in a non-vitalistic manner.[32]

However, while Lorenz in the 1940s was happily referring to Craig as "one of my most respected teachers," Craig was virtually unknown among the biologists and psychologists of his own country. Craig in fact had not been able to sustain a career studying animal behavior. As of the early 1920s, increasing deafness made it impossible for him to teach, and he was unable to locate a research position. The English ethologist W. H. Thorpe reports that when he lectured to a large audience at Harvard in 1951 and described the important contributions to behavior study made by the Americans Whitman, Wheeler, and Craig, only one or two people in the whole audience knew who Craig was.[33]

Craig was born in 1876 in Toronto, the son of a Scottish-born father and English-born mother. He attended high school in Hyde Park in Chicago and then went on to study at the University of Illinois, majoring in zoology and receiving the degrees of B.S. in 1898 and M.A. in 1901. One of his teachers at Illinois was Stephen Forbes, the noted ecologist. Another was Charles C. Adams, who came to Illinois in 1896 as an instructor when Craig was in his junior year. Adams became a good friend of Craig and an important source of support in Craig's later years.[34]

Stephen Forbes, like Whitman and many other investigators at the turn of the century, stressed the importance of studying living nature rather than just books or laboratory preparations. It was evidently Forbes's influence that led the University of Illinois to require graduate students in zoology to take field courses at a biological station as part of their degree requirements. Forbes established a biological field station at Havana, Illinois, in the 1890s. Craig worked there as an assistant in the winter and spring of 1898–1899, making systematic collections of fish at particular locations on the Illinois River and adjacent waters and then reporting the kinds and numbers of individuals of each species he found. In the course of his collecting, he wrote to Adams stating: "It's a very hard matter to tell much about the lives of fish by just catching them in nets, and not seeing them alive, and yet I'm getting more interested in it all the time." In words that would prove prophetic he also announced: "I have about decided that comparative psychology is the line for me. Whether there will be bread and butter in it or not, I don't know—that's a subordinate question."[35]

In 1901, after teaching in Harlan, Iowa and Fort Collins, Colorado, Craig decided to pursue a doctoral degree. While entertaining some hopes of going to Harvard on a Chicago Harvard Club scholarship, he was nonetheless of the opinion, as he told Adams, "that the best finish I could have to my education, that I know of now, would be working on pigeons under Whitman." Family considerations consolidated this opinion, and in the fall of 1901 Craig enrolled at Chicago.[36] He was a student at Chicago from 1901 to 1904 and then, after a three-year spell of teaching first in Ohio and then in North Dakota, he returned to Chicago for the 1907–1908 academic year to complete his dissertation. The summers of 1903 and 1906 he spent at Woods Hole. His Ph.D. was awarded from Chicago in 1908 for a dissertation entitled "The Expression of Emotions in the Pigeon."

Craig admired Whitman greatly, and he modeled his own animal behavior studies after those of the older biologist. His research was conducted on pigeons belonging to Whitman, and he evidently felt honored to be able to acknowledge the "constant, generous, and invaluable" aid Whitman had given him. On publishing his dissertation in 1909 Craig states: "Professor Whitman knows the emotions, the voices, and the gestures of the pigeons very much better than I do; he has told me a great many facts about the birds which my more limited experience has not afforded; and he has always given helpful answers to my questions as to what a bird is thinking about when it does a certain act." Craig felt indebted to Whitman not only for the factual information and interpretive insights his mentor had provided him but also "for the influence of his spirit of research." Craig hoped that he himself had gained some of his professor's "enthusiasm and steadiness of labor, sympathetic insight into the animal mind, patience with details, yet a constant reference to general problems."[37]

In his dissertation, Craig provided a detailed description of the sounds and body movements of one pigeon species, the blond ring dove. He described—and represented by musical notation—the voice of the ring dove from its time of hatching to its old age, and through its annual cycle and special brood cycle. He intended to follow this study with a comparison of the sounds and gestures of different species ("showing specific characteristics, homologies, and the possibility of voice and gesture throwing light on problems of phylogeny") and studies of inheritance, variation, selection, sociology, and psychology.[38]

Craig's first major scientific paper, "The Voices of Pigeons Regarded as a Means of Social Control," appeared in 1908. It was published not in a zoological or psychological journal but instead in the *American Journal of Sociology,* edited by the Chicago sociologist W. I. Thomas. His doctoral dissertation was published the following year in the *Journal of Comparative Neurology and Psychology.* This was followed in 1911 by companion pieces on the mourning dove and the passenger pigeon, published in the *Auk.*[39]

In "The Voices of Pigeons Regarded as a Means of Social Control," Craig acknowledged the great importance of sexual selection in accounting for bird song, but pointed out that the significance of voice in birds "is of a very much wider scope than has ever been suspected. The voice is a means of social control: that is to say, the voice is a means of influencing the behavior of individuals so as to bring them into co-operation, one with another." To understand social behavior on the part of lower animals, Craig maintained, it was not sufficient to regard the individual animals as being endowed with a set of social instincts. What was necessary, he said, was "to see that the instincts of the individual can effect their purposes only when they are guided and regulated by influences from other individuals. In a complete explanation of animal society, therefore, the account of the social instincts must be supplemented by an account of the social influences by which the instincts of many individuals are brought into harmonious co-operation."[40]

In good, Whitmanlike fashion, Craig indicated that bird song "ought never to be studied (as hitherto it has been studied) without reference to the whole system of vocal and gestural activity." He found that vocal utterances and song played significant roles in the food-begging behavior of young, the recognition by the young of their own kind, the various complex ceremonies involved in the stimulation and coordination of sexual behavior, the selection of a nesting site, the laying of eggs, the maintenance of the pair bond, the defense of territory, and so forth. He promised to give a more complete account of the forms of social control in pigeons in a later, larger work. He anticipated bringing out "within a year or so" an entire book on the subject of the functions of bird song as part of a new series of books on animal behavior edited by Robert Yerkes.[41]

In the fall of 1908 Craig went to the University of Maine to fill a vacancy in philosophy and psychology. Maine had some attractions for Craig, but it also had serious disadvantages. He was a one-man department of philosophy. As such, he gave the only courses on philosophy and psychology at the university. In his first semester he taught separate courses on elementary psychology, ethics, logic, and the history of philosophy, and he also offered a seminar and individually directed research.

Craig liked teaching, however, and the university allowed him to develop new courses according to his own interests. As a result he does not seem to have felt particularly overworked, at least not at first. He was pleased to be able to be his own master and happy to find that the local people were relatively liberal minded. In his second year at Maine he introduced a course on evolution, a course on social psychology, and a new "introduction to philosophy." In the University of Maine catalog he described his evolution course as "an elementary presentation of evolution in all its phases, cosmic, geological, organic, psychic, and social." His "introduction to philosophy" was intended as a sequel to the evolution course, taking up, among other issues, "the bearing of evolution upon ultimate philosophic problems." Craig believed that an introductory course on evolution—a course surveying evolutionary theories, the evolution of animal behavior, mental evolution, social evolution, and human evolution—was the proper introduction and foundation for further studies "in psychology, sociology, and allied fields."[42]

Although Craig began his teaching at Maine with enthusiasm, he concluded before long that the students at Maine were "the most unscholarly or antischolarly lot that I ever knew." He was also unhappy with the limited resources of the library and with not having another psychologist with whom to talk. In addition, he found that his salary was inadequate and that the university was unprepared to provide him funds for research.[43]

Like Whitman, Craig learned that financing a research program involving pigeons required all the resources he could muster. Three years of having to deal with a landlord convinced him that if he wanted to work on pigeons without interference he would have to buy a house and lot of his own. After he sank all his savings into a house and lot, borrowed on his life insurance, and secured a mortgage for what still remained to be paid on the house, however, he found it difficult to make ends meet. He told Adams in May, 1913: "We spend so little on clothes that we look hardly respectable. My research work suffers seriously. It was impossible for me to go to the meeting [of the American Psychological Association] at Cleveland last Christmas. I am too poor to buy the pigeon cages which I ought to have."[44]

Distressed by his financial situation, Craig asked himself what could be done in the future. Contemplating whether he should secure a better-paying position, leave comparative psychology to do research in abstract philosophy, or carry on with comparative psychology by making it pay for itself, he decided the last option was probably best. He described his plan to Adams: "We must keep hens; while I watch their behavior we can eat their eggs, and later we can put the specimens themselves in the pot. I must keep large pigeons as well as doves; we can eat the squabs." Desirous of studying birds that imitated others, he decided it would be best to choose parrots, rather than the mockingbirds or sparrows he would have actually preferred, because parrots might later be sold for a profit. "If I thus aim at the financial," he told Adams, "science must suffer somewhat; but I see no other way, for keeping birds and experimenting with them is a mighty expensive business. All scientific work is expensive, but this sort is more expensive than many branches of zoology."[45]

Craig's guarded optimism about making comparative psychology pay for itself did not last long. In December 1913 he wrote again to Adams indicating that the plan he had entertained nine months earlier no longer appealed to him because it would not allow him to do the research he wanted:

> Probably I could maintain a bird farm and make it pay, but it would take every minute of my time, and I should have not time for making observations & experiments, keeping records of them and writing up the results. I could make mass observations just as every pigeon breeder can do; but mass observations are not what are needed in comparative psychology to-

> day; what is needed is minute, exhaustiv[e] observation and carefully controlled experiment. I find that the farming business takes all a man's time and brings in just enough money to make it pay, and no more. And I am, after all, more in need of time than I am of money.[46]

With respect to money, Craig reported: "There is no hope, certainly not for a long time to come, of this university appropriating money for research in animal behavior. It gives no money for research in any field (except the Experiment Station, which is entirely separate)."[47]

Without adequate resources, Craig found it difficult to participate in the annual meetings of scientific societies. When he went in December 1913 to the meeting of the American Psychological Association in New Haven, where he delivered an early version of his paper on appetites and aversions as constituents of instincts, this was, he later told Adams, the first time he had been out of the state of Maine for two years and four months. "Hereafter," he resolved, "I shall go to the psychological meeting, or some such meeting, every year, for here I am, a psychologist, absolutely alone. You don't know how paralyzing such isolation is."[48]

Craig continued to develop new courses at Maine, including courses on anthropology, child psychology (soon renamed by him "genetic psychology"), applied psychology, experimental psychology (a laboratory course), and, among others, the history and philosophy of science, but he never taught a course on his own specialty, animal psychology (nor did he ever teach a course in zoology). Furthermore, with the exception of the courses he gave on intellectual history, his offerings over time tended to become more utilitarian in their orientation. As of 1917–18, he stopped giving his course on evolution. In that year his course on social psychology was renamed "social and economic psychology." The course, which promised to cover, among other things, the applications of psychology to "daily life, hygiene, education, art, advertising, [and] managing men," was required of seniors in mechanical engineering.[49]

At Maine, Craig never completed either the book on bird song that he had once contemplated for Yerkes's animal behavior series or the book he had hoped to write on the emotional or social life of pigeons. He had found it discouraging enough when a journal turned down one of his articles; worrying about whether a publisher would find a book manuscript acceptable made it difficult for him to put words to paper. He entertained for a while the idea of paying for the publication of the book he wanted to write—"a large book on pigeons, with many illustrations and musical notations"—once he had paid off his mortgage. He did not succeed in following through on this idea, however. His scientific productivity at Maine consisted solely of papers.[50]

By August 1920 Craig had decided he could not continue much longer at Maine. He wrote to C. B. Davenport at Cold Spring Harbor stating:

> The results of my study of pigeon behavior, which I have continued in one form or another up to the present time, should be brought together and treated in a monograph. They constitute one whole, the parts all intimately bound together, and they should be treated so. To do this work as it should be done is impossible at the University of Maine. At that institution I have not the time, nor any other conditions suitable for the work. In order to do much in research I must move from the University of Maine to some other university, where research in pure science is encouraged. But before taking up the work of teaching in a new university it would be well if I could devote one year to writing up the research on pigeon behavior. This letter is written to ask you if there is any possibility that within a year or a few years from now I might secure a position on the staff of the Station for Experimental Evolution, the position to be held for one year, and my work to consist solely in finishing and writing up my study of pigeon behavior. The product to be published in monograph form by the Carnegie Institution.[51]

Davenport could offer Craig no encouragement. He responded: "On account of depreciated buying power of a dollar the Institution was never so poor and the President's orders are plain; take on nothing new."[52]

Craig left Maine in 1922. In the spring of 1923 he was a visiting lecturer at Harvard, teaching animal psychology, but his prospects for further employment there or elsewhere, as he told Raymond Pearl and Robert Yerkes in May of 1923, were very uncertain. To Pearl he wrote:

> So far as I know now, there will be no opening for me at Harvard for next year. And I am too deaf to secure a position elsewhere. I shall probably have to give up teaching.
>
> I should be glad if I could secure a research position in zoology. I am willing to begin at the bottom, like a young man just out of college, and learn some specialty. If opportunity offered, I could well accept a half-time position, with half pay, or a position for only a part of each year.
>
> Since my getting such a position is problematic, I am making ready to leave college work and go into something purely mechanical, e.g., electrical work, in which case I hope to do a little with animal psychology in my spare time.[53]

As Craig explained to Yerkes, a research position in zoology was what he wanted most. "If I can find a research position in zoology that will enable me to devote my time to the studies of animal behavior on which I have worked all my life, that will be ideal." He was willing, however, to do something of a more practical nature. He thought he might continue his work on animal behavior by doing something on horses, perhaps for an agricultural experiment station, or for the War Department, or for some commercial concern.[54]

With Yerkes's support, Craig explored the possibility of getting a position with the Animal Husbandry Division of the United States Department of Agriculture. By the end of July, however, he had secured a quite different position. His new job was that of "Librarian in the Department of Biophysics of the Cancer Commission of Harvard University," working under Dr. W. T. Bovie. He told Yerkes that he very much enjoyed the job, which involved "indexing, abstracting and translating literature on biophysics." The pay, however, was pitifully low. At Maine his annual salary was $3,500. At Harvard it was $1,500.[55] Craig spent four years (1923–27) in Bovie's laboratory at the Harvard Medical School and then the following two years working in the Harvard College Library. He had no steady employment after that. He and his wife lived in England and Scotland from 1935 to 1937 and then moved to Albany, New York, in June 1937, where his friend C. C. Adams, then Director of the New York State Museum, promised to help Craig publish a monograph on the song of the wood pewee.[56]

Late in 1936 Konrad Lorenz told Oscar Heinroth that it was up to the two of them to establish "a truly comparative psychology" because the only other people who could do what was needed for the field were Jan Verwey in Holland, who had turned his attention to other things, and Wallace Craig in the United States, who was old and no longer publishing. Craig by this time seems indeed to have become convinced that not only his fortunes but also his talents were inadequate for making a major contribution to science. In a letter of 1936 to Leonard Carmichael, he explained why he had not published a paper he had read to the students in his Harvard seminar: "The manuscript which I read to you was written at Maine. When I was at the University of Maine I got an exaggerated idea of the importance of my theories. As soon as I came to Harvard I began to be deflated. I discovered that many of my students, even my undergraduate students, had better minds than my own. Since then I have given my research time to studies that are less ambitious and more exact. I shall keep on studying, and do the best I can."[57]

Problems of Disciplinary Identification and Institutional Support

Craig's case was unique in certain respects, but the general difficulties he experienced in trying to pursue a career as a student of behavior were not entirely atypical. Prominent among these were the problems of disciplinary identification and institutional support.

Disciplinary identification had not been so much of a problem for Whitman, at least not with respect to his behavior studies, because he did not begin to devote himself to these studies until after he was well-established professionally—and after he had clearly identified himself as a biologist. Furthermore, even then biological studies were for him but one part of the more comprehensive program of research that defined him in his own mind as a biologist. For Craig, in contrast, the problem of disciplinary identification proved to be more complicated. Craig, as we have seen, was trained as a zoologist, and he modeled his research after Whitman's. Whitman, however, was not the only professor at the University of Chicago interested in the evolution of behavior and mind. Craig's first scientific paper was read by the sociologist W. I. Thomas and the philosopher G. H. Mead before being published in the *American Journal of Sociology.* Craig took his first academic position, as a philosopher-psychologist at the University of Maine. If he had any feelings of dissonance with respect to the relation between his zoological training and the philosophical/psychological post he obtained, he may have helped dispel these feelings by the way he interpreted his research results. In his studies of mating and egg-laying behavior in birds, he tended to emphasize the influence of psychological factors on physiological states as much as the other way around. He joined the American Psychological Association and read a paper at one of its national meetings. After leaving Maine the only other academic position he ever held was as a psychologist—his brief appointment as visiting professor in comparative psychology at Harvard in 1923. However, when he came to look for a new position after that, his expressed desire was to secure something in zoology.[58]

Other students of animal behavior in Craig's day had occasion to ponder where their own professional training and particular interests left them with respect to the existing disciplines of biology and psychology, although they seem to have taken much less time than Craig did in deciding that their own fate ultimately rested with the discipline in which they were initially trained. For example, H. S. Jennings, who distinguished himself in the early years of the century by his work on the behavior of invertebrates, remarked in 1906 to Robert Yerkes: "while I am a sort of a homeless wanderer so far as my subject of investigation is concerned, it is certainly true that the Psychologists come nearer to taking up the matters in which I am interested than do any other of the Societies." Conversely, in 1909 John B. Watson expressed to Yerkes his frustrations with regard to "finding a proper place and scope for psychology." He wrote to Yerkes, "Am I a physiologist? Or am I just a mongrel? I don't know how to get on." But Jennings left his studies of behavior for studies of genetics in 1907, and Watson, though attracted by the methods of physiology and the status physiologists enjoyed, used his own animal behavior work not to develop ties with physiology but rather to launch a campaign to reform psychology as a whole. Although Yerkes had once thought that it did not matter whether psychology was considered part of biology or an independent science, by 1910 he had changed his mind on this score and now believed it was crucial for psychologists to develop faith in their own aims, methods, and abilities. As O'Donnell has shown, Yerkes's restructuring of his thinking on the proper aims and methods of psychology was intimately related to the insecurity of his position at Harvard. Evidently, how a student of animal behavior perceived the relations of biology and psychology depended very much on—and changed according to—the different pressures the individual experienced in the course of his or her career.[59]

An ongoing balancing act was required if one were simultaneously to carry on research, develop and maintain a constituency, see to the construction of institutional bases to support studies of behavior, and respond to the messages emanating from one's own discipline. Whitman devoted a substantial part of his own career to discipline- and institution-building, but in his latter years, the years in which he did his work on behavior, he retired into the seclusion of his own researches and did little to promote behavior studies. Craig seems never to have been disposed toward institution-building. Robert Yerkes, on the other hand, had both the energy and the organizational skills to match his own professional aspirations, and he recognized early in his career that his prospects of success as an animal psychologist were linked closely to the prospects of providing animal behavior studies with a secure institutional foundation.

Yerkes made a number of moves to give animal behavior studies increased visibility and social and institutional support. The most successful of these before the 1920s was his establishment of a journal devoted specifically to the field. This was not an accomplishment that came easily, however. It took eight years from the time Yerkes first raised the idea with Jennings to the time the first issue of the *Journal of Animal Behavior* appeared in 1911.

Concerned as he was in 1910 with representing himself, at least to his department and to the Harvard administration, as a proponent of a psychology independent from biology, Yerkes appreciated that a journal of animal behavior could only survive if psychologists, biologists, and even "nature lovers" joined in support of the enterprise. The prospectus he drew up in 1910 to announce the new journal specified that the journal would publish "field studies of the habits, instincts, social relations, etc., of animals, as well as laboratory studies of animal behavior or animal psychology." It was Yerkes's expressed hope that such a journal would "serve to bring into more sympathetic and mutually helpful relations the 'naturalists' and the 'experimentalists' of America." The *Journal of Animal Behavior*'s editorial board was chosen carefully, with an eye to geographical as well as disciplinary breadth. Over the course of the journal's seven-year existence, the papers contributed to the journal were divided almost equally between contributions by biologists and contributions by psychologists. It is therefore little wonder that a number of comparative psychologists of a later date, critical of the way their field had narrowed in the 1930s and 1940s into little more than a science of "rat learning," looked back upon the days of the *Journal of Animal Behavior* as a kind of golden age. However, a closer examination of the respective contributions of biologists and psychologists to the *Journal of Animal Behavior* reveals that already by the second decade of the twentieth century the psychologists differed significantly from the biologists in terms of the animals and problems they studied and the methods they employed.[60]

Significantly, for psychologists and biologists alike, laboratory work dominated the contributions to the *Journal of Animal Behavior.* This was overwhelmingly the case for the psychologists, for whom seventy-four out of seventy-five papers involved laboratory studies, with a full two-thirds of the total number of papers emanating from just four university laboratories (Harvard, Chicago, Hopkins, and Texas), all of which had been established between 1899 and 1908. Not a single one of the psychological papers was a field study. As for the biologists, slightly more than one-quarter of the seventy-two papers they contributed to the journal dealt primarily with fieldwork. The rest of their papers, however, resembled laboratory studies, and more than two-thirds were specifically identified as coming from university laboratories.

Despite the laboratory orientation they shared, however, the biologists and psychologists overlapped very little with respect to the different kinds of animals they used and the various problems they studied. The biologists worked primarily on invertebrates, paying particular attention to tropisms and other responses to stimuli. The psychologists worked almost exclusively on vertebrates, concerning themselves for the most part with problems of sensory discrimination and learning. While forty percent of the biologists' papers involved insects, none of the psychologists' papers did. Conversely, while seventy-four percent of the psychologists' papers involved mammals, only one (1.4 percent) of the biologists' papers did. Craig worked on the one class of animals where there was the greatest intersection of the interests of biologists and psychologists—thirteen percent of the biologists' papers and twenty percent of the psychologists' papers were on birds (Craig's papers have been counted here as biological contributions). As is quite obvious, however, birds did not constitute the preferred class of either discipline. While Craig was studying the interrelations of instinctive and learned elements in the complex behavior of pigeons, most behaviorally oriented biologists were studying the responses of invertebrates to stimuli and most comparative psychologists were studying the abilities of mammals to discriminate between stimuli or to learn to run through mazes.

In terms of his institutional setting, Craig was also out of the mainstream. Though doing laboratory work rather than fieldwork, he did not have at Maine the resources of a psychological or zoological laboratory at his disposal. With his own modest resources he tried to maintain a research program that would allow him to study behavioral phenomena exhibited over the course of his birds' lives. Unlike Whitman, unfortunately, Craig did not have a wealthy relative to help support his work, and unlike Yerkes, he was not an aggressive entrepreneur who knew how to further his studies by developing new institutions. Craig's pigeons would presumably have been no more costly to maintain than W. E. Castle's hooded rats, but Castle's long-term study of mammalian genetics was exceptional. It survived thanks to continuous support from the Carnegie Institution of Washington. Few other investigators, as it turns out, enjoyed such assistance.

Shortly after its founding in 1902, the Carnegie Institution of Washington looked as if it might help animal behavior studies as part of its general support for zoology. The Carnegie Institution's Advisory Committee on Zoology clearly had something on the order of Whitman's biological farm in mind when it called for the establishment of an experimental biological station that would concern itself with "experiments in heredity, in variation, in instinct, in modification, all of which should extend over a series of years and be planned systematically." What is more, the proposal that the Carnegie Institution soon endorsed—submitted by C. B. Davenport for a biological station at Cold Spring Harbor, New York—included as an attachment the paragraphs from Whitman's biological farm paper in which Whitman talked about the importance of studying "development, growth, life histories, species, habits, instincts, intelligence." Once Davenport's Station for Experimental Evolution was established, however, the station concentrated primarily on Davenport's own interests, which had moved from morphological and behavioral studies to genetics. Nothing came of the attempts, first by Jennings and then later by Yerkes and Watson, to secure Carnegie funds to establish a behavior station either in conjunction with Davenport's laboratory or independent of it. Oscar Riddle's work at Cold Spring Harbor in the 1910s included a study of the alteration of sex behavior through the injection of sex hormones (using Whitman's pigeons), but this work was not typical for the laboratory. Eventually, in 1921, Davenport's two primary enterprises at Cold Spring Harbor, the Department of Experimental Evolution and the Eugenics Records Office (the latter having come under the Carnegie Institution's wing in 1918), were united and named the Department of Genetics of the Carnegie Institution.[61]

More important for behavioral studies was the Marine Biological Laboratory established by the Carnegie Institution at Bird Key in the Dry Tortugas, Florida, with Alfred G. Mayer as its director. H. S. Jennings did some early work there, and it was there that John B. Watson conducted the studies on bird behavior that have led some writers to regard him as a pioneer ethologist in addition to being the father of American behaviorism.

The field studies Watson undertook in the Tortugas of the behavior of the noddy and sooty terns illustrate the high degree of contingency in animal behavior studies at a time when the institutional support for any behavior studies was not forthcoming. Watson's Tortugas studies were not the result of any long-term commitment on his part to fieldwork or even an application to do research at the Tortugas Laboratory. He was *invited* to do research there. Mayer, the director of the laboratory, was eager to justify the new laboratory's existence by attracting prominent scientists to work there in the summers. In 1907 he invited Watson to the Tortugas to study the behavior of sea gulls.

Watson had already tried twice to secure Carnegie funding for research projects of his own devising. In 1903 he had applied for support for a study of the sensory nerve fibers in the spinal roots of humans. In 1905 he had applied to study "the role which the various sense-organs play in the associations of animals." Both applications were rejected. Failing to receive funding to do research, Watson had to devote his summers to teaching summer school so that he could pay the debt he had accumulated in gaining an education.

To Watson, the opportunity to be paid in the summer for doing research instead of teaching was extremely attractive. Although the Carnegie Institution was able to provide him only with transportation and living expenses for the projected three months in the Tortugas, Mayer was able to come up with another $55 and the University of Chicago added $150, and this, plus $40 per month from the Audubon Society for being custodian of Bird Key, was enough for Watson to be able to make the trip. Before making the trip, Watson admitted to Yerkes that he did not know exactly what he was going to do in the Tortugas once he got there. He had read through the *Auk, Bird-Lore,* and other such journals in the hope of finding a model for his research, but he had not found anything he was prepared to call scientific. In a statement that suggests how low Whitman's profile at Chicago was in 1907, Watson told Yerkes: "Our library is vile on bird literature and the fellows in zoology here don't seem to know a whole lot about the behavior side." [62]

Watson's later work in the Tortugas under Carnegie auspices was supported more generously than his first work there. He was appointed a research associate of the Department of Marine Biology for three years (1912–14) with a yearly honorarium of $500 per year, and in 1915 he was awarded a $1,400 publication grant for his monograph with K. S. Lashley on "Homing and Related Activities of Birds." Significantly, however, this was the only money that the Carnegie Institution earmarked for a study of animal behavior in the first quarter of the twentieth century with the exception of its early support of the work of Jennings. In the first few years of the Carnegie Institution's existence, Jennings had received a grant, an assistantship, and the use of a research table at the Stazione Zoologica at Naples. The Carnegie Institution had also published Jennings's first monograph on invertebrate behavior. But the institution's policies with respect to minor grants had stiffened shortly after that. The institution's second president, Robert Woodward, decided that the institution's funds were better spent supporting a few large enterprises rather than a great many individual projects, and by its second decade the Carnegie Institution was making precious few awards to individual zoologists or biologists in support of ongoing research.

Indeed, as of 1910 the only money being given to zoology by the Carnegie Institution in its minor grants category was that given to W. E. Castle for his work on genetics and that given to the marine laboratory at Naples for the support of two tables. Castle had begun receiving support from the Carnegie

Institution in 1904. He continued to do so up into the 1920s. Between 1910 and 1918 he received from $1,000 to $2,500 per year. Frederic Clements began receiving Carnegie support in 1914, and T. H. Morgan in 1915.

There is no indication that Wallace Craig ever applied directly to the Carnegie Institution for funding, although as indicated above, he did explore the possibility of getting an appointment at Cold Spring Harbor in 1920. His aspirations were for the most part more modest, as for example when he applied in 1914 to the Elizabeth Thompson Science Fund for a grant of $150, but told the would-be benefactors that "even $50. would be a real aid." In 1923 he applied for a fellowship from the National Research Council but was turned down because the fellowship in question was supposed to be for young scientists just beginning their careers.[63]

Conclusion

The kind of work represented in Whitman's animal behavior lecture of 1898 did not become an integral part of American biology over the course of the next twenty-five years. Animal behavior studies were not adopted as a particularly fruitful means of elucidating phylogenetic relationships, evolutionary mechanisms, or the relations between instinct and mind. Nor for that matter did other biological approaches to behavior retain the vigor they displayed at the beginning of the century. The study of animal tropisms seems to have lost its enthusiasts as early hopes for a general theory of animal reactions foundered on the actual complexities of animal behavior and on the frustration of terminological wrangles over what the word "tropism" should be understood to mean.[64]

By the mid-1920s, some of the most prominent contributors to the study of animal behavior at the beginning of the century had died. Others had moved off into different fields. Whitman had been dead for more than a decade. So too had the talented amateur naturalist George Peckham (1845–1914), who with his wife, Elizabeth Peckham, had conducted such remarkable studies on the behavior of insects and arthropods, demonstrating as convincingly as anyone in the late nineteenth and early twentieth centuries the compatibility of the Darwinian theories of natural and sexual selection with the testimony of animal behavior. Recently dead was Charles H. Turner (1867–1923). Among the living were William Morton Wheeler, H. S. Jennings and Samuel J. Holmes, but none of them was campaigning for the development of behavior studies. Wheeler had continued his studies of ants and his somewhat idiosyncratic interpretation of instincts, but he had given up promoting the word "ethology" as the best designation of what modern biology ought to be about. Jennings and Holmes were devoting their attentions to other subjects. Jennings, as indicated above, had left behavioral studies for genetics in 1907, and Holmes since the 1910s had committed his energies to the cause of eugenics. With regard to a particular experiment on the behavior of bacteria, Jennings had noted to Yerkes in 1907 that "amid the great multitude of possible things, one has to choose for doing those that look most promising." This was an observation that applied not only to individual experiments but also to entire programs of research.[65]

The move from animal behavior studies to other areas was made not only by the biologists but also by the psychologists. In 1916 Jennings noted to Yerkes that Watson "[appeared] to be going into psychiatric work, and more or less into applied psychology in general," and that S. O. Mast was expressing the opinion that "there was no demand for work on animals in Psychological departments; that practically every man trained in that line was forced to go into human work . . . ; and that work on behavior was going to find its home, after all, in the Zoological departments."[66]

Mast, as it turned out, was wrong about the future of behavior studies in American zoology departments, at least for the short run. His perception of what was happening in American psychology departments, however, had some basis in fact. By the 1920s, most of Yerkes's students, having found little support from academic departments for animal psychologists, had moved into such fields as education, educational psychology, and vocational psychology. In 1921, of the 424 members of the American Psychological Association, only twenty-six (six percent)—a group including Craig, Holmes, and Wheeler—identified animal behavior studies among their research interests. This group was outnumbered by between three and four to one by members whose expressed research interests included education or educational psychology and by between four and five to one by the overlapping set of investigators who did research on testing. In 1921 a new journal entitled the *Journal of Comparative Psychology* was set up as a continuation of the *Journal of Animal Behavior* (1911–17) and *Psychobiology* (1917–20), but the editor, Knight Dunlap, was a man who was not a comparative psychologist himself and who was less excited about theoretical issues than about the possibilities of developing better experimental apparatus. As of 1925 Yerkes was just getting back into academia after having devoted two years to psychological testing for the Army and then five more (1919–24) to the activities of the National Research Council. Watson, having been forced out of academe five years earlier by the scandal of a love affair, was putting his interest in behavioral control to use as a consulting psychologist for the Madison Avenue advertising firm of J. Walter Thompson. The future of animal behavior studies as part of psychology was no more clear than the future of animal behavior studies as part of biology.[67]

Various problems associated with academic existence in general and with the study of animal behavior in particular inevitably contributed to a high degree of uncertainty in the careers of American students of animal behavior in the first quarter of the twentieth century. Wallace Craig's old friend, C. C. Adams, who had been unsuccessful in his own attempts "to secure facilities for the intensive study of animals in relatively natural conditions," expressed this problem quite aptly in a letter to Yerkes in 1932. Congratulating Yerkes on the primate laboratory Yerkes had finally succeeding in establishing, Adams noted:

> It is a strange fate that so many men must spend the best years of their lives before they have a real chance to do what they most desire to do. What would it have meant to you to have had the facilities you now have say, 20 years ago! Ritter had a long struggle to get his marine laboratory, and possibly this came too late for him, as I have heard that he lost interest in the project in later years. Whitman wanted his "biological farm" at Woods Hole, but it has not yet arrived in America, even today. It once looked as if the Desert Laboratory would make such a centre, but it has not developed as I had hoped. Not only do the men migrate but, as well, their interests. It is a severe strain on most men to have a sustained interest that lasts for many years. Of course external pressure has its influence as well.[68]

For a host of reasons—personal, professional, conceptual, methodological, and institutional—the American biologists concerned with animal behavior in the first quarter of the twentieth century did not succeed in establishing animal behavior studies as an area with special claims on the resources of the discipline of biology or on the broader American society. They were unable to bequeath to the next generation a conceptually integrated and institutionally secure set of theories, problems, and models of research. Animal behavior studies became neither an indispensible part of a generalized American biology nor a specialized subdiscipline. In the second quarter of the century a handful of American investigators including W. C. Allee, G. K. Noble, and Karl Lashley made important contributions to the study of animal behavior, but the conditions of existence for American studies of animal behavior in this period remained precarious. The particular approach to animal behavior study known as ethology was first established as a modern scientific discipline not in America but instead in Europe. Significantly enough, the first American to identify Lorenz's work as important and to take it upon herself to introduce Lorenz's ethological ideas to English-language readers was not a professional biologist but rather an exceptionally capable amateur, Margaret Morse Nice.

It should go without saying that in Europe as well as the United States the development of behavior studies was fraught with uncertainties and shaped by local situations and interests. The task of analyzing how animal behavior studies developed in Europe and how ethology was established as a discipline there, however, will have to be reserved for another study.

Acknowledgments

I am grateful to the American Philosophical Society, the Carnegie Institution of Washington, Western Michigan University, and Yale University Library for permission to quote from archival materials in their collections. I owe special thanks to Phyllis Burnham, Manuscript Curator of Regional History Collections at Western Michigan University, and Robert M. Croker for their assistance in locating Wallace Craig materials in the C. C. Adams Papers at Western Michigan University. I am also pleased to thank the participants in the conference on the history of American biology, held at Friday Harbor Laboratories in September, 1986, for their very helpful comments on an early draft of this paper. My work on Whitman and Craig began as part of a research project supported by the National Science Foundation, grant number SOC78-05922. My research on this paper was also supported in part by a grant from the Research Board of the University of Illinois at Urbana-Champaign.

Notes

1. On the history of ethology see W. H. Thorpe, *The Origins and Rise of Ethology* (London: Heinemann, 1979); Richard W. Burkhardt, Jr., "On the Emergence of Ethology as a Scientific Discipline," *Conspectus of History,* 1981, *1:* 62–81, and "The Development of an Evolutionary Ethology," in D. S. Bendall, ed., *Evolution from Molecules to Men* (Cambridge: Cambridge University Press, 1983), pp. 429–444; and John R. Durant, "Innate Character in Animals and Man: a Perspective on the Origins of Ethology," in Charles Webster, ed., *Biology, Medicine and Society 1840–1940* (Cambridge: Cambridge University Press, 1981), pp. 157–192. For Lehrman's comments, see Daniel S. Lehrman, "A Critique of Konrad Lorenz's Theory of Instinctive Behavior," *Quarterly Review of Biology,* 1953, *28:* 337–363, on p. 337. Lorenz's comments on the founders of ethology are in Konrad Lorenz, *Studies in Animal and Human Behaviour,* vol. 1 (Cambridge, Mass.: Harvard University Press, 1970), p. xv.

2. See C. O. Whitman, "Specialization and Organization, Companion Principles of all Progress.—The Most Important Need of American Biology," *Biological Lectures Delivered at the Marine Biological Laboratory of Wood's Holl,* 1891, *1890:* 1–26; "Some of the Functions and Features of a Biological Station," *Biol. Lect.,* 1898, *1896–1897:* 231–242, on p. 241; "Animal Behavior," *Biol. Lect.,* 1899, *1898:* 285–338; "A Biological Farm. For the Experimental Investigation of Heredity, Variation and Evolution and for the Study of Life-histories, Habits, Instincts and Intelligence," *Biological Bulletin,* 1902, *3:* 214–224. Whitman's "Animal Behavior Lecture" has been reprinted in Jane Maienschein, ed., *Defining Biology. Lectures from the 1890s* (Cambridge, Mass.: Harvard University Press, 1986). William Morton Wheeler, "'Natural History,' 'Oecology' or 'Ethology'," *Science,* n.s., 1902, *15:* 971–976.

3. See Whitman, "Some of the Functions and Features of a Biological Station," pp. 240–241. On the differences between Whitman and Loeb, see Philip J. Pauly's excellent study, *Controlling Life: Jacques Loeb and the Engineering Ideal in Biology*

(New York: Oxford University Press, 1987), pp. 80–81. Whitman, "A Biological Farm," p. 222.

4. On the work of the animal psychologists, see especially John M. O'Donnell, *The Origins of Behaviorism. American Psychology, 1870–1920* (New York: New York University Press, 1985), pp. 179–208.

5. On Whitman, see Maienschein, in this volume; Frank R. Lillie, "Charles Otis Whitman," *Journal of Morphology,* 1911, *22:* xv–lxxvii; Charles B. Davenport, "The Personality, Heredity, and Work of Charles Otis Whitman, 1843–1910," *American Naturalist,* 1917, *51:* 5–30; Edward S. Morse, "Charles Otis Whitman," *National Academy of Sciences Biographical Memoirs,* 1912, *7:* 269–288; and Ernst Mayr, "Whitman, Charles Otis," *Dictionary of Scientific Biography,* vol. 14 (N.Y.: Charles Scribners Sons, 1976), pp. 313–315.

6. On the childhood interest in animals on the part of ethologists, see Konrad Lorenz, "Introduction; the Study of Behavior," in Jurgen Nicolai, *Bird Life* (New York: G. P. Putnam's Sons, 1974), p. 15. On Whitman's youth, see Davenport, "Whitman," p. 20, and Lillie, "Whitman," p. xvi. Lorenz's quote is from Konrad Z. Lorenz, "The Comparative Method in Studying Innate Behavior Patterns," *Symposia of the Society for Experimental Biology,* 1950, *4:* 221–268, on p. 222.

7. Although Whitman appears to have had an interest in behavior throughout his career, his serious studies of behavior seem to have been concentrated in two periods, 1895 to 1898 and 1903 to 1907. The first period resulted in his publication of two papers, "Animal Behavior" and "Myths in Animal Psychology." In the second period he concentrated on the reproductive behavior of pigeons. Some of his published and manuscript writings were incorporated in Harvey A. Carr, ed., *Posthumous Works of Charles Otis Whitman,* vol. 3, *The Behavior of Pigeons* (Washington, D.C.: The Carnegie Institution of Washington, 1919), a volume put together with Wallace Craig's assistance. Whitman's definition of biology is from "Myths in Animal Psychology," *The Monist,* 1899, *9:* 524–537, on p. 524. Mary Alice Evans and Howard Ensign Evans, *William Morton Wheeler, Biologist* (Cambridge, Mass.: Harvard University Press, 1970), p. 9, report the anecdote of how Whitman's students greeted one another.

8. Lillie, "Whitman," p. lvii.

9. Whitman, "Animal Behavior," pp. 288, 296.

10. Ibid., p. 302 (Whitman's italics), and Whitman, "A Biological Farm," p. 221.

11. Whitman, "Animal Behavior," pp. 299, 311.

12. Ibid., pp. 309, 332.

13. Ibid., pp. 310n., 328. On p. 310n., Whitman objected to Loeb's idea of "unorganized chemical substances" in the egg just as he objected to the ahistorical nature of Loeb's view of tropisms. On Whitman's use of behavioral characters as well as color patterns to reconstruct the evolutionary history of the pigeons, see Lillie, "Whitman," p. lxiv. On the retrospective appreciation of Whitman's insight on using behavior patterns for phylogenetic purposes, see for example Konrad Lorenz, "Part and Parcel in Animal and Human Societies." in Lorenz, *Studies in Animal and Human Behaviour,* vol. 2 (Cambridge, Mass.: Harvard University Press, 1971), pp. 130–131. Additional similarities between Whitman's work and that of later ethologists include acknowledgment that instincts could "run down," his experiments on what different races of pigeons did when their eggs were moved just outside of their nests, his description of

the phenomenon that would later be called "imprinting," and his view of how the instinctive behavior of domestic and feral races differed from one another. On this last point, Whitman's perspective was more positive than Lorenz's. While Whitman suggested that "domestication lets down the bars to choice and at the same time gives more opportunities for free action" (p. 334; see also pp. 335–336), Lorenz tended to regard the effects of domestication as by and large detrimental. The political dimensions of Lorenz's views on "domestication" in the human race are discussed by Theodora J. Kalikow, "Konrad Lorenz's Ethological Theory: Explanation and Ideology, 1938–1943," *Journal of the History of Biology,* 1983, 16: 39–73.

14. Whitman, "Animal Behavior," pp. 322, 323. See also Whitman's comments on the incubation instinct in birds, p. 328.

15. Ibid., pp. 306, 308–309.

16. Ibid., p. 321.

17. Ibid., p. 329. Earlier in his paper (p. 311), Whitman displayed a view quite typical of his time in observing: "In the human race instinctive actions characterize the life of the savage, while they fall more and more into the background in the more intellectual races." Whitman also discussed the relations of instinct and intelligence in a letter of 25 September 1906, to the Chicago *Tribune* (responding to an article by E. Ray Lankester on the origins of intelligence).

18. Whitman, "Animal Behavior," pp. 335–336.

19. Ibid., pp. 336, 338.

20. Whitman, "Myths in Animal Psychology," pp. 537, 538.

21. Ibid., pp. 524–525.

22. In the last decade of his life, Whitman published only eight papers: three brief notices regarding his research on the heredity and evolution of the color patterns of pigeons, two articles on the Marine Biological Laboratory, his biological farm article, the address he delivered at the St. Louis exposition of 1904 "On the Problem of the Origin of Species," and another, shorter discussion of the problem of the origin of species. See Lillie, "Whitman," pp. lxxvi–lxxvii, for the complete references. On Whitman's use of imprinting in his hybridization experiments, see Carr, *The Behavior of Pigeons,* p. 28. I am indebted to Kevin MacNeil for calling Carr's observation to my attention.

23. Lillie, "Whitman," p. lxxi.

24. Quoted by Mrs. Emily Whitman in a letter to R. S. Woodward, President of the Carnegie Institution of Washington, 27 September 1911, Archives, Carnegie Institution of Washington, D.C. (hereafter Carnegie Archives).

25. Lillie, p. lxxi; see also p. xxxix.

26. Mrs. Emily Whitman to Woodward, 27 September 1911. In a later letter to Woodward (undated, but apparently from late February 1912), Emily Whitman stated: "In the purchase of the expensive birds we were aided by my brother Mr. L. L. Nunn—who helped us with these as with other enterprises—as the Morphological Journal and the Woods Holl Laboratory." Whitman had kept herons and flickers as well as pigeons, but his wife disposed of the herons and flickers after his death because they were too expensive and too difficult to keep. Carnegie Archives.

27. Riddle to Woodward, 26 September 1911; Woodward to Davenport, n.d.; Davenport to Woodward, 4 October 1911 (replying to a letter from Woodward, 30 September 1911); Carnegie Archives.

28. Mathews to Woodward, 27 November 1911; Carnegie Archives.

29. The letter from Woodward to Riddle, 19 February 1912, and the articles of agreement are in the Carnegie Archives. Riddle edited volumes 1 (*Orthogenetic Evolution in Pigeons*) and 2 (*Inheritance, Fertility and the Dominance of Sex and Color in Hybrids of Wild Species of Pigeons*) and Carr edited volume 3 (*The Behavior of Pigeons*).

30. Margaret Morse Nice, *Research is a Passion with Me* (Toronto: Consolidated Amethyst Communications, 1979), pp. 140–141; Konrad Lorenz, "Der Kumpan in der Umwelt des Vogels," *Journal für Ornithologie,* 1935, *83:* 137–413; idem, "Über die Bildung des Instinktbegriffes," *Die Naturwissenschaften,* 1937, *25:* 289–300, 307–318, 324–331; idem, *Studies in Animal and Human Behaviour,* vol. 1, p. xix.

31. Wallace Craig, "Appetites and aversions as constituents of instincts," *Biol. Bull.,* 1918, *34:* 91–107.

32. Lorenz acknowledges that it was with the help of Craig and Erich von Holst that he "at last arrived at the clear conceptualization of appetitive behaviour, innate releasing mechanism and innate motor pattern"; see Lorenz, *Studies in Animal and Human Behaviour,* vol. 1, p. xx. Lorenz defined instinctive and appetitive behavior more narrowly than Craig had originally done, but noted in his paper of 1942, "Inductive and Teleological Psychology," that Craig was "in agreement down to the last detail with my more narrow and precise definition of the two concepts" (ibid., p. 361). In his paper "The Comparative Method in Studying Innate Behaviour Patterns," Lorenz maintained that "Craig's great discovery" was that "it is the discharge of consummatory actions and not [their] survival value which is the goal of appetitive behavior."

33. Konrad Lorenz, "Inductive and teleological psychology," in Lorenz, *Studies in Animal and Human Behaviour,* vol. 1, 361. W. H. Thorpe, *The Origins and Rise of Ethology* (London: Heinemann, 1979), pp. 48–49. Thorpe reports that he had thought Craig was dead, and was thus astonished to learn that Craig was alive and in the audience.

34. A brief obituary of Craig, written by A. W. Schorger, appeared in the *Auk,* 1954, *71:* 496. More information on Craig's life up to 1918 appears in *The Alumni Record of the University of Illinois, Semi-Centennial Edition, 1918.* Additional information on Craig's activities can be found in correspondence of his at the American Philosophical Society, Philadelphia (in the C. B. Davenport and the Leonard Carmichael papers), in the C. C. Adams papers at Western Michigan University, Kalamazoo, Michigan (hereafter Adams Papers), and in the Robert M. Yerkes papers at Yale University, New Haven. I am grateful to Professor Robert A. Croker for providing me with information about the Craig correspondence in the C. C. Adams papers.

35. George W. Bennett, "A Century of Behavior Research: Aquatic Biology," *Illinois Natural History Survey Bulletin,* 1958, *27:* 163–178 (information cited on p. 165); Stephen Alfred Forbes and Robert Earl Richardson, *The Fishes of Illinois* (Urbana: Illinois State Laboratory of Natural History, 1908), p. xi; Craig to Adams, 12 December 1898, Adams Papers.

36. Craig to Adams, 24 April 1901; see also Craig to Adams, 2 May 1901, and 11 May 1901, Adams Papers.

37. Wallace Craig, "The Voices of Pigeons Regarded as a Means of Social Control," *American Journal of Sociology,* 1908, *14:* 86n.; and "The expressions of

emotion in the pigeons. I. The Blond Ring Dove (*Turtur risorius*)," *Journal of Comparative Neurology and Psychology,* 1909, *19:* 29–80, on p. 31.

38. Craig, "The Expressions of Emotion in the Pigeons, I," 30–31.

39. Ibid.; Craig, "The Expressions of Emotion in the Pigeons. II. The Mourning Dove (*Zenaidura macroura* Linn.)," *Auk,* 1911, *28:* 398–407, and "The Expressions of Emotion in the Pigeons. III. The Passenger Pigeon (*Ectopistes migratorious* Linn.)," *Auk,* 1911, *28:* 408–427.

40. Craig, "The Voices of Pigeons Regarded as a Means of Social Control," pp. 86, 87.

41. Ibid., pp. 86, 100. On pp. 89–90 of this paper, Craig described Whitman's experiences with what would later be called "imprinting." Craig wrote that "we must believe that young doves have no inherited tendency to mate with birds of a particular kind; they learn to associate with a particular kind during the period when they are being fed, when the characteristics of their nursing-parents are vividly impressed upon their young minds." On Craig's thoughts on writing a book at this time, see Craig to Yerkes, 13 September 1907, Yale University, New Haven, Yerkes Papers, Sterling Library Archives, (hereafter Yerkes Papers).

42. On Craig's feelings about teaching at Maine, see Craig to Adams, 24 May 1913, Adams Papers. On the courses he offered at Maine, see the *University of Maine Catalog,* 1909–10, p. 106, and 1913–14, p. 169.

43. Craig to Adams, 24 May 1913, Adams Papers.

44. Ibid.

45. Ibid.

46. Craig to Adams, 18 December 1913, Adams Papers.

47. Ibid.

48. Craig, "Attitudes of Appetition and of Aversion in Doves," *Psychological Bulletin,* 1914, *11:* 56–57. Craig to Adams, 4 February 1914, Adams Papers.

49. *University of Maine Catalogs,* 1914–20.

50. For Craig's plans with regard to publishing, see Craig to Yerkes, 22 February 1909, Yerkes Papers; Craig to Adams, 10 September 1911, and 4 February 1914, Adams Papers. In the 1940s, Craig worked on a book manuscript that he entitled "A Study of the Space System of the Perceiving Self," for which he received support from the American Philosophical Society. On this subject, see the correspondence relating to Craig in the Leonard Carmichael Papers, Archives, American Philosophical Society, Philadelphia (hereafter Carmichael Papers).

The papers Craig published while at Maine included "Oviposition Induced by the Male in Pigeons," *Journal of Morphology,* 1911, *22:* 299–305; "Observations on Doves Learning to Drink," *Journal of Animal Behavior,* 1912, *2:* 273–279; "The Stimulation and the Inhibition of Ovulation in Birds and Mammals," *J. Anim. Behav.,* 1913, *3:* 215–221, "Male Doves Reared in Isolation," *J. Anim. Behav.,* 1914, *4:* 121–133; "Appetites and Aversions as Constituents of Instincts," *Biol. Bull.,* 1918, *34:* 91–107; and "Why Do Animals Fight?" *International Journal of Ethics,* 1921, *31:* 264–278. Together with Oscar Riddle, Craig also prepared the chapter on "Voice and Instinct in Pigeon Hybridization and Phylogeny" for vol. 3 of the *Posthumous Works of Charles Otis Whitman.* Craig's later work included "A Note on Darwin's Work on the Expression of the Emotions in Man and Animals," *Journal of Abnormal and Social Psychology,* 1921–22, *16:* 356–366; and "The Song of the Wood Pewee *Myio-*

chanes virens Linnaeus: A Study of Bird Music," *New York State Museum Bulletin,* 1943, no. 334.

51. Craig to C. B. Davenport, 8 August 1920, Davenport Papers, Archives, American Philosophical Society, Philadelphia (hereafter Davenport Papers).

52. C. B. Davenport to Craig, 16 August 1920, Davenport Papers.

53. Craig to Pearl, 13 May 1923 (copy in Yerkes Papers). Pearl wrote to Yerkes 16 May 1923 stating "it seems to me that in this case specifically we owe a certain duty to pure science to see that the activity of a valuable brain is not lost to science. . . . Craig does not need a great deal of money, indeed never has had much. His requirements in every way to keep him at useful work are very modest" (Yerkes Papers).

54. Craig to Yerkes, 15 May 1923, Yerkes Papers.

55. Craig to Yerkes, 30 July 1923, Yerkes Papers.

56. See copy of letter from Craig to Bovie, dated 7 October 1943, Adams Papers.

57. Lorenz is quoted in Katharina Heinroth, *Oskar Heinroth, Vater der Verhaltensforschung* (Stuttgart: Wissenschaftliche Verlagsgesellschaft, 1971), p. 160; Craig to Carmichael, 3 February 1936, Carmichael Papers.

58. In addition to belonging to the American Psychological Association, Craig belonged to the American Ornithologists' Union, Sigma Xi, and the American Philosophical Society. He does not appear, however, to have joined the American Society of Naturalists or any other society besides the Ornithologists' Union that was specifically oriented toward biological concerns. For Craig's thoughts on psychological influences on physiological processes, see his "Oviposition Induced by the Male in Pigeons."

59. Jennings to Yerkes, 21 January 1906, and Watson to Yerkes, 29 October 1909, Yerkes Papers; Robert M. Yerkes, "Psychology in Its Relations to Biology," *Journal of Philosophy,* 1910, *7:* 113–124. On Yerkes's experiences at Harvard, see O'Donnell, *The Rise of Behaviorism,* pp. 191–200.

60. Wartime dislocations and skyrocketing publication costs caused the *Journal of Animal Behavior* to discontinue publication in 1917. On the journal, see Richard W. Burkhardt, Jr., "The *Journal of Animal Behavior* and the Early History of Animal Behavior Studies in America," *Journal of Comparative Psychology,* 1987, *101:* 223–230.

61. *Carnegie Institution of Washington. Yearbook,* 1902, 1: 167, 274–283, and subsequent *Yearbooks.*

62. On the Tortugas Laboratory, see Alfred Goldsborough Mayer, "Marine Biological Laboratory at Tortugas, Florida," *Carnegie Institution of Washington. Yearbook,* 1904, *3:* 50–54, and subsequent reports in the *Yearbook.* Information on Watson's early applications to the Carnegie Institution is to be found in the Carnegie Institution of Washington archives. For Watson's early plans to go to the Tortugas, see Watson to Yerkes, 7 February 1907, and 29 March 1907, Yerkes Papers. See Watson, "The Behavior of Noddy and Sooty Terns," *Papers from the Tortugas Laboratory of the Carnegie Institution of Washington,* 1908, *2:* 187–255; and Watson and K. S. Lashley, "Homing and Related Activities of Birds," *Papers Tort. Lab. Carn. Inst. Wash.,* 1915, *8:* 1–104.

63. See Craig to Adams, 4 February 1914, Adams Papers; and Craig to Pearl (copy), 13 May 1923, Yerkes Papers.

64. On the decline of the study of tropisms, see Philip J. Pauly, "The Loeb-Jennings Debate and the Science of Animal Behavior," *Journal of the History of the Behavioral Sciences,* 1981, *17:* 504–515.

65. Jennings to Yerkes, 30 October 1907, Yerkes Papers.

66. Jennings to Yerkes, 10 June 1916, Yerkes Papers. The failure of animal behavior studies to take deep root in American zoology departments is illustrated by the fate of behavior courses in several major departments in the first half of the century, notably Columbia, Johns Hopkins, and Berkeley. At Columbia, T. H. Morgan, the Professor of Experimental Zoology, introduced in 1906 a course on tropisms (in addition to his other courses on experimental zoology, experimental embryology, and regeneration), and by 1910 he was also offering a course entitled "the experimental study of instincts." The Columbia catalog continued to list Morgan's courses on tropisms and instincts up into the mid 1920s. How often he actually taught these courses is not clear, however, and the courses were dropped either just after he left Columbia for the California Institute of Technology or shortly before. Morgan, needless to say, had identified other, more rewarding research problems. Samuel J. Holmes, upon his arrival at Berkeley in 1912, took over the course on animal behavior that had been introduced by Harry B. Torrey the previous year, and he continued to teach it until his retirement in the 1930s. When Holmes retired, however, the course was dropped from the books and as of 1950 no behavior course had found its way back into the biological curriculum at Berkeley. At Johns Hopkins, H. S. Jennings turned his behavior course over to S. O. Mast around 1910. Mast continued to teach animal behavior until he retired in the 1930s, but after Mast's retirement, behavior ceased to be listed among the topics offered by the Hopkins biologists.

Particularly interesting is what happened to behavioral studies in the zoology department at Chicago. Before Whitman's death, no undergraduate courses on animal behavior were offered by the department (although animal "habits" were treated in the field zoology and animal ecology courses taught by Victor Shelford and the ornithology course taught by Reuben Strong). In 1910–11, Shelford taught a special course on animal behavior as well as his courses on field zoology, ecology, and geographic zoology, and behavior was taught regularly thereafter, either in conjunction with ecology or independently of it. The courses Shelford initiated were later taught by Morris Wells, from 1915 to 1919, and then, as of 1921, by Warder Clyde Allee, who had been a Shelford student. Interestingly enough, and appropriately enough as well, Allee did not see his own behavioral work, which was arguably as important a contribution to behavior study as that of any other American zoologist in the second quarter of the century, as a continuation of a tradition begun at Chicago by Whitman. Allee indeed seems to have had little feeling that Whitman's behavioral work was particularly important for him. When F. R. Lillie read the material Allee had prepared for the historical introduction to Allee's forthcoming, coauthored *Principles of Animal Ecology,* Lillie chided Allee for not having done justice to Whitman's contribution to the study of behavior. Unsure of himself on this score, Allee wrote to Yerkes asking "How significant was the contribution of Professor Whitman to the work in animal behavior?" Allee to Yerkes, 21 June 1943, Yerkes Papers. For Allee's revised comments on Whitman, see W. C. Allee, et al., *Principles of Ecology* (Philadelphia: W. B. Saunders Company, 1949), p. 24.

67. On the fate of Yerkes's students, see O'Donnell, *The Rise of Behaviorism,* 195–197. On the research interests of American Psychological Association members, see its *Yearbook,* 1921. See also Burkhardt, "*The Journal of Animal Behavior.*"

68. C. C. Adams to Yerkes, 16 February 1932, Yerkes Papers.

Ronald Rainger

7 Vertebrate Paleontology as Biology: Henry Fairfield Osborn and the American Museum of Natural History

In late nineteenth-century America, the study of paleontology flourished. After the Civil War, a multitude of geological and geographical surveys of the western states and territories provided new occupational opportunities that enabled scientists to uncover a wealth of fossil material. Work by vertebrate paleontologists Joseph Leidy, O. C. Marsh, and Edward Drinker Cope yielded a spate of discoveries that brought them international fame and placed America at the forefront of that field of science. Cope and Alpheus Hyatt, an invertebrate paleontologist, were the principal architects of a non-Darwinian theory of evolution and the nominal heads of an American school of neo-Lamarckians. Paleontology, it has been claimed, was among the first disciplines in which American scientists made empirical and theoretical contributions that were not merely derivative of European science.[1]

Yet despite the efflorescence of research in that field, paleontology remained largely peripheral to the developments occurring in American biology. Institutionally, vertebrate and invertebrate paleontology generally had no place in the new centers established for biology. As American biology became increasingly based in specialized departments, laboratories, and societies (see Appel, Benson, Maienschein, and Pauly, in this volume), paleontology found homes elsewhere. Prominent new biology departments at Johns Hopkins, Chicago, Harvard, and other universities included few courses or instructors in paleontology. Expanding natural history museums most often placed paleontology in departments of geology, or on occasion in its own separate department. Although Marsh maintained a large paleontology laboratory in the Yale Peabody Museum of Natural History, the new Field Museum in Chicago and United States National Museum in Washington defined paleontology as a subdivision of geology and organized it accordingly. In disciplinary and institutional terms, paleontology was rarely considered a part of biology.[2]

Conceptually, too, paleontology had only indirect and rather indefinite ties to biology. Although Marsh and Leidy produced numerous important studies of fossil vertebrates, they concentrated on describing and classifying specimens and only rarely addressed questions of development, evolution, or inheritance.[3] Cope, on the other hand, was a theorist, a neo-Lamarckian who constructed a comprehensive evolutionary theory based on the use and disuse of parts and the inheritance of acquired characteristics. Yet he was also a maverick. Despite his ingenious speculations on morphological problems, Cope was without a home, a scientist who had no secure and permanent institutional foundation from which to sustain his program of research. Forced by political and economic difficulties to sell his large collections of fossil vertebrates in the mid 1890s, he thus effectively terminated his research in vertebrate paleontology. Although Cope's views on variation, evolution, and inheritance had a significant impact in the 1880s and early 1890s, circumstances ultimately provided him with little opportunity to maintain his research or to perpetuate his influence on the analysis of biological questions pertaining to fossil vertebrates.[4]

Biology also had little relevance to work in invertebrate paleontology. Hyatt, like Cope, had an active interest in the whole range of questions pertaining to evolution, and in the early twentieth century a group of students and disciples applied his methods, views, and even his terminology to a wide variety of fossil invertebrates. Scientists such as Robert Tracy Jackson, Amadeus Grabau, and Joseph Augustus Cushman maintained an active interest in morphological questions.[5] By the 1920s, however, they were a rapidly dwindling minority. Always closely associated with the effort to identify mineral resources, invertebrate paleontology in the early twentieth century increasingly became a subset of geology. In the context of a rapidly expanding industrial economy, made evident most clearly in the dramatic development of the oil industry, the study of fossil invertebrates came more and more to serve strictly economic interests, as markers or indicators for mineral deposits. Looking back on the development of invertebrate paleontology in that period, Percy E. Raymond claimed that "in this commercial age, few invertebrate paleontologists are interested in anything except the description of new species."[6] Geology, not biology, offered opportunities for invertebrate paleontologists, and consequently geological questions dominated training and research in the discipline. As geology threatened to subsume invertebrate paleontology, the study of fossil vertebrates increasingly became a diversion and social concern of the rich, a manifestation of the interests of an Andrew Carnegie, Marshall Field, or J. Pierpont Morgan who, in their support of natural history museums, viewed vertebrate paleontology as a means to educate and entertain the urban masses.[7] In the early twentieth century, neither vertebrate nor invertebrate paleontology demonstrated much likelihood of becoming a foundation for serious programs of research in biology.

The Department of Vertebrate Paleontology (DVP) at New York's American Museum of Natural History, however, constitutes a significant exception to that rule. Under the direction of Henry Fairfield Osborn, a biologist, administrator, and entrepreneur, the American Museum developed a large, dynamic department grounded in the study of biological questions pertaining to fossil vertebrates. "Paleontology," according to Osborn, "is not geology, it is zoology; it succeeds only in so far as it is pursued in the zoological and biological spirit."[8] For Osborn problems of biogeography, the process and pattern of evolution, or the relationship between inheritance and development, more than traditional issues in systematics and stratigraphy, were the heart and soul of vertebrate paleontology. Not only did he publish prolifically on such subjects, he also influenced and directed the work of others. Pursuing topics of investigation that Osborn considered interesting and important, his principal associates, William Diller Matthew and William King Gregory, investigated biological problems and appreciably affected research in American vertebrate paleontology.

Moreover, Osborn was able to develop an institutional base for biological research on fossil vertebrates. Interested from an early age in establishing new foundations for American biology, Osborn had the requisite economic resources, the social and political connections, and the administrative abilities. In contrast to Cope or even Marsh, whose paleontological empire was crumbling by the mid 1890s,[9] Osborn developed and maintained a viable program of research on fossil vertebrates. By virtue of his conception of the science and the important infrastructures he created, vertebrate paleontology at the American Museum flourished as a field of biology.

Osborn's Social and Scientific Background

Osborn's role in developing a center for biological research in vertebrate paleontology derived from his academic training as well as the social and economic opportunities at his command. Originally attracted to science by the romance of fossil hunting, he chose the study of morphology as the means for structuring a scientific career. While his education defined the biological problems he would address, the influence and assistance provided by his family enabled Osborn in the 1880s to establish a career in biology. From the outset he combined an interest in biological research with a desire to organize and administer institutions in his chosen line of work.

Osborn was drawn to science while an undergraduate at Princeton University (then the College of New Jersey). Having read the exploits of Marsh and Cope, Osborn and two companions organized a fossil hunting expedition to the Bridger Basin of Wyoming in 1877. The experience encouraged Osborn and his colleague William Berryman Scott to choose careers in science, specifically work associated with vertebrate paleontology. At the time academic opportunities in paleontology or biology were only beginning in this country, and Princeton had only begun to offer graduate work in science. But through their close association with the college president James McCosh, Osborn and Scott embarked on a course of graduate study and soon secured academic positions. McCosh, interested in promoting science and graduate education, allowed Osborn and Scott to use the college's E. M. Museum to prepare the fossil materials obtained on their summer expeditions. More important, he carefully laid out plans for the two students to pursue graduate work in morphology. With advice and recommendations from McCosh, Osborn in 1878 took courses in anatomy and physiology with William Henry Welch in New York. The next year he joined Scott to study comparative anatomy with T. H. Huxley in London, and embryology with Francis Maitland Balfour in Cambridge. Taught by two of the leading morphologists of the late nineteenth century, Osborn developed an interest in the problems of development, evolution, and inheritance.[10]

Osborn's graduate training shaped his initial scientific investigations. Although he completed a study in vertebrate paleontology for his Sc.D. from Princeton in 1881, much of his early research focused on the morphological problems considered important by his mentors. Fascinated by the embryological work he had done with Balfour, Osborn in the early 1880s published several studies on the development of the brain in reptiles and amphibians.[11] At Princeton he did research in physiological psychology. Working with McCosh, Osborn and Scott took leading roles in teaching science courses, developing faculty clubs, and promoting research in psychology. Osborn also published studies on memory, imagery, and other topics that combined work in comparative anatomy, neurophysiology, and psychology. Those studies, however, required a facility for fieldwork and experimentation that Osborn, by his own admission, did not possess. Despite having done research in several areas of biology, Osborn in 1885 returned to the study of vertebrate paleontology.[12]

Previously Osborn had avoided devoting himself to vertebrate paleontology, in part because of the hostilities between Cope and Marsh, and in part because of the wider range of opportunities available in morphology and comparative anatomy.[13] Now, in 1885, aware of his limitations in both the laboratory and the field, he concentrated on the study of fossil vertebrates and spent six months working in the Munich laboratory of Karl Alfred von Zittel. He did so, however, not with the intention of teaching the science at Princeton; Scott already had that job.[14] Rather Osborn would direct his efforts to pursuing, organizing, and supporting research in vertebrate paleontology. Interested in the study of fossil vertebrates as a means for examining morphological questions, Osborn relied on financial assistance from his family to further his work.

Osborn's social and economic background no doubt influenced him in making the commitment to vertebrate paleontology. Born in 1857, he grew up in a family of considerable wealth and status. His mother, Virginia Reed Osborn, was the daughter of Jonathan Sturges, a prominent New York merchant. Although Osborn hardly knew Sturges, on turning twenty-one in 1878 he received $20,000 from his maternal grandfather.[15] Likewise his father, William Henry Osborn, was a wealthy and powerful businessman. Having made a fortune at a young age, William in 1855 became president of the Illinois Central Railroad. Bringing that corporation out of serious financial straits to a position of strength and prosperity within ten years, Osborn made the Illinois Central one of the country's leading railroads and earned a substantial fortune and reputation for himself. Such wealth and the social and political connections that accompanied it aided the young Osborn in his desire to pursue a scientific career. Money provided by his family enabled him to undertake his earliest paleontological expeditions and to pursue graduate research in biology. His father's financial assistance may have influenced Osborn's appointment at Princeton, and clearly his father provided considerable support to improve the college library holdings, to develop faculty and student discussion clubs, and to expand facilities for scientific research and education.[16]

More than a successful and wealthy businessman, the elder Osborn was also an entrepreneur who wanted to make his son head of the railroad empire he had created. Balking at his son's interest in fossil hunting and later cutting short his graduate studies, Osborn's father hired him to help administer the recent acquisition of the Chicago, St. Louis, and New Orleans Railroad. Despite his father's efforts to groom him for a business career, however, Osborn retained a strong interest in science. Family friction resulted. Eventually acceding to his son's wishes to establish a career in science, the elder Osborn nonetheless remained involved in his son's activities. Having failed to convince the young Osborn to take up business, William Osborn helped to provide facilities that would enable his son to operate as an entrepreneur in biology and vertebrate paleontology. Relying on the money, the model, and advice provided by his father, Osborn combined scientific research with efforts to order and organize the newly emerging field of biology.[17]

Osborn and New Foundations for Biology

In his earliest years at Princeton, Osborn and his colleague Scott identified themselves as biologists and actively sought association with others in the field. In 1883 both men joined the new Eastern Society of Naturalists (see Appel, in this volume), and by 1884 Osborn was helping to run that organization and the biological section of the American Association for the Advancement of Science.[18] His scientific interests and financial resources also led Osborn to inaugurate efforts to establish a biological organization, journal, and laboratory.

In the 1880s Osborn was part of an expanding group of American biologists who sought new outlets for their science. At the time none of the principal science periodicals seemed entirely appropriate to Osborn's morphological interests. *The Journal of the American Microscopical Society* was a specialized, technical publication, while the *American Journal of Science* and *Science* were general periodicals that did not allow for lengthy biological studies. Neither did *American Naturalist,* a periodical owned and run by Cope that was in considerable financial trouble.[19] In 1884 several biologists including Osborn felt the time was ripe for a more specialized, professional journal; others sought to develop a biological society. Osborn had a hand in both efforts.

In March 1884 Osborn wrote the English biologist E. B. Poulton that he was considering starting a journal of morphology. With financial backing from his father, Osborn proposed a journal that would publish articles on comparative anatomy, plant and animal morphology, paleontology, histology, and embryology, "but will exclude all systematic, historical, and purely descriptive work."[20] The journal embodied Osborn's conception of the most important fields and issues in biology; accordingly, he wanted himself, Scott, Poulton, E. B. Wilson, and Benjamin Sharp as editors. Apparently Osborn also wanted to keep the venture secret, although Wilson made it known to Harvard biologist E. L. Mark who was independently attempting to establish an embryological society and publication. Although Mark had contacted Osborn and Scott about his idea for "cooperation in embryological work between a few persons who have its interests most at heart," Osborn's journal plans presented difficulties for Mark, and both efforts soon foundered[21] (see Appel in this volume). Yet at this early date, Osborn, drawing on the financial resources and opportunities at his disposal, aspired to promote biological research through the development of appropriate infrastructures.

Osborn had considerably greater success in expanding research opportunities at Princeton. Having committed himself to the study of vertebrate paleontology, Osborn in 1886 sought to establish a center for the science. The morphological laboratory represented, albeit on a small scale, Osborn's commitment to financing and directing a program of shared objectives on biological research pertaining to fossil vertebrates. Drawing on advice and financial assistance from his father, he pressed for a reduced teaching load and improved laboratory facilities at Princeton as a basis from which to begin. Osborn arranged for his "salary" to go to an assistant who had responsibility for the practical (laboratory) work for his courses on anatomy, morphology, and physiological psychology. Similarly, he personally hired a preparator and an artist, Rudolph Weber, with whom he had worked in Munich. Osborn's interest in enhancing his own research opportunities in vertebrate paleontology led him to finance a morphological laboratory that would provide new and better facilities than existed at the older zoological and geological museums on campus.[22]

In addition to providing financial backing, Osborn worked to coordinate and organize research in vertebrate paleontology. Although he and Scott had previously studied, taught, and even published together, the arrangements made in 1886 provided for a closer and more well-defined working relationship. In one respect, that relationship seemed to entail some separation of tasks. Although Princeton sponsored fieldwork and numerous expeditions to western fossil beds, Osborn did not engage in those activities. Collecting and preparing specimens are crucial components of vertebrate paleontology, and the fact that Osborn did not do fieldwork may have reflected certain practical decisions. Since 1882 Scott, as professor of geology, had led the university expeditions and had done so with considerable success.[23] There was little need for Osborn to participate in such work, particularly because he claimed to have little facility for fieldwork. Socioeconomic factors may also have played a role. Osborn had provided the money for additional personnel and improved facilities for research in vertebrate paleontology, and it is conceivable that, as an entrepreneur, he chose not to engage in fieldwork. In later years at the American Museum, Osborn had only a very limited involvement in fieldwork, although he was well aware of its importance to research in vertebrate paleontology and oversaw the expeditions by his staff. On those rare occasions when he did participate, it was largely a formal affair: to oversee the work or to promote publicity for spectacular finds at places like Bone Cabin Quarry or the Gobi Desert.[24] Osborn's lack of fieldwork in the 1880s does not mean that there existed a well-defined division of labor or delegation of work at Princeton more generally, but it does suggest that in addition to doing research he paid close attention to managing and administering research in a manner that foreshadows his later role at the American Museum.

Aside from fieldwork, Osborn and Scott collaborated in all other aspects of research in a way that was productive and mutually reinforcing. In 1886, for example, Osborn made plans with Scott to write a jointly authored textbook on American fossil vertebrates.[25] They also collaborated on several papers. For a study on fossil mammals from the White River Miocene, they journeyed together to Cambridge and Philadelphia to examine specimen collections. Dividing up materials for study, Osborn focused his attention on perissodactyls (odd-toed hoofed mammals) while Scott examined fossil carnivores and artiodactyls (even-toed hoofed mammals). They then analyzed and reviewed each others' work for joint publication. Similarly, they authored a joint monograph on the fossil mammalia of the Uinta formation.[26] Each man also continued to publish separately, although here too there was extensive exchange of ideas.

Osborn and Scott both focused their attention on problems not typically addressed by vertebrate paleontologists—that is, biological questions of variation, evolution, and inheritance. Whereas Marsh, Leidy, and others had generally ignored such issues, Osborn and Scott recognized the study of fossil vertebrates as a means for examining problems central to late-nineteenth-century morphology. Certainly Huxley, Balfour, and McCosh had helped to generate their interests in such issues. Even more influential was Cope, whose neo-Lamarckian theory of evolution had attracted Osborn and Scott. In a number of studies published in the late 1880s, both men contended, with Cope, that the fossil record demonstrated evolution to be an orderly, regular process that could not be explained by the natural selection of random variations. In line with Cope's theory, Osborn and Scott maintained that the use or disuse of parts resulted in structural changes, and the inheritance of those acquired changes constituted evolution. Osborn and Scott were preoccupied with morphological questions, and at first they accepted Cope's neo-Lamarckian interpretation of variation, evolution, and inheritance.[27]

By 1890, however, Osborn and Scott had rejected certain central features of Cope's theory, and in its place defined their own views on morphological issues. Both responded to new hereditary theories, notably August Weismann's work, and Osborn in particular abandoned the neo-Lamarckian principle of the inheritance of acquired characteristics.[28] Both men also called into question Cope's emphasis on use and disuse as the means for explaining variation and evolution. Continued research on fossil mammals led them to claim that evolution is an irreversible process, one that proceeds in definite directions and yields rectilinear, virtually predictable trends. According to Scott,

> the relatively fixed direction taken by variations, which has been insisted upon by so many observers (e.g., Askenasy, Eimer, Geddes, Thompson, and Osborn) comes out most clearly in the series of fossil mammals. Granting that unlimited variation is no necessary part of the selection theory, it seems strange that new facets on the bones and new cusps on the teeth should appear only in such definite ways and that there should not be many tentative attempts and false starts before the proper development is hit upon. In the structure of the carpus and tarsus we find that in any given phylum very definite lines of evolution are early established and closely adhered to, and the changes are just those called for by the operation of dynamical influences.[29]

Impressed by the highly regular nature of variation and the occurrence of widespread evolutionary parallelism, Scott by 1891 rejected the concept that evolution was an adaptive process based on the use and disuse of parts. Instead he adopted the ideas of Wilhelm Waagen, a German invertebrate paleontologist, and put forth an orthogenetic interpretation, claiming that the impact of external factors on an internal directing force produced "mutations"—gradual, cumulative changes that resulted in linear, parallel lines of descent. Though Scott first presented such an interpretation, it was Osborn in the mid 1890s who drew upon Waagen's concept of mutations and the principle of evolutionary irreversibility to fashion a comprehensive theory of evolution.[30]

Osborn's and Scott's commitment to orthogenesis was reinforced by the methods they adopted for classifying organisms and constructing phylogenies. Unlike Cope, who had followed traditional methods, Osborn and Scott employed vertical schemes of classification, an approach whereby they separated specimens, even among related organisms, into distinct lines of descent at the first sign of morphological difference. Scott, in two highly suggestive papers published in 1890–91, described mammalian evolution in terms of numerous parallel and separate lines of descent. Discussing camels and other closely related mammals, he claimed that "all the great groups of selenodont artiodactyls when traced back are seen to arise independently from the abundant and widespread Eocene type, which may be called the Buno-Selenodont, and which forms, as it were, a lake from which several streams, flowing in partly parallel, partly divergent directions, are derived." For Scott, and later for Osborn, evolution was not a diverging, branching tree of life, but rather a bush with numerous, parallel and independent lines.[31]

Certainly Osborn and Scott were not alone in their commitment to non-Darwinian explanations of the process and pattern of evolution. For years Hyatt had explained evolution as a process that paralleled individual development and resulted in racial senescence and species extinction. Similarly, Theodor Eimer and a host of others argued that various internal biological mechanisms or guiding principles directed evolution along definite, linear paths. Non-Darwinian, orthogenetic evolutionary theories abounded in the late nineteenth century.[32] Yet the views of Osborn and Scott were distinctive, particularly in the context of American vertebrate paleontology. Few of their colleagues examined evolutionary questions, and virtually none, including Cope, relied on vertical schemes of classification or adopted Waagen's understanding of variations, mutations, and evolution. Interested in the study of fossil vertebrates as a means for investigating biological questions, Osborn and Scott had worked together to develop their own interpretation of evolution and related questions. Collaboration and the dynamic interplay of ideas made possible through the establishment of a center for research at Princeton enabled Osborn and Scott to secure a place for themselves in the community of biologists and paleontologists.

For Osborn, those developments were particularly important. Unsure in 1885 of what program of research to pursue, uncertain of his abilities and perhaps even his position, Osborn had achieved some status in vertebrate paleontology through the work done at Princeton. In part he did so by drawing on his own resources to establish a well-funded and organized center for pursuing scientific research. In addition, he drew upon Scott. Concerned with biological questions and knowledgeable about vertebrate paleontology, Osborn nonetheless relied a good deal on Scott's greater experience in the laboratory and the field as well as his insightful views on evolution. By virtue of Osborn's facility for raising money, organizing research, and drawing on the talents of others, Princeton became a small but active research center. Such achievements met his scientific and administrative aspirations, and when he moved on to the American Museum Osborn worked to develop vertebrate paleontology along similar lines.

Osborn, Columbia, and the American Museum

In 1891 Osborn left Princeton for New York City, specifically to take positions at Columbia University and the American Museum of Natural History. At the American Museum he quickly developed a program in vertebrate paleontology that, in terms of specimen collections, personnel, and economic support, far surpassed anything at Princeton. Osborn, by virtue of a network of powerful political, economic, and social connections, established a large, diversified department of vertebrate paleontology devoted to the study of problems in evolution, biogeography, and functional morphology that he considered significant.

Osborn's opportunity to develop such a program came first from Columbia. In 1890 Seth Low, the newly appointed president of Columbia, inaugurated a campaign to transform "the College into a complete University adapted to maximum service to American needs."[33] After establishing a central administration, a coordinated curriculum, and a coherent educational philosophy that emphasized graduate education and research in all fields, Low took the lead in selecting faculty for the new university. Osborn was among those he solicited. Though originally interested in Osborn for a position as a professor of physiological psychology, Low quickly changed his offer when Osborn indicated that his primary field of interest was biology and that he would consider a position at Columbia if Low and the university could guarantee the rapid development of a first-rate department of biology.[34] Over the next year the two worked closely together on such a project. Osborn laid out his plans for a biology program and requirements for building and equipping such a department, while Low shared with Osborn his concerns for fitting biology in with other programs and his overall vision for the university. Osborn's program, presented to the Columbia board of trustees in April 1891, embodied his conception of the foundations of biology: a program that offered courses in vertebrate and invertebrate zoology, morphology, and embryology. Based on his views, as well as Low's negotiations with the medical school and other departments, botany, physiology, and bacteriology would be taught as related but separate fields of study through other university departments.[35] Osborn's program reflected what for many was the "core" of biology (see Introduction to this volume). Negotiating with Low, Osborn was not only able to make such a program a reality, he also established vertebrate paleontology as a part of that program.

While Osborn's plan for biology at Columbia did not explicitly include vertebrate paleontology, he nonetheless developed an associated program in that field and made sure it had close ties to the biology department at Columbia. In August 1890 Low, with Osborn's prompting, wrote Morris K. Jesup, president at the American Museum, and inquired about the possibility of establishing cooperative programs between the two institutions. Low, seeking to develop a university that would serve the needs of New York City, no doubt recognized the value of the American Museum, by that time the largest science museum in the country. Connections with the museum would promote the interchange of speakers, collections, and other resources between the two institutions, and Low suggested to Jesup that "in the direction of biology especially, . . . the collections of your Museum might be made of great value to our students and instructors."[36] A businessman and politician by background, Low had little knowledge of biology and its needs, but recognizing the opportunity to establish a cooperative relationship with a major museum, he negotiated for Osborn a curatorship in vertebrate paleontology at the American Museum. He also established an arrangement whereby Columbia's biology department offered courses and research in vertebrate paleontology, although students would do such work primarily at the American Museum.[37] At the time such an explicit association of vertebrate paleontology with biology existed only in New York, and that arrangement helped to shape the research done there in vertebrate paleontology.

Although Osborn inaugurated a new program in biology at Columbia, one that soon achieved national and international status, the American Museum became the primary arena for his scientific and administrative activities. Osborn's principal interests lay in vertebrate paleontology, a discipline that required extensive fieldwork, collections, and preparatory laboratories—none of which Columbia could provide. The American Museum possessed those facilities and resources, and by 1890 had developed important collections in mammology, ornithology, and geology.[38] Moreover, the American Museum possessed extensive financial resources, another necessity for vertebrate paleontology, and indeed Osborn's close personal connections with important administrators at the museum enabled him to obtain substantial support for the development of a department of vertebrate paleontology. Albert Bickmore, the founder and supervisor of the museum, was a close family friend, and others, including Osborn's uncle, J. Pierpont Morgan, and William E. Dodge, his brother's father-in-law, sat on the museum's board of trustees. Those men and others provided the financial and political backing that enabled the American Museum to establish a large, well-staffed department capable of doing the fieldwork, preparation, and research necessary in vertebrate paleontology.[39]

Osborn, of course, also played a crucial role in that development. In New York, as in Princeton, he spent a good deal of his own money to build a successful program. Originally committed to subscribing $1,500 annually to the department's expenses, by 1900 he was contributing more than $2,000 per year.[40] Moreover, he actively supervised and managed the department's programs. He devoted considerable effort and attention to soliciting the money, personnel, and materials necessary to run a multifaceted department. Despite his infrequent participation in fieldwork, Osborn nonetheless defined for his collectors the principal sites for fieldwork, the kinds of specimens they should particularly search for, and the geographic and stratigraphic information they should obtain.[41] Similarly, he personally hired and supervised the work of numerous artists, illustrators, and preparators. Combining the powerful support provided by the museum administration with his own scientific knowledge, personal financial contributions, and managerial skills, Osborn developed a department that by 1920 had transformed the methods for preparing and displaying fossil specimens and outstripped all other centers for the science in size of staff, size and quality of specimen collections, and number of expeditions.[42] Most important, Osborn drew upon the resources, opportunities, and facilities available through the American Museum to develop a program that emphasized research on biological questions pertaining to fossil vertebrates.

W. D. Matthew and Vertebrate Paleontology at the American Museum

The move to the American Museum and Columbia not only enabled Osborn to establish valuable infrastructures for the institutional development of biology and vertebrate paleontology but also provided greater opportunities to pursue the line of biological research he had begun at Princeton. Osborn continued to emphasize the study of fossil vertebrates, and, drawing on the fieldwork, preparation, and research done by his museum staff, he published a vast number of taxonomic and descriptive papers in vertebrate paleontology. He also remained preoccupied with evolution and, building on his earlier work with Scott, developed in the mid 1890s a comprehensive evolutionary theory. Furthermore, his positions at Columbia and the American Museum now enabled Osborn to influence and coordinate the work of others, notably Matthew and Gregory.

Osborn's scientific research and influence are particularly evident in his work on evolution and geographical distribution and its impact on Matthew. Throughout the early 1890s Osborn sought to develop a theory of evolution that could account for variation, inheritance, and the relationship between ontogeny and phylogeny. Drawing on historical analysis and even more the earlier work done at Princeton, Osborn constructed a theory that relied heavily on the ideas of Waagen, ideas that Scott had found so suggestive. Distinguishing between Waagen's concepts of variations and mutations, Osborn claimed that changes in environment or development produced variations that had taxonomic importance, but were of little long-term evolutionary significance. For Osborn, changes that affected an organism's hereditary constitution were especially noteworthy.

> While the environment and activity of the organisms may supply the stimuli in some manner unknown to us, definite tendencies of variation spring from certain very remote ancestral causes; for example, in the middle Miocene the molar teeth of the horse and the rhinoceros began to exhibit similar variations; when these are traced back to the embryonic and also to the ancestral stages of tooth development of an earlier geological period, we discover that the six cusps of the Eocene crown, repeated today in the embryonic development of the jaw, were also the centers of phylogenetic variation; these centers seem to have predetermined at what points certain new structures would appear after these two lines of ungulates had been separated by an immense interval of time. In other words, upper Miocene variation was conditioned by the structure of a lower Eocene ancestral type.[43]

Evolution for Osborn constituted the unfolding of a predetermined plan that lay latent in an ancestral germ plasm. External factors triggered an internal hereditary potential or predisposition that in turn gave rise to gradual, continuous evolutionary changes. For Osborn the hereditary factor was the causal mechanism for evolutionary change; it also controlled evolutionary trends by producing mutations that were cumulative, linear, and irreversible.[44]

Having developed a theory of evolution, Osborn began to apply it to questions of geographical distribution. His analysis in 1897 of the evolution of rhinoceroses immediately made him aware of important similarities among widely separated species. While such evidence confirmed his belief in widespread evolutionary parallelism, it also led him to recognize a serious problem that underlay any interpretation of the evolution or geographical distribution of vertebrates—namely, the fact that there existed no well-defined system of biostratigraphy and correlation. The lack of a coherent, worldwide system of correlation for the Cenozoic era motivated Osborn to address the problem. He did so in part through research and publication; more characteristically, he became an advocate, publicist, and organizer who inaugurated an international vanguard for reform. Soliciting support and information from a number of vertebrate paleontologists, Osborn also relied extensively on Matthew to do research on Cenozoic biostratigraphy and correlation. He recruited Matthew to examine questions that he considered important, and thereby provided the means for them both to contribute to the analysis of questions in correlation, biostratigraphy, and biogeography.[45]

Matthew was in many respects a highly valued asset to Osborn. Trained primarily as a geologist, he became interested in vertebrate paleontology only in his last year as a graduate student at the Columbia University School of

Mines. Taking Osborn's courses on the evolution of vertebrates and mammalian morphology, he soon became an assistant in the new and rapidly expanding DVP. There he quickly gained a reputation for doing accurate and important work. With a knowledge of geology and stratigraphy that most vertebrate paleontologists of the day did not possess, Matthew offered new insights on a variety of issues. Working with the museum's large collection of Eocene mammals, Matthew in 1897 reformulated the understanding of the stratigraphy of the Puerco, a formation that had long proved troublesome to American vertebrate paleontologists. In other early studies he developed a valuable classification of mammals in the western states and, with others, reinterpreted the understanding of the geological processes responsible for the rich deposits of fossil mammals in that region. His abilities in the laboratory and the field demonstrated early on that Matthew was an intelligent, resourceful scientist.[46]

Not surprisingly, Matthew came to play an important part in Osborn's work in correlation and geographical distribution. For Osborn's 1909 monograph on Cenozoic mammal horizons of North America, Matthew not only contributed the faunal lists of tertiary mammalia but also much of the data on geological processes and faunal life zones. Similarly he assisted in Osborn's organized efforts to reform the understanding of correlation and biostratigraphy. Matthew translated studies by Louis Dollo and Santiago Roth; he also provided detailed criticisms of the works by Roth and Florentino Ameghino on South American paleontology and geology that disputed the interpretations of numerous European and North American paleontologists. Matthew's intimate knowledge of fossil mammal specimens and his understanding of stratigraphy and geology made him an invaluable resource to Osborn.[47]

In pursuing such research under Osborn's direction, Matthew soon moved beyond strictly stratigraphic questions to deal with related biological problems that intrigued Osborn. Most noteworthy were studies on the radiation and geographical distribution of vertebrates. Osborn's concern with such questions, derived largely from his interest in evolution, set him apart from his colleagues; at that time virtually no American vertebrate paleontologists, including Cope and Scott, had examined the subject of biogeography.[48] Osborn did, and conseqently developed the principle of adaptive radiation that fit nicely with his theory of evolution. While aware of the occurrence of widespread adaptation and divergence, Osborn emphasized that "the modifications which animals undergo in this adaptive radiation are largely of a mechanical nature, they are limited in number and kind by hereditary, stirp, or germinal influences."[49] Work on geographical distribution led him to speculate on the sources of radiation and distribution and the similarities and differences of strata, flora, and fauna in different parts of the world. In the early 1900s Osborn suggested that Africa and Antarctica had served as major centers for radiation and distribution.[50] Although he offered intriguing hypotheses, Osborn did not fully pursue the subject; however, Matthew did. Studying geological processes, comparing extensive lists of flora and fauna, and analyzing changes in ancient climate, land mass, and ocean levels, Matthew sought to determine the origin and distribution of the principal families of vertebrates.

Matthew's earliest researches on geographical distribution reflected Osborn's influence. Osborn's work in 1900 included a restoration of Antarctica, and during the next several years he supervised departmental work on a series of paleogeographic maps that defined changes in land masses and animal migrations. Matthew, in his first analysis of geographical distribution, included a number of those maps and concentrated on describing the hypothetical outlines of the continents in earlier geological epochs. Although Matthew embraced Alfred Russel Wallace's theory of a Holarctic origin and distribution for vertebrates, he also followed Osborn in suggesting that some families of mammals, like the creodonts, had radiated out from Antarctica. He likewise adopted Osborn's interpretation that Antarctica, though a source of distribution, had no connection to Africa.[51] Osborn initially interested and influenced Matthew in the study of paleobiogeography, but Matthew soon delved into the subject in considerably greater detail and advanced a new and different interpretation, one that Osborn eventually came to accept.

Despite his early belief that some families of mammals had migrated from a homeland in Antarctica, Matthew soon became a staunch advocate of the northern origin and dispersal of all vertebrates. His own research on the evolution and distribution of fossil rodents, along with new studies on marsupials by Robert Broom and Pauline Dederere, provided evidence of a North American-European origin for some of the most troublesome groups.[52] Moreover, in the years after 1906 Matthew became increasingly committed to a particular interpretation of the geological history of the earth. Whereas sunken continents, land bridges, and other extensions were becoming commonplace in the contemporary literature on paleogeography, Matthew steadfastly adhered to an understanding of the earth based on the relative permanence of continents and ocean basins and the concept of isostacy. Influenced by the views of Wallace and the geological theory put forth by Thomas Crowder Chamberlin, Matthew maintained that alternating conditions, in effect a balancing of geological processes within a stable framework, characterized the history of the earth. "The conclusion appears unavoidable," he wrote, "that in a broad way the present distribution of land and shallow water on the one hand, of deep water on the other, has been substantially unchanged." Based on his knowledge of the fossil record and committed to such an understanding of the earth's structure, Matthew roundly attacked hypotheses of drifting continents, extended land bridges, and other devices developed to account for the geographical distribution of vertebrates from centers other than Holarctica. His book *Climate and Evolution* (1915) was an impassioned argument for geological permanentism and the Holarctic theory of geographical distribution.[53]

Matthew's study of geographical distribution was also the basis for his analysis of other biological problems emphasized by Osborn, in particular the process and pattern of evolution. As early as 1902 Matthew took up the question of "the influence which changes of climate have had on the evolution of life, especially the higher land animals." Matthew, relying on the interpretations of Chamberlin, argued that periods of land elevation resulted in the development of cold, arid climates that promoted the rapid evolution and extensive migration of mammals.

> The point of most interest in reviewing thus the progress of the mammals during Tertiary time is that their main trend of evolution was not a predeterminate one, carried out on certain lines inherent in the organism, but was an adaptation to changing external conditions. As the climate changed and with it the conditions of their habitat, the animals changed correspondingly, and those races which were best fitted for the new conditions, either because of natural adaptability or because in their original habitat the new conditions were earlier reached than elsewhere and they thus had a handicap over their competitors, these races spread widely and became dominant. . . . The evolution of the mammal race was chiefly a response to changed conditions of life.[54]

From the outset Matthew explicitly rejected Osborn's interpretation of evolution and defined changes in external, environmental conditions as the principal causal factor for organic change.

In developing his interpretation of evolution, Matthew not only disputed Osborn's theory, but even ignored many of the questions that Osborn had considered. Whereas Cope and Osborn had devoted much attention to defining the internal biological mechanisms by means of which organisms responded to external conditions, Matthew did not. Claiming that environmental factors produced migrations and evolutionary change, he did not delve into a detailed examination of the how and why of such a process. In one respect Matthew's failure to investigate the factors responsible for evolution may have reflected a relative disinterest in biological questions and a continued emphasis on geological, environmental concerns. On the other hand, however, Matthew's interest in examining the process of evolution was itself manifestly a biological concern. Matthew would in fact advance a Darwinian interpretation of evolution, arguing that "change is due to the pressure of the environment acting through selection upon individual variations," and building on his interest in evolution he would later address a number of other related biological questions.[55]

Matthew's commitment to the theory of evolution by natural selection is evident in a number of studies. His interpretation of sabre-toothed cats is a good example. According to several early twentieth-century paleontologists, including Scott and Frederick B. Loomis, the history of those cats demonstrated that evolution followed predetermined paths that, once begun, could not be reversed and in fact took on an increased momentum leading to extinction. Among sabre-tooths, it was argued, the canine teeth were so affected and eventually reached a stage of over-specialization so that the animals could not open their jaws wide enough to eat and thus died out.[56] To Matthew that interpretation was outlandish. Early felids were well adapted to prey upon larger hoofed mammals, and as the latter became larger and thicker skinned, the sabre-tooths became "progressively larger, more powerful and developed longer and heavier weapons to cope with and destroy them." Denying any role for evolutionary momentum and overspecialization in the evolution of those cats, Matthew claimed that the utilitarian value of any character was the determinant of its evolution, and "the moment the harmfulness of a character outbalanced its usefulness" natural selection ensured its elimination.[57]

Likewise a commitment to Darwinian views characterized Matthew's interpretation of the evolution of the horse. The horse was the classic example of vertebrate evolution, although most paleontologists explained its history in orthogenetic terms. For Matthew the evolution of the horse went hand in hand with the evolution of the plains environment, and in a number of studies he described the disappearance of the side toes, the increase in the length of the leg and foot, and the change from short-crowned to high-crowned molar teeth that characterized the evolutionary history of that family. Working with an outstanding collection of horse material at the American Museum, he held that natural selection operated on changes produced by "climate, physical and geographic conditions, and associated fauna." Matthew then explained that natural selection produced a diverging pattern of "several branch lines more or less clearly defined, running into parallel or partly diverse adaptations." Matthew thus rejected the orthogenetic and neo-Lamarckian interpretations that dominated much of early twentieth-century evolutionary theory.[58]

Matthew's study of evolution, particularly his work on fossil horses, also led him to investigate other biological questions—namely, the nature of variations and the species concept. For years Osborn had maintained that variations were continuous and followed certain predetermined paths. Matthew also thought variations continuous, although more than Osborn he recognized and emphasized the importance of individual differences and variations within a species. Possibly his commitment to Darwinism—his belief that natural selection operated on individual differences—influenced Matthew's views. More likely, his fieldwork and the large amount of fossil material at his disposal made him increasingly aware of extensive variation among fossil specimens and increasingly hesitant to define taxonomic categories or construct phylogenies on the basis of any one or a few specimens. Certainly the wealth of horse material collected by American Museum expeditions in the 1910s and

1920s led him to question the typological assumptions that underlay the views of other vertebrate paleontologists. Material from one location, the Snake Creek deposits of western Nebraska, he claimed

> represents many thousands of individuals, no two of them exactly alike in the complex details of tooth construction. If the standards of species distinction that have been accepted by most American students of fossil Equidae were applied conscientiously to this great collection, the result would be to place upon record scores if not hundreds of "new species" from this one locality. But the thousands of isolated teeth or other fragmentary specimens would clearly show that there are no really constant and uniformly associated distinctions between such "species." They are merely individual differences and it is the scanty or fragmentary character of the material or a failure to make a thorough and impartial study of all materials available for comparison, the natural tendency to compare only the types or best preserved specimens, or to use drawings in place of the originals, that have been responsible for maintaining many of these species as distinct.[59]

Matthew thus directly challenged the typological views, the definition of species, and the understanding of variation that characterized the work of Osborn and others. He also sought to lay the foundation for a new interpretation by drawing on the contemporary work of biologists.

In the late 1920s Matthew worked to establish rules and regulations for defining species that were based in biology. Paleontologists, he argued, needed to recognize the wide range of variations within a species and establish their understanding of fossil species on the range and limitations of modern species. Calling for a species concept "defined by a group of associated fixed (hence inherited) characters, and including many interbreeding strains with a tinge of hybridism all around the contacts of its distribution range with that of other related species," Matthew pointed to the relevance of field studies, ecology, and even genetics for paleontology.[60] Indeed, he corresponded with geneticists T. H. Morgan, F. B. Sumner, and others on questions of genetics and the relationship of work in experimental evolution to paleontology. By 1930 Matthew could state that the geneticists' concept of mutations "is much better understood than it was in Darwin's time, thanks especially to the researches of T. H. Morgan and his school. Some of them are inherited according to definite laws. They are 'mutations' of the same nature as the larger more conspicuous and more occasional mutations which geneticists have studied in detail. Others are the non-heritable differences between individuals due to slight or considerable differences in their environment and growth history." Matthew understood the basics of classical genetics and recognized a striking similarity between the mutations geneticists studied in the laboratory and the continuous variations he found in the field. Certainly he never systematically applied the findings of genetics to the fossil record; however, his interest in biological problems, specifically variation, evolution, and the species concept, led Matthew to realize the important consequences of biological research for paleontology and to begin to incorporate that research into his own work.[61]

William King Gregory and Vertebrate Paleontology at the American Museum

In much the same way that Osborn's interest in biological questions provided a basis for Matthew's work, so too did it influence the researches of another associate, William King Gregory. Whereas Matthew, following Osborn's lead, took up questions pertaining to geographical distribution and evolution, Gregory examined problems concerning the relationship between form and function. Combining research on extinct and modern animals in a way that was unique among American vertebrate paleontologists, Gregory posed questions and developed techniques that offered new insights into the study of the fossil record. Drawing on his researches in functional morphology, Gregory, like Osborn and Matthew, would also investigate variation, evolution, and inheritance among fossil vertebrates.

In developing a program of research on functional morphology, Gregory drew on both the institutional and intellectual resources provided by Osborn. As Osborn's hand-picked successor, he became professor of vertebrate paleontology at Columbia in 1910. Ten years later he also took on the role of curator of a new Department of Comparative Anatomy at the museum. That department, which served as the training ground for numerous graduate students in biology from Columbia, emphasized research in comparative myology and osteology—the study of muscle and skeletal systems in fossil and recent vertebrates. For that work Gregory and his students drew on facilities and materials not only from Columbia and the DVP, but also the Bronx Zoo, whose president was Osborn. Clearly the institutional networks for biology that Osborn had established played a crucial role in Gregory's research.[62]

Gregory's interest in questions pertaining to habit, function, and adaptation also stemmed in part from Osborn's biological emphases. Osborn's concern with a wide spectrum of biological issues led him to recognize that issues concerning the relationship between form and function were central to the work of the vertebrate paleontologist. In 1892 he referred to the work of the Russian paleontologist Vladimir Kovalevsky as "a model union of the detailed study of form and function with theory and the working hypothesis. It reflects the fossil not as a petrified skeleton, but as moving and feeding; every joint and facet has a meaning, each cusp a certain significance."[63] In his own research, Osborn on occasion examined questions of function and adaptation, explaining, for example, how foot and limb structure among hoofed mammals resulted from changes in weight and pressure.[64] Moreover, his interest in such problems influenced the work of the DVP staff.

In the early 1900s, when preparators and scientists at the American Museum attempted to mount gigantic vertebrates like *Brontosaurus,* concerns about form and function came to the fore. The sheer size and bulk of those creatures raised serious questions about their habits and functions: How did they stand? Did they walk or crawl like other reptiles? What kind of skeletal and muscular structure enabled them to move, eat, and undertake other functions? Those questions led Matthew and others to branch out in new directions, specifically to combine studies of vertebrate musculature (myology) with the paleontologists' traditional emphasis on the study of skeletal hard parts. Gregory later described that innovative research.

> During the years before 1905, while the skeleton of the huge *Brontosaurus* was being removed from the matrix and restored and mounted by Adam Hermann and his assistants, Doctor Matthew studied the problems involved in the reconstruction and mounting. Before the limbs and girdles were mounted, he and Walter Granger dissected an alligator, marked the areas of origin and insertion of the limb muscles on the girdle and limb bones, and then identified as far as possible the corresponding areas on the bones of *Brontosaurus,* and finally by means of scale drawings and paper strips representing the muscles, they endeavored to determine the course and direction of the principal muscle masses in so far as they would influence the posture of the limbs and girdles, especially the angulation of the elbows and knees. This was apparently the first application of the data of comparative myology to the mounting of an extinct animal, and the studies that Matthew and Granger made at this time led eventually to further developments of those subjects of comparative myology and osteology by other workers in the Museum and elsewhere.[65]

Matthew did some additional investigations in functional morphology, but it was Gregory who was most strongly influenced by the studies of the DVP staff. As a Columbia graduate student working in the DVP, he participated in the researches on the relationship between form and function in *Brontosaurus.* In later years, questions of functional morphology became the cornerstone of his research in vertebrate paleontology.[66]

Following the approach inaugurated by Matthew and Granger, Gregory relied heavily on the comparative analysis of fossil and recent vertebrates. Biological questions, notably the need to understand how the interaction of muscle and skeleton operated to affect bodily functions and adaptation, lay at the heart of Gregory's program of research. He, more than other vertebrate paleontologists, combined the traditional study of skeletal hard parts with myology as a means for understanding questions of form, function, and adaptation. Gregory and his students not only examined fossil specimens but also dissected and analyzed recent vertebrates as a means for inferring the musculature, the relationship between muscle and skeletal structure, and the habits of extinct organisms. One of Gregory's students, Roy W. Miner, best defined the assumptions that guided such research. "The vertebrate endoskeleton may, therefore, be said to have been extensively molded by muscle activity, and its form to the smallest detail often shows indications of this fundamental influence. If, therefore, the skeleton of vertebrate types is studied in conjunction with the muscular and connective tissue systems and interpreted from this light, much can be inferred, conversely, regarding the original musculature of fossil types by an intelligent study of their skeletal remains, especially if careful comparisons are made with the skeletons of the most nearly related living forms."[67] Reliance on biological techniques and an intimate knowledge of recent as well as extinct vertebrates characterized and distinguished Gregory's work.

Gregory's researches also considerably expanded the understanding of fossil vertebrates. Combining comparative myology and osteology, he moved beyond traditional systematic and phylogenetic studies to offer explanations of how organisms functioned, adapted, and evolved. This approach is particularly evident in his analyses of pose and motion. A study by Gregory and his student Charles L. Camp, ostensibly a reconstruction of the primitive reptile *Cynognathus,* was in fact a detailed analysis of the relationship between structure and function that explained how animals stand and move. According to Gregory, the reason why functional differences existed between *Cynognathus* and more modern reptiles

> was related to the fact that in this stage of evolution the adaptations for preventing the body from settling down of its own weight on one side, when one foot was raised from the ground, were less perfectly developed than in later types which can raise the body far from the ground and even hold it for some time in a true standing pose. When a quadruped lifts, for example, the hind left limb off the ground, the right hind foot being truly on the ground, the backbone and sacral region are only prevented from sagging toward the left by the action of those muscles in the right side of the acetabulum which tend to pull the pelvis and backbone toward the head of the right femur. In this way also, the sacral ribs are being forced downward and outward against the inner side of the ilium. As long as the animal remains small and the body is not raised high above the ground, especially if the connection with the sacral ribs is loose, the lifting effect can be obtained by muscles running from the neural spines to the backbone in the knee joint as well as by the ventral muscles running from the pubis and ischium to the femur; but when the size becomes very great and when the body is lifted well above the ground, the muscles are reinforced

> by the greatly expanded gluteal mass, running from the outer surface of the ilium to the upper part of the femur. Hence, at this more advanced stage, the area on the blade of the ilium for the gluteal muscles is correspondingly expanded.[68]

Analyzing how muscles operated in relationship to skeletal structure, Gregory and Camp had provided an explanation of motion among extinct vertebrates. That question, how to explain animal motion, was central to their study, and relying on work in comparative myology and osteology Gregory and Camp offered important new insights into the biology of fossil vertebrates.

Their work also provided the means for explaining adaptation and evolutionary divergence. According to Gregory and Camp, different functional needs and requirements gave rise to changes in skeletal and muscular structure. Without adopting a Lamarckian interpretation that explained evolution explicitly as a result of habit, Gregory's researches nonetheless provided adaptational reasons for the differences in animal structure and function. Gregory and Camp's study in 1918 served as the model that influenced Roy Miner's work on the primitive reptile *Eryops;* Alfred Sherwood Romer's efforts to explain the function, adaptation, and evolution among dinosaurs; and even G. K. Noble's studies on recent reptiles.[69]

Biological questions and the work in functional morphology likewise played a central role in Gregory's research on the adaptation and evolution of primates. Osborn's decision to take up the study of humans, based largely on his social and political interests, no doubt had an impact on Gregory. Equally important were scientific developments—specifically, the discovery of the Piltdown specimens in 1911 and 1913, and a request from Matthew and Walter Granger for Gregory to assist in the study of new specimens of *Notharctus,* an Eocene lemur.[70] Gregory approached the study of primates much as he had the study of reptiles and amphibians: he emphasized questions pertaining to function, adaptation, and evolution. The phylogenetic relationship between humans and apes was a primary issue, and by comparing types and casts of fossil hominids made available through the DVP with specimens of recent primates, Gregory documented the common heritage and close evolutionary relationship between humans and anthropoids. He thus became a leading proponent of the thesis of the anthropoid ape ancestry of humans.[71] Gregory's work also indicated that profound changes occurred in structure among those related organisms, and he relied on analyses of changes in habit and function to explain such adaptation and evolutionary divergence.

For Gregory changes in habit, particularly as they related to motion, explained the marked differences between humans and apes. Drawing on his own researches as well as the studies of the British anthropologists Arthur Keith and Grafton Eliot Smith, Gregory maintained that the change from an arboreal way of life based on brachiation (swinging in trees) to bipedal life on the ground had profound consequences. For example, in explaining

> the transformation of a gorilloid form of foot into the human foot we have to do with widely differing conditions, namely the transformation of erectly sitting, brachiating quadrumana into erectly walking plantigrade bipeds. When the Miocene ancestors of the Hominidae began to spend more time on the ground and less time in the trees, it was perfectly natural that their powerful hallux should have been utilized as the main axis of the foot, instead of the weaker digits II, III, IV; because on rough forest ground the strongly grasping hallux with its powerful flexors and adductors would be almost as useful in maintaining the balance in the upright pose as it would be in the trees.[72]

Changes in the structure of the hand, he claimed, "are no doubt associated with the marked differences in the modes of locomotion and habitual usage in the hands of gorilla and man." So too were the evolution of upright posture; the improvement of the brain, eyes, and balancing mechanism; and the evolution of the human arrangement of the viscera.[73] For Gregory changes in habit and function, specifically motion, had led ape-like organisms to evolve what we recognize as human structural and functional characteristics. Only by examining primate form, function, and change of function could Gregory explain the evolution of human feet, limbs, and dentition. Only by research in comparative myology and osteology, by combining the analysis of function and structure among fossil and recent primates, could he establish accurate phylogenies and a meaningful understanding of primate evolution.

Gregory's researches on the habits, functions, and adaptations of fossil vertebrates also laid the foundation for his views concerning the nature of the evolutionary process. It is not surprising that Gregory, by virtue of his close association with Osborn and Matthew, developed an interest in evolutionary issues and, on the basis of his work in functional morphology, sought to explain variation, inheritance, and the process and pattern of evolution. Convinced by his study of primates that changes in habit produced major transformations in function and structure, Gregory stressed the significance of adaptation and evolutionary divergence, and challenged the ruling orthogenetic interpretations of the day. At first, in the 1910s, he directed his criticism toward such British scientists as Guy Pilgrim and Frederick Wood-Jones.[74] Later, however, Gregory directly attacked Osborn's ideas. Although he focused his criticism on Osborn's interpretation of primate evolution, Gregory also spread his net more widely and eventually challenged the entire foundation of Osborn's understanding of variation, evolution, and inheritance.

Gregory was motivated to define his views as the result of his reaction to Osborn's "Dawn Man" theory of human evolution. In part an attempt to contain the controversy that followed the Scopes Trial, the theory also embodied Osborn's commitment to orthogenesis and extensive taxonomic splitting. Osborn's views were built on his belief that evolution is an irreversible process. Applying that idea to the construction of phylogenies, Osborn maintained that the remote ancestors of any given type, in order to be designated as such, had to exhibit unmistakable signs of the characteristics evident in their descendants; those that did not he placed in separate lineages. Drawing on that interpretation and emphasizing the morphological, behavioral, and functional differences between humans and other primates, Osborn claimed that humans had not evolved from apes but rather from a hypothetical "Dawn Man" ancestor of the Eocene epoch. Osborn's theory thus constituted a direct challenge to Gregory's interpretations, and, despite personal reservations, Gregory debated and criticized Osborn's views in the halls of the American Museum, in scientific meetings, and in scholarly journals as well as newspapers.[75]

From Gregory's perspective, Osborn's theory embodied a number of serious misinterpretations. In the first place, Osborn had misunderstood Louis Dollo's law of evolutionary irreversibility. According to Gregory, Osborn supported "a sort of *emboitment* hypothesis in which the visible characters of the late forms are mentally imputed even to their remote ancestors."[76] Osborn in effect had read the present into the past, and thus failed to realize that Dollo's law allowed for changes in form and function among organisms in a line of descent. Indeed functional changes—transformations that produced modifications and on occasion even reversals in evolutionary trends—lay at the heart of Gregory's understanding of evolution. For him evolution was not a predetermined, irreversible process and, rejecting Osborn's belief in orthogenesis, Gregory also repudiated Osborn's commitment to polyphyleticism, extensive parallelism, and taxonomic inflation. In contrast to his mentor who had interpreted the evolution of primates and virtually all families of mammals as bushlike phylogenies of parallel, rectilinear lines of descent, Gregory understood evolution as an ever-branching, ever-diverging tree of life.[77]

Gregory also sought to explain the causes of evolution, and here too his conclusions differed significantly from Osborn's. To Gregory the evidence of fossil and recent vertebrates demonstrated adaptation, basic patents both in the individual characters of organisms and their overall design. Whereas Osborn held that such evidence bespoke planning and purpose, for Gregory adaptation was the consequence of the natural selection of random variations. Writing to the South African paleontologist Robert Broom, who like Osborn understood evolution as a deterministic process, Gregory claimed:

> 'Planning' is simply the result of experience read backward and projected into the future. To me the 'purposive' action of a beehive is simply the summation and integration of its units, and Natural Selection has put higher and higher premiums on the most 'purposeful' integration. It is the

> same way (to me) in the evolution of the middle ear, the steps in the Cynodonts (clearly shown by me in 1910 and by you later in Oudenodon) make it easier to see how such a wonderful device as the middle ear could arise without any predetermination or human-like planning, and in fact in the good old Darwinian way, if only we admit that as the 'twig is bent the tree's inclined' and that each stage conserves the advantages of its predecessors. . . . The simple idea that planning is only experience read backward and combined by selection in suitable or successful combinations takes the mystery out of Nature and out of man's minds.[78]

Although he preferred documenting evolution to explaining its causes, Gregory nonetheless was a Darwinian whose studies "were submitted in evidence of the power of Natural Selection to produce wide secular differentiation among the descendants of a never entirely stable germ plasm."[79]

Gregory's concern with evolution likewise led him to consider questions of variation and inheritance, and he incorporated the findings of early twentieth-century biology. His own researches on vertebrates led Gregory to recognize the extent and importance of variability among vertebrates. As he noted, "my studies on the phylogeny of fossil and recent mammals lead me to visualize evolution as follows: (a) a growing and very plastic process with occasional change or reversal in the trend of reduction or enlargement of parts; (b) with occasional change in the direction of function; (c) successive ancestors with growing or changing potentiality, opening out or restricting itself at each horizon; (d) variability irregular, sometimes very large; (e) chance universal and primordial, predetermination derived and secondary, solely a result of historical growth."[80] Evolution is thus a plastic, highly variable process based on the selection of random variations. Moreover, Gregory identified random variations as genetic mutations and thus as subject to the laws of Mendelian inheritance. Gregory was by no means fully conversant with the technical details of work in genetics, and he, like Osborn, had personal and professional difficulties with T. H. Morgan. Nevertheless, Gregory understood the tenets of classical genetics, embraced genetics as the means for understanding inheritance, and could "not see that there is any conflict between sound work in genetics, paleontology, taxonomy, and comparative anatomy." For Gregory, the study of the fossil record also required a knowledge and understanding of the work being done in biology, including experimental biology.[81]

Conclusion

By 1930, then, both Gregory and Matthew had rejected Osborn's theory of evolution and presented their own, quite different interpretations. Both men set aside Osborn's belief in determinate variation, evolutionary irreversibility, and orthogenesis and argued instead for a theory of evolution based on the natural selection of random variations. Although Matthew and Gregory only rarely laid out their understanding of the process and pattern of evolution in a systematic, comprehensive manner, both embraced Darwin's theory of evolution and turned to the work in classical genetics to interpret their findings from the fossil record.

Although they challenged and repudiated Osborn's understanding of variation, evolution, and inheritance, in a deeper sense Gregory and Matthew sustained and extended the work of their mentor. The biological questions that preoccupied Osborn were central to the work of Gregory and Matthew, and it was from that foundation that they advanced their evolutionary interpretations. Osborn's study of geographical distribution directly involved Matthew, and by pursuing that suggestive line of inquiry Matthew later developed a sustained, powerful argument for vertebrate biogeography. Moreover, keeping in mind the evolutionary questions that Osborn always emphasized, Matthew drew upon fieldwork and research in biogeography to define his understanding of the species concept, evolution, and inheritance. Osborn's interest in problems of form and function also had an impact on Matthew, although its influence on Gregory was more important. Osborn's own researches as well as the work done by the DVP staff led Gregory to pursue questions pertaining to structure, function, and adaptation and eventually he developed a well-defined, systematic program of research in comparative myology and osteology. Drawing on an approach and set of questions derived from his interest in functional morphology, Gregory looked to animal habits, functions, and changes of function to explain adaptation and evolutionary divergence. From that basis he eventually confronted Osborn, but in so doing addressed the very biological problems that Osborn deemed significant. Despite profound differences in methodology and interpretation, Osborn's interest in evolution, geographical distribution, and functional morphology defined the lines of inquiry that both Matthew and Gregory pursued.

The interest in such biological questions did not transform the discipline of paleontology. In the 1920s, just as in the 1880s, many students of the fossil record remained preoccupied with descriptive, taxonomic questions, and vertebrate paleontology was still primarily a museum science. At the American Museum, however, where Osborn's interests and influence extended to Gregory and Matthew, biological research on fossil vertebrates flourished. Furthermore, Osborn not only had a powerful intellectual influence on his students but also established an institutional center that supported, sustained, and promoted such research. In contrast to Leidy, Cope, or even Marsh, Osborn created a department of vertebrate paleontology where the work of his principal students and associates would have an impact on later generations. Indeed, Matthew's views on geographical distribution became the dominant theory in vertebrate biogeography, and as such he influenced the interpretations of many biologists and paleontologists.[82] Gregory, building on the valuable institutional connection between Columbia and the American Museum, developed a program that, through the work of students such as Camp, Romer, Percy Butler, Charles Breder, and others, shaped much of the important research done in vertebrate paleontology in the next generation.[83] Furthermore, the work of Matthew and Gregory had a powerful impact on later evolutionary investigations. Setting aside older views of Osborn and others on the nature of evolution and inheritance, Gregory and Matthew pointed out the relevance and importance of work in zoology, genetics, and even embryology and physiology for paleontologists attempting to interpret the fossil record. Their work thus demonstrated that paleontology was a field of biology. It is no coincidence that George Gaylord Simpson, who in the 1940s would establish an influential new interpretation of the fossil record based on the findings of population genetics, worked in the institutional and intellectual context for vertebrate paleontology that Osborn had established at Columbia and the American Museum.[84]

Acknowledgments

I would like to thank the members of the Friday Harbor conference as well as Bobb Schaeffer and Richard H. Tedford of the Department of Vertebrate Paleontology of the American Museum of Natural History for their valuable comments and criticisms on an earlier draft of this paper. The staffs of the Library and the Department of Vertebrate Paleontology of the American Museum of Natural History, the New-York Historical Society, the University of California at Berkeley, Columbia University, and Princeton University provided valuable assistance. A National Science Foundation grant SES 85-12626 and a grant from the Texas State Organized Research Fund supported this work.

Notes

1. William H. Goetzmann, *Exploration and Empire: The Explorer and the Scientist in the Winning of the American West* (New York: Norton, 1978); and Nathan Reingold, ed., *Science in Nineteenth-Century America: A Documentary History* (New York: Hill & Wang, 1964), pp. 236–240. On Cope and Hyatt's neo-Lamarckism, see Edward J. Pfeifer, "The Genesis of American Neo-Lamarckism," *Isis*, 1965, *56:* 156–167; and Peter J. Bowler, *The Eclipse of Darwinism: Anti-Darwinian Evolution Theories in the Decades Around 1900* (Baltimore: Johns Hopkins University Press, 1983), pp. 118–140.

2. Several universities that offered courses in paleontology, such as Chicago, Kansas, Nebraska, and Princeton, did so through geology departments. See Elizabeth Noble Shor, *Fossils and Flies: The Life of a Compleat Scientist Samuel Wendell*

Williston (1851–1918) (Norman: University of Oklahoma Press, 1971), pp. 205–209. On museums, see Charles W. Gilmore, "A History of the Division of Vertebrate Paleontology in the United States National Museum," *Proceedings of the United States National Museum*, 1941, *90:* 305–377; Charles Schuchert and Clara Mae LeVene, *O. C. Marsh, Pioneer in Paleontology* (New Haven: Yale University Press, 1940); Helen J. McGinnis, *Carnegie's Dinosaurs* (Pittsburgh: Carnegie Institute, 1982), pp. 11–26; and "An Historical and Descriptive Account of the Field Columbian Museum," *Field Columbian Museum, Historical Series*, 1894, *1:* 1–91.

3. O. C. Marsh, *Odontornithes, A Monograph on the Extinct Toothed Birds of North America* (Washington, D.C.: Government Printing Office, 1880); idem, *Dinocerata: A Monograph on an Extinct Order of Gigantic Mammals* (Washington, D.C.: Government Printing Office, 1886); U.S. Department of Interior, United States Geological Survey, "The Dinosaurs of North America," by O. C. Marsh, *Annual Reports* 16 (Washington, D.C.: Government Printing Office, 1896), pp. 133–244; Joseph Leidy, "The Ancient Fauna of Nebraska," *Smithsonian Contributions to Knowledge*, 1854, *6:* 1–126; idem, *The Extinct Mammalian Fauna of Dakota and Nebraska* (1869; reprint, New York: Arno, 1974); U.S. Department of the Interior, Report of the United States Geological Survey of the Territories, "Contributions to the Extinct Vertebrate Fauna of the Western Territories," by Joseph Leidy, vol. 1 (Washington, D.C.: Government Printing Office, 1873), pp. 1–358.

4. Cope's chief works on evolutionary theory include *The Origin of the Fittest. Essays in Evolution* (New York: Macmillan, 1887); and *The Primary Factors of Organic Evolution* (Chicago: Open Court, 1896). On Cope, see Peter J. Bowler "Edward Drinker Cope and the Changing Structure of Evolution Theory," *Isis*, 1977, *68:* 249–265; Reingold, *Science in Nineteenth-Century America*, pp. 237–240; and Henry Fairfield Osborn, *Cope, Master Naturalist. The Life and Letters of Edward Drinker Cope* (Princeton: Princeton University Press, 1930).

5. The influence of Hyatt's work is discussed in Robert Tracy Jackson, "Alpheus Hyatt and His Principles of Research," *American Naturalist*, 1913, *47:* 195–205; Percy E. Raymond, "Invertebrate Paleontology," in Geological Society of America, ed., *Geology, 1888–1938. Fiftieth Anniversary Volume* (New York: Geological Society of America, 1941), pp. 90–94; and Ronald Rainger, "The Continuation of the Morphological Tradition: American Paleontology, 1880–1910," *Journal of the History of Biology*, 1981, *14:* 129–158, on pp. 152–157.

6. Raymond, "Invertebrate Paleontology," p. 90.

7. See n. 2. Also Helen Lefkowitz Horowitz, *Culture and the City: Cultural Philanthropy in Chicago from the 1880s to 1917* (Lexington: University of Kentucky Press, 1976); John Michael Kennedy, "Philanthropy and Science in New York City: The American Museum of Natural History, 1868–1968" (Ph.D. dissertation, Yale University, 1968); and Donna Haraway, "Teddy Bear Patriarchy: Taxidermy in the Garden of Eden, New York City, 1908–1936," *Social Text*, 1984, *5:* 20–64.

8. Henry Fairfield Osborn, "The Present Problems of Paleontology," *Popular Science Monthly*, 1905, *46:* 226–242, on p. 227.

9. Reingold, *Science in Nineteenth Century America*, p. 239. See also Thomas G. Manning, *Government in Science: The U.S. Geological Survey, 1867–1894* (Lexington: University of Kentucky Press, 1967), pp. 204–216; and Schuchert and LeVene, *Marsh*, pp. 248–313.

10. On Princeton in the 1870s, see Thomas Jefferson Wertenbaker, *Princeton: 1747–1896* (Princeton: Princeton University Press, 1946); and J. David Hoeveler, Jr., *James McCosh and the Scottish Intellectual Tradition* (Princeton: Princeton University Press, 1981), pp. 284–294. On the careers of Osborn and Scott, see Henry Fairfield Osborn, *Fifty Two Years of Research, Observation, and Publication 1877–1929* (New York: Charles Scribner's Sons, 1930), pp. 55–73; William King Gregory, "Henry Fairfield Osborn," *National Academy of Sciences Biographical Memoirs,* 1938, *19:* 53–119; and William Berryman Scott, *Some Memories of a Palaeontologist* (Princeton: Princeton University Press, 1939), pp. 1–150.

11. Henry Fairfield Osborn, "Preliminary Observations upon the Brain of Amphiuma," *Proceedings of the Academy of Natural Sciences of Philadelphia,* 1883, *35:* 177–186; idem, "Observations upon the Foetal Membranes of the Opposum and other Marsupials," *Quarterly Journal of Microscopical Science,* 1883, *23:* 473–484; and idem, "Preliminary Observations upon the Brain of Menopoma," *Proc. Acad. Nat. Sci. Phil.,* 1884, *36:* 262–274.

12. The Osborn papers in the Archives, American Museum of Natural History (hereafter Osborn Papers, American Museum), document his activities in McCosh's classes, library meetings, and discussion groups. See also Hoeveler, *McCosh,* pp. 291–295, and Scott, *Some Memories,* pp. 146–147. Osborn and McCosh jointly published one article, "A Study of the Mind's Chamber of Imagery," *Princeton Review,* 1884, *13:* 50–72. Osborn also wrote "Illusions of Memory," *North American Review,* 1884, 476–486; and "Visual Memory," *The Journal of Christian Philosophy,* 1884, *3:* 439–450. Osborn noted his lack of facility for fieldwork and embryological technique in *Fifty-Two Years,* p. 65.

13. Henry Fairfield Osborn in a letter to William Berryman Scott, 11 July 1880, Box 2, Henry Fairfield Osborn Papers, New-York Historical Society (hereafter Osborn Papers, Historical Society). Osborn noted that he turned down a request from Albert Bickmore to take a position in paleontology at the American Museum because, in his opinion, Cope and Marsh stood like Scylla and Charybdis over the field.

14. Scott's position in geology and paleontology at Princeton is discussed in *Some Memories,* pp. 123, 133–134, 150–195; and George Gaylord Simpson "William Berryman Scott," *Natl. Acad. Sci. Biog. Mems.,* 1948, *25:* 175–203, esp. pp. 175–180.

15. Virginia Reed Osborn to Henry Fairfield Osborn, 28 July 1878, Box 2, Osborn Papers, Historical Society. "Sturges, Jonathan" *National Cyclopaedia of American Biography* vol. 3 (New York: James T. White, 1893), p. 350.

16. "Osborn, William Henry" *Dictionary of American Biography,* vol. 14 (New York: Scribners, 1934), pp. 72–73. Carlton J. Corliss discusses his role in the development of the Illinois Central Railroad in *Mainline of Mid-America: The Story of the Illinois Central Railroad* (New York: Creative Age Press, 1950). References to the elder Osborn's donations to scientific clubs, laboratory facilities, and library collections at Princeton are in Henry Fairfield Osborn to William Henry Osborn, 15 May 1882, Box 3; William Henry Osborn to Henry Fairfield Osborn, 16 May 1882 and 15 June 1882, Box 2; and Virginia Reed Osborn to Henry Fairfield Osborn, October 1887, Box 4, Osborn Papers, Historical Society. That Osborn's father may have paid his son's salary or provided monetary support to the college for his son's position is

suggested in Henry Fairfield Osborn to William Henry Osborn, 11 June 1883, Box 3, Osborn Papers, Historical Society.

17. In letters to Scott, 29 July 1880, 7 October 1880, and 8 October 1880, Box 2, Osborn Papers, Historical Society, Osborn discussed his interest in obtaining a degree and an academic position from Princeton and his negotiations with William Keith Brooks for a fellowship to Johns Hopkins. Although the elder Osborn objected to his son's interest in pursuing a scientific career, Henry Osborn continued embryological research and kept up with current events in science. Henry Fairfield Osborn to William Berryman Scott, 19 June 1880, Box 2, Osborn Papers, Historical Society.

18. Henry Fairfield Osborn to William Berryman Scott, 12 April 1884, Box 3, Osborn Papers, Historical Society. On Osborn's role in the Eastern Society of Naturalists, see Henry Fairfield Osborn to Edward Drinker Cope, 28 July 1884, Edward Drinker Cope Papers, Archives, Department of Vertebrate Paleontology, American Museum of Natural History, New York (hereafter, Cope Papers).

19. Edward Drinker Cope to Henry Fairfield Osborn, 4 April 1886, 19 October 1887, and 4 October 1889, Cope Papers.

20. Henry Fairfield Osborn to E. B. Poulton, 20 March 1884, Box 3, Osborn Papers, Historical Society. The quote is from a flier of 5 February 1885 entitled "Plan of Organization for a Proposed Biological Journal subject to alteration by the Board of Editors when chosen," Box 3, Osborn Papers, Historical Society.

21. E. L. Mark to Henry Fairfield Osborn, 24 March 1884, Box 3, Osborn Papers, Historical Society. The issue concerning keeping the journal secret is discussed in E. B. Wilson to Henry Fairfield Osborn, 23 May 1884, and E. L. Mark to Osborn, 17 May 1884 and 26 May 1884, Box 3, Osborn Papers, Historical Society. Mark decided to abandon the idea of embryological club because of questions concerning membership and because he feared the club might appear to be a Harvard-Princeton enterprise designed to exclude others.

22. Henry Fairfield Osborn to William Henry Osborn, 29 November 1885, Box 5, Osborn Papers, Historical Society. Osborn communicated his views to the Princeton administration in Henry Fairfield Osborn to William H. Roberts, 11 May 1886, Box 4, Osborn Papers, Historical Society; and Henry Fairfield Osborn to M. Taylor Pyne, 11 May 1886, Box 4, Osborn Papers, Historical Society. In the letter to Roberts, Osborn noted that in order to concentrate on research in vertebrate paleontology with Scott, he wanted to "confine my work in the Department to the general supervision of the laboratory and the delivery of (advanced) lectures." Osborn also used the morphological laboratory as a center to coordinate and promote graduate student studies on neurological problems that intrigued him. Those studies, most of which appeared in the *Journal of Morphology,* include: Henry Fairfield Osborn, "The Foetal Membranes of the Marsupials," *Journal of Morphology,* 1887, *1:* 373–380; idem, "A Contribution to the Internal Structure of the Amphibian Brain," *J. Morph.*, 1888, *2:* 51–96; Henry Orr, "A Contribution to the Embryology of the Lizard," *J. Morph.*, 1887, *1:* 311–372; Isaac Nakagawa, "The Origin of the Cerebral Cortex and the Homologies of the Optic Lobe Layers in the Lower Vertebrates," *J. Morph.*, 1890, *4:* 1–10; Charles F. W. McClure, "The Segmentation of the Primitive Vertebrate Brain," *J. Morph.*, 1890, *4:* 35–56; Oliver S. Strong, "The Cranial Nerves of Amphibia," *J. Morph.*, 1895, *10:* 101–230; and Alvin Davison, "A Contribution to the Anatomy and Physiology of *Amphiuma means,*" *J. Morph.*, 1895, *11:* 375–410.

23. Scott, *Some Memories,* pp. 150–188.

24. Evidence for Osborn's minimal participation in fieldwork derives from the fact that the Department of Vertebrate Paleontology possesses no Osborn field notebooks, and that Osborn's personal papers and diaries include field notes only for work in the Huerfano Basin of Colorado in 1897, a geological tour of the dinosaur beds of Colorado and Wyoming in 1897–1900, a general reconnaissance of lower Tertiary beds in 1906, and some notes and diagrams of deposits at Agate Springs and Bitter Creek, Nebraska, Box 24, Osborn Papers, Historical Society. Concerning his much-publicized trips to Bone Cabin Quarry, the Fayum of Egypt, and the Gobi Desert, Osborn went out well attired, often with his family, and generally after staff parties had done extensive work.

25. Matters concerning the publication of that work are the subject of numerous letters between Osborn and Scott in Boxes 4, 5, and 6, Osborn Papers, Historical Society.

26. Henry Fairfield Osborn and W. B. Scott, "Preliminary Account of the Fossil Mammals from the White River and Loup Fork Formations, contained in the Museum of Comparative Zoology," *Bulletin of the Museum of Comparative Zoology,* 1887, *13:* 151–171, and 1890, *20:* 65–100; and Henry Fairfield Osborn and W. B. Scott, "The Mammalia of the Uintah Formation," *Transactions of the American Philosophical Society,* n.s., 1890, *16:* 461–572.

27. Cope, *The Origin of the Fittest* and *The Primary Factors of Organic Evolution.* Scott and Osborn adopted Cope's interpretations in such essays as William Berryman Scott, "On the Osteology of *Mesohippus* and *Leptomeryx,* with Observations on the Modes and Factors of Evolution in the Mammalia," *Journal of Morphology,* 1891, *5:* 301–406; Henry Fairfield Osborn, "On the Structure and Classification of the Mesozoic Mammalia," *Journal of the Academy of Natural Sciences of Philadelphia,* 2nd Ser., 1888, *9:* 186–265; and idem, "The Evolution of Mammalian Molars to and from the Tritubercular Type," *Am. Nat.,* 1888, *22:* 1067–1079.

28. Henry Fairfield Osborn, "Evolution and Heredity," *Biological Lectures delivered at the Marine Biological Laboratory of Wood's Holl,* 1891, *1890:* 130–140; and "Difficulties in the Heredity Theory," *Am. Nat.,* 1892, *26:* 537–567.

29. Scott, "Osteology of *Mesohippus* and *Leptomeryx,*" p. 384.

30. William Berryman Scott, "On Variations and Mutations," *American Journal of Science,* 3rd ser., 1894, *48:* 355–374. Also see n. 44.

31. Scott, "*Mesohippus* and *Leptomeryx,*" p. 371.

32. Alpheus Hyatt, "On the Parallelism between the Different Stages of Life in the Individual and those in the entire Group of the Molluscous Order Tetrabranchiata," *Memoirs of the Boston Society of Natural History,* 1866, *1:* 193–209; and "Phylogeny of an Acquired Characteristic," *Proceedings of the American Philosophical Society,* 1893, *32:* 349–367. See also Bowler, *The Eclipse of Darwinism.*

33. James Martin Keating, "Seth Low and the Development of Columbia University 1889–1901" (Ph.D. dissertation, Columbia University, 1973), p. 100.

34. Following Low's original overture of 23 April 1890, Osborn responded in a letter to Seth Low, 1 May 1890, Osborn Papers, American Museum. The subsequent negotiations between Osborn and Low concerning the development of a department of biology at Columbia are in the Seth Low Correspondence, Central Record Files, Low Memorial Library, Columbia University, New York (hereafter Low Correspondence).

35. Letters that particularly concern the program, buildings, and money for a department of biology include Low to Osborn, 30 December 1890, 19 February 1891, and 21 February 1891, Low Correspondence. Osborn's original plan for the program, including exhibits outlining the relationship of biology to other fields of study and a budget for the proposed department, are in the Minutes of the Columbia University Trustees, vol 9. (1890–1891): 1–7, Central Record Files, Low Library. Osborn's views were somewhat similar to Charles Otis Whitman's, and in fact the two had quite a close working relationship. Osborn was a supporter of Whitman's *Journal of Morphology,* and in the late 1880s and early 1890s Osborn and several of his Princeton students contributed articles to the *Journal*. See n. 22. Similarly, Osborn was a powerful proponent of the Marine Biological Laboratory: in the early 1890s he delivered several lectures at the MBL, raised money to support a Columbia table at the MBL, and from 1896 to 1902 served as president of the MBL's Board of Trustees. References to his efforts to obtain Columbia's support for biological research at the MBL are in the Low Correspondence. Though Osborn, in a letter to Frank Lillie, 27 June 1919, Osborn Papers, American Museum, claimed that he had destroyed most of the records pertaining to his association with the MBL, there is some correspondence between Osborn and Whitman in the Osborn Papers, American Museum, and in the Charles Otis Whitman Correspondence, Archives, Department of Vertebrate Paleontology. A brief outline of Osborn's activities as president of the MBL is in Box 19, Osborn Papers, Historical Society.

36. Seth Low to Morris K. Jesup, 20 August 1890, Low Correspondence.

37. Keating, "Seth Low," pp. 19–41. Osborn's negotiations with Jesup for the establishment of the DVP are defined in a letter from Osborn to Jesup, 18 April 1891, Department of Vertebrate Paleontology, Annual Reports, 1891, Box 1: 1, Carton 2, Archives, Department of Vertebrate Paleontology. The arrangements between the DVP and the biology department at Columbia are referred to in Henry Fairfield Osborn, "Zoology at Columbia," *Columbia University Bulletin,* 1897: 1–10, on pp. 3, 9. See also Douglas Sloan, "Science in New York City, 1867–1907," *Isis,* 1980, *71:* 35–76.

38. Kennedy, "Philanthropy and Science," pp. 66–68, 76–109.

39. For references to Osborn's connections to Bickmore, see n. 13, as well as Virginia Reed Osborn to Henry Fairfield Osborn, 7 October 1875, Box 1, Osborn Papers, Historical Society, and Henry Fairfield Osborn to Virginia Reed Osborn, 28 July 1878, Box 2, Osborn Papers, Historical Society. Morgan, at the time one of the premier financiers and businessmen in the country, had originally gone into business with William Henry Osborn, and together they married the Sturges sisters. Following the death of his first wife, Morgan remained close to Osborn and the Osborn family. Box 1, Osborn Papers, Historical Society. See also Andrew Sinclair, *Corsair: The Life of J. Pierpont Morgan* (Boston: Little Brown and Co., 1981). Morgan's many contributions to Osborn's department, including assistance for the purchase of Cope's collections of fossil vertebrates, payment for the purchase and remounting of the Warren Mastodon, and reduced railroad rates for Osborn's collectors and fossil specimens, are referred to in the Department of Vertebrate Paleontology, Annual Reports, 1896, 1897, 1899, and 1906, Archives, Department of Vertebrate Paleontology. The connection to the Dodge family is also indicated in the Osborn family genealogy, Box 1, Osborn Papers, Historical Society. William E. Dodge, Sr., a prominent figure in the New York State Chamber of Commerce, also had railroad interests and was an original member

of the museum's board. His granddaughter Alice married Osborn's brother William Church Osborn in 1886. "Dodge, William Earl," *Natl. Cycpd. Am. Biog.*, vol. 3, pp. 174–175.

40. Osborn to Jesup, 18 April 1891. Detailed information on Osborn's financial contributions is in the Department of Vertebrate Paleontology, Annual Reports, 1897–1901, Box 1: 1, Carton 2.

41. For example, Henry Fairfield Osborn to O. A. Peterson, 18 November 1893, and 2 December 1893; Henry Fairfield Osborn to G. R. Wieland, 24 September 1900, Department of Vertebrate Paleontology, Field Correspondence, Box 2: 3, Carton 1, Archives, Department of Vertebrate Paleontology; and Henry Fairfield Osborn to William Diller Matthew, 31 July 1902, Department of Vertebrate Paleontology, Field Correspondence, Box 2: 3, Carton 2. Also W. D. Matthew, "The Carnivora and Insectivora of the Bridger Basin, Middle Eocene," *Memoirs of the American Museum of Natural History,* 1909, *9:* 289–567, on pp. 293–297.

42. Among those who Osborn hired and personally supervised were the artists Charles R. Knight, Alistair Brown, Margaret Flinsch Buba, Roger Bullard, Helen Cox, Elizabeth Fulda, Lindsay M. Sterling, and Rudolph Weber; photographer A. E. Anderson; research assistants Helen Warren Brown, Christina D. Matthew, and Lucille Merriman; personal secretaries (who worked as research and editorial assistants) Florence Milligan, Johanna K. Mosenthal, Mabel R. Percy, H. Ernestine Ripley, and Ruth Tyler; and numerous preparators. As indicated by the department's annual reports and personnel records, the DVP maintained a staff of at least ten to fifteen members each year for the years after 1897. By 1900 the department was fielding three to four collecting expeditions per year, and its collection of specimens was by far the largest in the country. Under the direction of the chief preparator Adam Hermann, the DVP inaugurated new means for mounting and displaying fossil specimens in active, lifelike poses. See Adam Hermann, "Modern Laboratory Methods in Vertebrate Paleontology," *Bulletin of the American Museum of Natural History,* 1909, *26:* 283–331; Schuchert and LeVene, *Marsh,* p. 91; and Sylvia Massey Czerkas and Donald F. Glut, *Dinosaurs, Mammoths, and Cavemen: The Art of Charles R. Knight* (New York: Dutton, 1982). On the preeminence of the DVP, see William Berryman Scott, "Development of American Vertebrate Paleontology," *Proc. Am. Phil. Soc.,* 1927, *66:* 409–429; and Joseph T. Gregory, "North American Vertebrate Paleontology, 1776–1976," in Cecil J. Schneer, ed., *Two Hundred Years of Geology in America: Proceedings of the New Hampshire Bicentennial Conference on the History of Geology* (Hanover, N.H.: University Press of New England, 1979), pp. 305–335.

43. Henry Fairfield Osborn, "The Hereditary Mechanism and the Search for the Unknown Factors of Evolution," *Biol. Lect.,* 1895, *1894:* 79–100, on p. 95.

44. Osborn's later papers developing that interpretation include "The Four Inseparable Factors of Evolution," *Science,* 1908, *27:* 148–150; "Evolution As It Appears to the Paleontologist," *Sci.,* 1907, *26:* 744–749; and "Tetraplasy, the Law of the Four Inseparable Factors of Evolution," *J. Acad. Nat. Sci. Phil.,* 2nd ser., 1912, *15:* 275–309.

45. Henry Fairfield Osborn, "Correlation between Tertiary Mammal Horizons of Europe and America," *Annals of the New York Academy of Sciences,* 1900, *13:* 1–72. On Matthew, see n. 47.

46. Henry Fairfield Osborn, "Memorial of William Diller Matthew," *Bulletin of*

the Geological Society of America 1931, *42:* 55–95; and William King Gregory, "William Diller Matthew's Contributions to Mammalian Paleontology," *American Museum Novitates,* 1931, 473: 1–23. Matthew's early studies include "A Revision of the Puerco fauna," *Bull. Am. Mus. Nat. Hist.,* 1897, *9:* 259–323; "A Provisional Classification of the Freshwater Tertiary of the West," *Bull. Am. Mus. Nat. Hist.,* 1899, *12:* 19–75; and "Is the White River Tertiary an Aeolian Formation?" *Am. Nat.,* 1899, *33:* 403–408.

47. Louis Dollo, "The Fossil Vertebrates of Belgium," trans. W. D. Matthew, *Ann. N.Y. Acad. Sci.,* 1909, *19:* 99–119; and W. D. Matthew, "Patagonia and the Pampas Cenozoic of South America: A Critical Review of the Correlations of Santiago Roth," *Ann. N.Y. Acad. Sci.,* 1910, *19:* 149–160. Matthew's work was done under the auspices of a grant that Osborn in 1908 obtained from the National Science Foundation to coordinate international research in biostratigraphy and correlation. In addition to translating and criticizing the work of other scientists, Matthew and Osborn solicited from vertebrate paleontologists throughout the world observations on the appearance and abundance of species and genera; their range, time, and distribution; and data on geographical and geological conditions. Although Osborn administered the project, Matthew did much of the detailed work. Henry Fairfield Osborn and W. D. Matthew, "Geological Correlation through Vertebrate Paleontology by International Cooperation," *Ann. N.Y. Acad. Sci.,* 1909, *19:* 41–44.

48. Scott, of course, did considerable research on vertebrate biogeography, although his classic work, *A History of Land Mammals in the Western Hemisphere* (New York: Macmillan, 1913), did not appear until later.

49. Henry Fairfield Osborn, "The Law of Adaptive Radiation," *Am. Nat.,* 1902, *36:* 353–363, on pp. 355–356.

50. Henry Fairfield Osborn, "The Geological and Faunal Relations of Europe and America during the Tertiary Period and the Theory of the Successive Invasions of an African Fauna," *Sci.,* 1900, *11:* 561–574.

51. W. D. Matthew, "Hypothetical Outlines of the Continents in Tertiary times," *Bull. Am. Mus. Nat. Hist.,* 1906, *22:* 353–383. Alfred Russel Wallace, *The Geographical Distribution of Animals,* 2 vols. (New York: Hafner Publishing Co., 1962).

52. W. D. Matthew, "Osteology of *Blastomeryx* and Phylogeny of the American Cervidae," *Bull. Am. Mus. Nat. Hist.,* 1908, *24:* 535–562; idem, "Osteology and Relationship of Paramys and Affinities of the Ischyromyidae," *Bull. Am. Mus. Nat. Hist.,* 1910, *28:* 43–71. For references to the works by Dederer and Broom, see W. D. Matthew, *Climate and Evolution* (New York: New York Academy of Sciences, 1939), pp. 95–103.

53. Matthew, *Climate and Evolution,* p. 4. Matthew drew his views particularly from Chamberlin's article, "A Group of Hypotheses Bearing on Climatic Changes," *Journal of Geology,* 1897, *5:* 653–683.

54. W. D. Matthew, untitled paper presented to the Linnaean Society of London, 14 January 1902, William Diller Matthew Papers, vol. 4, Earth Sciences Library, University of California at Berkeley. Also see Ronald Rainger, "Just Before Simpson: William Diller Matthew's Understanding of Evolution," *Proceedings of the American Philosophical Society,* 1986, *130:* 453–474.

55. W. D. Matthew, "Time Ratios in the Evolution of Mammalian Phyla: A Contribution to the Problem of the Age of the Earth," *Sci.,* 1914, *60:* 232–235, on p. 233.

56. Frederick B. Loomis, "Momentum in Variation," *Am. Nat.*, 1905, *39:* 839–843. Scott did not hold to this ridiculous belief, but he did maintain that the validity of the law of irreversibility of evolution ruled out Matthew's suggestion that the teeth of the sabre-tooths may have increased and later decreased in size; see Scott, *A History of Land Mammals*, pp. 535–541.

57. W. D. Matthew, "Phylogeny of the Felidae," *Bull. Am. Mus. Nat. Hist.*, 1910, *28:* 289–316, on pp. 306–307.

58. W. D. Matthew to F. A. Bather, no date, Matthew Papers, vol. 26; and W. D. Matthew, "The Pattern of Evolution," *Scientific American*, September 1930: 192. Matthew's most important study of horses is W. D. Matthew, "The Evolution of the Horse: A Record and its Interpretation," *Quarterly Review of Biology*, 1926, *1:* 139–185.

59. W. D. Matthew, "Third Contribution to the Snake Creek Fauna," *Bull. Am. Mus. Nat. Hist.*, 1924, *50:* 59–210, on p. 154.

60. W. D. Matthew to F. B. Sumner, 8 May 1925, Sumner Correspondence, Archives, Department of Vertebrate Paleontology. Matthew defined his views in "Range and Limitations of Species as seen in Fossil Mammal Faunas," *Bull. Geol. Soc. Am.*, 1930, *41:* 271–274; and "Critical Observations on the Phylogeny of the Rhinoceroses," *University of California Publications in Geology*, 1931, *20:* 1–9. See also Rainger, "Just Before Simpson," pp. 468–469.

61. Matthew, "Pattern of Evolution," p. 193. See also Matthew's correspondence in 1925 with T. H. Morgan. T. H. Morgan Correspondence, Archives, Department of Vertebrate Paleontology.

62. Osborn's role in promoting Gregory's position at Columbia is the subject of much correspondence, including letters to Columbia President Nicholas Murray Butler dated 15 April 1916, 31 January 1921, and 9 April 1924, Osborn Papers, American Museum. Gregory's appointment to the curatorship of the Department of Comparative Anatomy is noted in *Annual Reports of the American Museum of Natural History*, 1921, *53:* 33–34. Osborn's powerful position at the Bronx Zoo is discussed in William Bridges, *A Gathering of Animals: An Unconventional History of the New York Zoological Society* (New York: Harper and Row, 1974); and Helen L. Horowitz, "Animal and Man in the New York Zoological Park," *New York History*, 1975, *55:* 426–453.

63. Henry Fairfield Osborn, "Rise of the Mammalia in North America," *Proceedings of the American Association for the Advancement of Science*, 1894, *42:* 189–227, on p. 191.

64. Henry Fairfield Osborn, "The Angulation of Limbs of Proboscidea, Dinocerata, and other Quadrupeds in adaptation to Weight," *Am. Nat.*, 1900, *34:* 89–94; "Oxyaena and Patriofelis Restudied as Terrestrial Creodonts," *Bull. Am. Mus. Nat. Hist.*, 1900, *13:* 270–271; and *The Titanotheres of Ancient Wyoming, Dakota, and Nebraska* 2 vols. (Washington, D.C.: Government Printing Office, 1929), vol. 2, pp. 731–759.

65. William King Gregory, "William Diller Matthew, 1871–1930," *Natural History*, 1930, *30:* 664–666, on p. 664.

66. Matthew's studies on functional morphology include: "The Pose of Sauropod Dinosaurs," *Am. Nat.*, 1910, *44:* 547–560; "The Ground Sloth Group," *American Museum Journal*, 1911, *11:* 113–119; and "A Tree Climbing Ruminant," *Am. Mus. J.*, *11:* 162–163. I have discussed Gregory's work in greater detail in "What's the Use:

William King Gregory and the Functional Morphology of Fossil Vertebrates," *J. Hist. Biol.* (forthcoming). See also Edwin H. Colbert, "William King Gregory," *Natl. Acad. Sci. Biog. Mems.*, 1975, *46:* 91–133.

67. Roy W. Miner, "The Pectoral Limbs of *Eryops* and other Primitive Tetropods," *Bull. Am. Mus. Nat. Hist.*, 1925, *51:* 147–312, on pp. 149–150.

68. William King Gregory and C. L. Camp, "Studies in Comparative Myology and Osteology, No. III," *Bull. Am. Mus. Nat. Hist.*, 1918, *36:* 447–563, on p. 519.

69. Miner, "The Pectoral Limbs of *Eryops,*" Alfred Sherwood Romer, "The Locomotor Apparatus of Certain Primitive Mammal-like Reptiles," *Bull. Am. Mus. Nat. Hist.*, 1922, *46:* 517–606; idem, "Crocodilian Pelvic Muscles and their Avian and Reptilian Homologues," *Bull. Am. Mus. Nat. Hist.*, 1923, *48:* 533–552; idem, "The Pelvic Musculature of Saurischian Dinosaurs," *Bull. Am. Mus. Nat. Hist.*, 1923, *48:* 605–617; idem, "Pectoral limb Musculature and Shoulder-Girdle Structure in Fish and Tetrapods," *Anatomical Record,* 1924, *27:* 119–143; and idem, "The Pelvic Musculature in Ornithischian Dinosaurs," *Acta Zoologica,* 1927, *8:* 225–275. On Noble, see G. K. Noble, "The Phylogeny of the Salientia," *Bull. Am. Mus. Nat. Hist.*, 1922, *46:* 1–89.

70. William King Gregory, "The Dawn Man of Piltdown England," *Am. Mus. J.*, 1914, *14:* 189–200; and idem, "Studies on the Evolution of Primates II. Phylogeny of Recent and Extinct Anthropoids with special reference to the Origin of Man," *Bull. Am. Mus. Nat. Hist.*, 1916, *35:* 258–355.

71. William King Gregory, "The Origin of Man from the Anthropoid Stem—When and Where," *Proc. Am. Phil. Soc.*, 1927, *66:* 439–463; "How Near is the Relationship of Man to the Chimpanzee-Gorilla Stock," *Qrtly. Rev. Biol.*, 1927, *2:* 549–560; "Were the Ancestors of Man Primitive Brachiators," *Proc. Am. Phil. Soc.*, 1928, *67:* 129–150; "The Upright Posture of Man: A Review of its Origin and Evolution," *Proc. Am. Phil. Soc.*, 1928, *67:* 339–377; and "Is the Pro-Dawn Man a Myth," *Human Biology,* 1929, *1:* 153–165.

72. Gregory, "The Evolution of Primates," p. 333.

73. Ibid. Also Gregory, "Were the Ancestors of Man Primitive Brachiators," p. 136.

74. Much of Gregory's criticism of Pilgrim is in "The Evolution of Primates." On Gregory's criticism of the work of Wood-Jones, see William King Gregory, "A Critique of Professor Frederick Wood-Jones's Paper: 'Some Landmarks in the Phylogeny of Primates,'" *Hum. Biol.*, 1930, *2:* 99–108; and idem, *Man's Place Among the Anthropoids* (Oxford: The Clarendon Press, 1934), a full-length study structured around Gregory's opposition to Wood-Jones's hypothesis that humans had evolved from an advanced tarsioid, not from monkeys or apes.

75. Osborn's more important articles on the "Dawn Man" theory include "Fundamental Discoveries of the last decade in Human Evolution," *Bulletin of the New York Academy of Medicine,* 2nd ser., 1927, *3:* 513–521; "Recent Discoveries Relating to the Origin and Antiquity of Man," *Sci.*, 1927, *65:* 481–488; "The Influence of Habit in the Evolution of Man and the Great Apes," *Bull. N.Y. Acad. Med.*, 2nd ser., 1928, *4:* 216–230; and "Is the Ape-Man a Myth," *Hum. Biol.*, 1929, *1:* 2–16. See Rainger, "What's the Use"; John G. Fleagle and William L. Jungers, "Fifty Years of Higher Primate Phylogeny," in Frank Spencer, ed., *A History of American Physical Anthropology 1930–1980* (New York: Academic Press, 1982), pp. 187–230, esp.

p. 195; and Peter J. Bowler, *Theories of Human Evolution: A Century of Debate, 1844–1944* (Baltimore: Johns Hopkins University Press, 1986).

76. William King Gregory, "A Critique of Professor Osborn's Theory of Human Origin," *American Journal of Physical Anthropology,* 1930, *14:* 133–164, on p. 155.

77. See the following articles by Gregory: "The Origin, Rise, and Decline of *Homo Sapiens,*" *Scientific Monthly,* 1934, *39:* 481–496; "The Roles of Undeviating Evolution and Transformation in the Origin of Man," *Am. Nat.*, 1935, *69:* 385–404; and "On the Meaning and Limits of Irreversibility of Evolution," *Am. Nat.*, 1936, *70:* 517–528. On Dollo, see Stephen Jay Gould, "Dollo on Dollo's Law: Irreversibility and the Status of Evolutionary Laws," *J. Hist. Biol.*, 1970, *3:* 189–212.

78. William King Gregory to Robert Broom, n.d. [1933], William King Gregory Papers, Box 3, Archives, American Museum of Natural History, New York (hereafter Gregory Papers). Gregory employed similar explanations in "On Design in Nature," *The Yale Review,* 1924, *13:* 334–345; and "Basic Patents in Nature," *Sci.*, 1933, *78:* 561–566.

79. William King Gregory, "Fish Skulls: A Study of the Evolution of Natural Mechanisms," *Trans. Am. Phil. Soc.*, 1933, *23:* 75–481, on p. 449.

80. William King Gregory, "Is the Pro-Dawn Man a Myth," *Hum. Biol.*, 1929, *1:* 153–165, on p. 162.

81. Gregory, "Design in Nature," pp. 337–338; and idem, "The Transformation of Organic Designs: A Review of the Origin and Deployment of the Earlier Vertebrates," *Cambridge Philosophical Society,* 1936, *11:* 311–344, on p. 339. The quotation comes from a letter from Gregory to W. P. Pycraft, 8 June 1936, Box 6, Gregory Papers.

82. Karl P. Schmidt, "Animal Geography," in Edward L. Kessel, ed., *A Century of Progress in the Natural Sciences* (New York: Arno, 1974), pp. 777–781. See also Philip J. Darlington, Jr., "The Origin of the Fauna of the Greater Antilles, with Discussion of Dispersal of Animals Over Water and Through the Air," *Qtrly. Rev. Biol.*, 1938, *13:* 274–300; idem, "The Geographical Distribution of Cold-Blooded Vertebrates," *Qtrly. Rev. Biol.*, 1948, *23:* 1–26, 105–123; George S. Myers "Fresh-Water Fishes and West Indian Zoogeography," *Annual Reports of the Smithsonian Institution,* 1938: 339–364; idem, "Fresh-Water Fishes and East Indian Zoogeography," *Stanford Ichthyology Bulletin,* 1949, *4:* 11–21; and esp. George Gaylord Simpson, "Mammals and Land Bridges," *Washington Academy of Sciences,* 1940, *30:* 137–163.

83. For references to Camp, Miner, and Romer, see n. 69. On Percy M. Butler, see "Studies of the Mammalian Dentition—Differentiation of post-canine Dentition," *Proceedings of the Zoological Society of London,* 1939, *109:* 1–36; and idem, "A Theory of the Evolution of Mammalian Molar Teeth," *Am. J. Sci.*, 1941, *239:* 421–450. See also Charles Breder, "The Locomotion of Fishes," *Zoologica,* 1926, *4:* 159–297. I am grateful to Joseph T. Gregory, personal communication, for pointing out William King Gregory's influence on the work of those individuals.

84. Biographical and autobiographical accounts of Simpson emphasize that his proximity and access to biologists at Columbia, particularly to the work in population genetics by Theodosius Dobzhansky, were crucial in leading him to refashion the understanding of the fossil record. George Gaylord Simpson, *Concession to the Improbable: An Unconventional Autobiography* (New Haven: Yale University Press, 1978), pp. 33–45, 108–120; Ernst Mayr, "George Gaylord Simpson," in Ernst Mayr

and William B. Provine, eds., *The Evolutionary Synthesis: Perspectives on the Unification of Biology,* (Cambridge, Mass.: Harvard University Press, 1980), pp. 452–463; Stephen Jay Gould, "G. G. Simpson, Paleontology, and the Modern Synthesis," in Mayr and Provine, *The Evolutionary Synthesis,* pp. 153–172; and Leo F. Laporte, "Simpson's *Tempo and Mode in Evolution* Revisited," *Proc. Am. Phil. Soc.,* 1983, *127:* 365–416. Rainger, "Just Before Simpson," emphasizes that the work being done in vertebrate paleontology at both Columbia and the American Museum also predisposed Simpson to embrace the contemporary research in experimental biology and a Darwinian understanding of evolutionary problems, and thus significantly influenced his reinterpretation of the fossil record.

Joel B. Hagen

8 Organism and Environment: Frederic Clements's Vision of a Unified Physiological Ecology

In 1899 Henry Chandler Cowles, a young geologist turned botanist, described the complex life history of the sand dunes bordering Lake Michigan.[1] From the variety of dunes that he observed, Cowles envisioned a developmental process leading from transient heaps of sand, or embryonic dunes, through a series of stages to mature, stabilized dunes covered with dense vegetation. As the small, lifeless embryonic dunes formed on the beach, they were captured by grasses whose massive, fibrous roots were well adapted to trap and hold sand. These sand binders produced stable centers for further colonization. Consequently, both the dune and a colony of plants grew in what Cowles described as a symbiotic partnership. Given proper environmental conditions, the dune and plant community would continue to grow and develop in a fairly predictable manner. The terminus of this process was a mature or climax community: in the case of Cowles's study site, a large sandy hill covered with deciduous forest. This developmental phenomenon, ecological succession, became a major focus of botanical research during the early twentieth century.

Cowles's elegant study of succession had been foreshadowed by the earlier research of the Danish botanist, Eugenius Wärming. However, during the first two decades of the twentieth century, the study of succession became something of an American specialty, and Cowles, assisted by his students at the University of Chicago, became recognized as the "great pioneer" in this area of ecological research.[2] America provided a fertile environment both physically and intellectually for the study of succession. The process of succession often requires hundreds of years to complete. Although the frontier may have been coming to a close in 1900, American botanists could still find pristine habitats to serve as natural laboratories for tracing the long-term development of vegetation. The general intellectual environment in America may have suggested thinking of succession as a developmental process. Organismal metaphors and developmental analogies were the "common property" of American intellectuals during the late nineteenth and early twentieth centuries.[3]

Cowles's paper—a masterpiece of careful observation and accurate description—remains a classic study in dynamic plant ecology. However, he aspired to more than descriptive natural history. For Cowles, the goal of ecology was to discover causal laws. Vegetation was in a constant state of flux and only by determining the causes of change could the ecologist bring order to an otherwise chaotic body of data. In the introduction to his study, Cowles stated, "The ecologist, then, must study the order of succession of the plant societies in the development of a region, and he must endeavor to discover the laws which govern the panoramic changes."[4]

Cowles clearly believed that succession was a law-governed process. The parallels that he drew between ontogeny and succession suggested that the laws governing the development of individual plants and the laws governing the development of plant communities might be very similar. Cowles never developed this suggestive analogy; however, his contemporary and sometimes competitor, Frederic Clements, adopted these ideas and developed them into an elaborate theoretical system. For Clements, the plant community was a kind of organism in the sense that it arose, grew, matured, and died. Much of Clements's early professional career was devoted to discovering and formalizing laws of succession.

Clements's brilliant, but controversial, comparison between ontogeny and succession has been a focus for ecological discussion throughout the twentieth century. Despite a strong reaction against this idea, Clements's organismal concept continues to attract some ecologists.[5] The objective of this essay is not to evaluate the present standing of Clementsian ecology, but to place Clements's early theoretical writings within the context of the developing discipline of ecology and of American biology.

At the turn of the century many young ecologists complained that their nascent discipline lacked a coherent conceptual and methodological focus. These ecologists were impressed by the laboratory botany of the late nineteenth century, and they frequently turned to experimental plant physiology rather than to natural history as a source of inspiration. Clements's theoretical writings can be properly understood only within this intellectual context. Clements was attempting to define the conceptual boundaries of the new field of ecology and to distinguish it from what he considered to be a moribund natural history.

Given this physiological perspective, the idea of the plant community as a kind of organism was logical. Indeed, organismal language was frequently employed in the literature of early ecology. What set Clements's theoretical work apart was the explicitness of his organismal ideas and the broad generalizations that he drew from them. Clements intended to construct a dynamic ecology similar in methods and theory to experimental physiology. Ecological succession was the paradigmatic example of a quasi-physiological process, and Clements's organismal ideas in this narrow context are well known. I shall argue, however, that Clements also used his organismal concept in a much more general way. By comparing plant communities to individual organisms, Clements attempted to create a unified physiological foundation for ecological theory.

The Emergence of an Amorphous Discipline

Ecology was one of a number of biological disciplines to emerge at the beginning of the twentieth century.[6] By 1900 there was a recognizable group of biologists who referred to themselves as ecologists. Several influential ecological treatises appeared between 1895 and 1905. Ecologists formed organizations that eventually developed into professional ecological societies. By 1920, journals devoted exclusively to ecological topics were established in the United States and Britain. This pattern of development was typical of many biological disciplines around 1900. In some important ways, however, the development of ecology as a scientific discipline was unique.

Institutionally, ecology was centered not at the elite private universities or research laboratories in the East, but rather at the rapidly expanding universities of the Midwest. Several historians, most notably Ronald Tobey, have emphasized the importance of the Universities of Chicago and Nebraska for the early development of ecology.[7] These two schools produced an impressive list of influential ecologists: Frederic Clements, H. C. Cowles, C. C. Adams, Victor Shelford, H. L. Shantz, Arthur Vestal, W. S. Cooper, and Paul Sears. Many of these scientists went on to establish other midwestern centers of ecological research at schools such as the Universities of Minnesota and Illinois. This western ecology also flourished outside of academe. For example, supported by the Carnegie Institution of Washington, Frederic Clements and his associates pursued ecological research at an alpine research station located in the Rocky Mountains of Colorado.

That early ecological research should have been centered in the American heartland is perhaps not surprising. Here was a favorable environment for applied science, as well as pure research; from the beginning, ecology had close ties with agriculture and forestry. In his book, *Saving The Prairies,* Ronald Tobey has demonstrated that grassland ecology in America was significantly shaped by the agricultural context within which it developed.[8] Of 307 members of the Ecological Society of America in 1917, forty-three listed forestry and twelve listed agriculture as their primary areas of interest.[9] But early ecologists were not simply "dusty-boots botanists of the West," to borrow a phrase from Joseph Ewan.[10] Ecologists, as Eugene Cittadino has argued, also were passionately interested in the broader biological problems of adaptation, development, and distribution.[11] Most early ecologists believed that these problems could be studied only in natural laboratories—forests, prairies, lakes, and dunes—and such natural laboratories were readily accessible from Chicago, Minneapolis, and Lincoln.

Ecology was also unique in the degree to which it was shaped by botanical problems and ideas. According to Cittadino, "ecology was first recognized and consciously pursued during the 1890s as a specialization within botany."[12] Many of the most influential figures during the formative years of ecology—Eugenius Wärming, Andreas Schimper, Alfred Tansley, Cowles, and Clements—were botanists. This botanical orientation persisted well into the twentieth century and had a profound intellectual impact on the field. Perhaps the most fruitful concepts in early ecology, community and succession, were essentially botanical ideas. In later years, animal ecologists would be forced to confront these ideas: adopting them, modifying them, reacting against them, but rarely ignoring them. Looking back on the early history of ecology in 1949, a group of leading zoologists noted that plant ecology had acted as a catalyst for animal ecology. Plant ecology not only had provided basic concepts and methods, but also "it gave psychological stimulus at the turn of the century by showing the zoologist that first-rate botanists were investigating ecological problems and getting results."[13]

A later generation of zoologists may have been impressed by the accomplishments of early plant ecologists, but during the period from 1890 to 1910 botanists struggled to define this new area of research. Robert McIntosh has characterized early ecological theory as a "loose amalgam of ideas,"[14] and this figure of speech accurately captures the sense of uncertainty that early twentieth-century botanists felt toward ecology. Some botanists questioned whether ecology was a legitimate field or simply a passing fad. Indeed, according to Paul B. Sears, the Ecological Society of America was established in 1915 only after ecologists failed for several years to convince botanists to allow an ecological section within the larger Botanical Society of America.[15] Perhaps recognizing that it was rather nebulous, early plant ecologists themselves tended to be self-consciously critical of their nascent discipline. For example, William Francis Ganong deplored the "pretentiousness of statement" and "weakness of logic" that characterized much of the ecological literature of his day.[16] Cowles and Clements denounced the dilettantes and tyros who dabbled in ecology.[17] However, even the small cadre of professional ecologists seemed unsure of the direction of their discipline. When Cowles was asked to comment on the current status of ecological research at the 1903 meeting of the American Association for the Advancement of Science, he remarked, "the field of ecology is chaos. Ecologists are not agreed even as to fundamental principles or motives; indeed, no one at this time, least of all the present speaker, is prepared to define or delimit ecology."[18]

Cowles may have been reticent about defining ecology, yet he did present a fairly coherent outline of the ecological research of the day. According to Cowles, this research fell into two broad categories. First, ecologists attempted to "unravel the mysteries of adaptation."[19] The study of adaptation reflected a keen interest among ecologists in understanding the relationship between plants and the physical environment. Ecologists were particularly interested in discovering how plants survived under severe conditions. Much of this interest had been stimulated by earlier advances in microscopic anatomy and physiology, carried out primarily in the botany laboratories of German universities. Although this research might loosely be termed Darwinian because it assumed that adaptation was a product of evolution, early twentieth-century ecologists tended to be quite ambivalent toward specific theories of evolution. Cowles cautioned that there was little scientific basis to choose among natural selection, mutation, or inheritance of acquired traits. He cited with approval the recent studies of Hugo de Vries and the French neo-Lamarckian Gaston Bonnier, not because their evolutionary conclusions were necessarily valid, but because they had employed innovative experimental methods in their research. For Cowles experimentation was the sine qua non of the study of adaptation. This enthusiasm for experimentation and skepticism toward evolutionary speculation accurately reflected the views of many of his contemporaries.[20]

The second broad area of research that Cowles identified might be termed ecological plant geography.[21] A primary focus of this type of research, particularly in America, was the development of plant communities. As I have discussed, Cowles had pioneered this area of research with his study of succession on sand dunes. However, ecological plant geographers were also concerned with classifying communities and explaining the distribution of these communities in terms of environmental factors. This ecological approach to plant geography had antecedents in the work of the nineteenth-century botanists Alexander von Humboldt, August Grisebach, and Oscar Drude, even though ecologists often gave only grudging acknowledgment to these illustrious predecessors. While nineteenth-century botanists had drawn correlations between groups of plants and the general features of the environment, early twentieth-century ecologists believed that they could go further and discover the specific environmental causes of distribution, an optimism evident in the title of the early ecological treatise, *Plant Geography Upon a Physiological Basis,* written by the German botanist, Andreas Schimper.[22] American ecologists, including Cowles and Clements, were equally enthusiastic about a physiological approach to plant geography.

Physiology was more than an academic discipline for plant ecologists; it was a general approach to biological research.[23] Unlike traditional natural history, physiology suggested the study of processes and structural-functional relationships. It also suggested the laboratory with connotations of instrumentation and rigorous experimentation. Ecologists not only intended to transfer these methods from the laboratory to the field, but also to study "nature's own experiments"—for example, by investigating the process of succession as it actually ocurred in undisturbed habitats.[24] Finally, for some ecologists physiology suggested both organismal concepts and mechanistic explanations. As I shall discuss, both were central to Frederic Clements's vision of a unified physiological ecology.

The close relationship between physiology and ecology is not so evident today. Nor do ecologists today necessarily see a close relationship between the two broad areas of ecological research that Cowles discussed in 1903. Early twentieth-century ecologists, however, believed that the study of individual adaptation ought to be unified with the study of plant communities. They also believed that these areas of research ought to be unified using physiological concepts. The chaos that Cowles perceived in ecology was due to the fact that no one had systematically drawn the connections between these areas of research.

If Cowles was reluctant to draw such a connection himself, Clements certainly was not. Unlike the perceptive but cautious Cowles, Clements was possessed of a restless imagination and a penchant for sweeping generalizations. In his first major ecological work, *Research Methods in Ecology* (1905), the young botanist at the University of Nebraska presented a striking vision of a unified plant ecology. This vision, though controversial, brought Clements international recognition as a gifted scientist. It also presented a broad research program, a program that Clements would pursue during his tenure as chairman of the Botany Department at the University of Minnesota (1907–17) and after 1917 as research associate with the Carnegie Institution of Washington (1917–41).

A superficial reading of Clements's *Research Methods* gives the impression that it is a manual of ecological techniques. The book is replete with illustrations of scientific instruments and instructions for their use. Instrumentation was a major focus of *Research Methods,* for Clements believed that measurement and experimentation were the keys to scientific ecology. However, Clements's book was more than a laboratory manual; it was also a manifesto for the new ecology. In the opening chapter, "The Foundation of Ecology," Clements delivered a blistering critique of traditional natural history, particularly as it was practiced by descriptive botanists. For Clements, these descriptive botanists were amateurs, dilettantes, or, perhaps worst of all, herbarium taxonomists whose knowledge of plants was limited to dessicated remains in musty museum cabinets.[25] This polemical critique was accompanied by an enthusiastic endorsement of experimental physiology. Indeed, according to Clements, ecology was "nothing but a rational field physiology."[26] The rest of the text was devoted to demonstrating that all ecological phenomena could be reduced to simple stimulus-response relationships between the physical environment and ecological organisms, which included both individual organisms and plant communities.

Clements's enthusiasm for experimental botany was a natural consequence of his training at the University of Nebraska, where he studied under the eclectic botanist and gifted educator, Charles Bessey. While not primarily an experimentalist, Bessey was a leading exponent of the new botany.[27] This pedagogical approach, a manifestation of a more general trend in American biological education, placed heavy emphasis on laboratory investigation, particularly microscopy.[28]

Despite his introduction to the new botany during the early 1890s, Clements's vision of a unified physiological ecology crystallized only after he had completed his Ph.D. in 1898. His graduate research at the University of Nebraska was neither experimental nor physiological. Inspired by the geographical studies of Drude, Clements and his fellow graduate student, Roscoe Pound, had produced an impressive but fairly conventional descriptive account of regional vegetation. Pound's and Clements's joint doctoral dissertation, *The Phytogeography of Nebraska,* cataloged species, described plant formations, and correlated these formations with general features of the environment.[29] Yet they made no consistent attempt to analyze the environment into discrete factors, much less to measure these factors. Indeed, *The Phytogeography of Nebraska* is a transitional work. In it the authors use ecology as only one of several useful perspectives from which to study plant distribution. By 1905, Clements would reverse the relationship; plant geography was but one aspect of a more inclusive science of ecology. In the jargon of his 1905 book, Clements's graduate research had been *reconnaissance,* and Clements was adamant about the status of this type of descriptive study vis-à-vis ecological *investigation:* "The objects of such a survey are to obtain a comprehensive general knowledge of the topography and vegetation of the region. . . . Standing by itself, it is not ecological research: it is preparation for it."[30] *The Phytogeography of Nebraska* does contain many ecological insights. Yet in contrast to *Research Methods in Ecology,* Pound's and Clements's early book barely mentions experimental physiology. Nor is there any hint that plant communities are living organisms.

Clements's Organismal Concept

The exact origin of Clements's organismal ideas is obscure. Ronald Tobey has presented a plausible argument for an indirect influence from the sociological tradition of Auguste Comte, Herbert Spencer, and Lester Frank Ward, all of whom used organismal metaphors in discussing human societies.[31] Tobey argues that Clements could have been exposed to these ideas at the University of

Nebraska, perhaps through his friend Pound. One might also cite Cowles as a likely source. Clements did not frame his organismal concept until after he had read Cowles's suggestive study of the sand dunes on Lake Michigan.[32] Whatever the exact source of his organismal ideas, Clements was using an explanatory device commonly employed by intellectuals of the late nineteenth and early twentieth centuries. It was a concept that fit well with Clements's physiological approach to ecology.

The organismal concept played two important roles in early Clementsian ecology, one methodological and the other explanatory. Methodologically, it suggested experimental analogues to techniques used in more traditional laboratory botany. Because the plant community was a kind of organism, Clements reasoned, "methods essentially similar" to those used in the study of individual organisms could be used to investigate communities.[33] For example, in Clementsian ecology the quadrat was to be used in the same way that laboratory botanists had used microscopical cross-sections of tissue. The microscope had allowed botanists to look inside the plant, study its fine structure, and correlate changes in structure with various physical causes. The arrangement of cells in a leaf might be greatly influenced by the intensity of a physical factor such as sunlight (Figure 8.1). Clements was obviously impressed by this microscopical technique; he included numerous microscopical illustrations in *Research Methods in Ecology.*

The ecologist could investigate the fine structure of the community in much the same way using the quadrat.[34] As its name implied, a quadrat was a square area, usually one meter on a side, marked off within a plant community. The plants within this area were identified, counted, and recorded on a chart (Figure 8.2). A permanent quadrat could be studied for several years and any changes in vegetation could be compared with the original chart. Just as the laboratory biologist could correlate changes in the fine structure of a leaf with changes in experimental conditions, so the ecologist could correlate changes within the quadrat with measured changes in the environment. These natural experiments on plant communities that the ecologist studied using the quadrat were analogous to the controlled experiments performed on individual plants by the laboratory botanist.

The organismal concept also played an important explanatory role in Clementsian ecology. By comparing plant communities to individual organisms, Clements believed that ecology could be unified within a single conceptual framework. Certainly the best example of this attempted unification was the comparison that Clements drew between individual development and ecological succession. Succession was the life cycle of a complex organism, the community. For Clements this was more than simply a figure of speech.

Because organismal concepts are often assumed to be antimechanistic, and Clements's ideas, in particular, have been identified with an antimechanistic tradition in biology,[35] it is important to consider carefully what Clements meant by "organism." In *Research Methods in Ecology,* I believe that Clements, far from being antimechanistic, set forth a rather mechanical theory of ecology. Although this theory was later modified, and although Clements made vague references to philosophical holism in his later writings, Clementsian ecology never totally lost its early mechanistic bias. Organisms, in Clements's view, were physicochemical machines controlled by discrete physical forces. In essence, Clements reduced causation to simple stimulus-response reactions between the physical environment and organisms. "Ecology," Clements concluded, "sums up this relation of [physical] cause and [biological] effect in a single word."[36] This mechanistic scheme fit well with Clements's physiological perspective on ecology.

The environment consists of a complex of factors, both physical and biological; however, according to Clements, this complex could be analyzed into a few discrete causes. Indeed, earlier field work by such pioneer experimentalists as Bonnier had been vitiated by the fact that specific causal factors had not been isolated.[37] For Clements, the most important or *direct* environmental factors were light, humidity, and the water content of soil.[38] Other physical factors acted only indirectly. For example, temperature exerted an influence primarily through changes in humidity. Similarly, the effects of soil texture could be reduced to differences in water content. Finally, the environment contained *remote* factors—for example, biotic factors. In *Research Methods,* Clements did not completely ignore the biotic environment, but he certainly relegated it to a secondary role. The impact of animals on community structure and function was barely mentioned. Biotic interactions such as competition

Figure 8.1. Cross-sections of oak leaves (*Quercus novimexicana*): (1) sun leaf, and (2) shade leaf. (From Frederic Edward Clements, *Research Methods in Ecology* [Lincoln, Neb.: University Publishing Company, 1905], p. 142.)

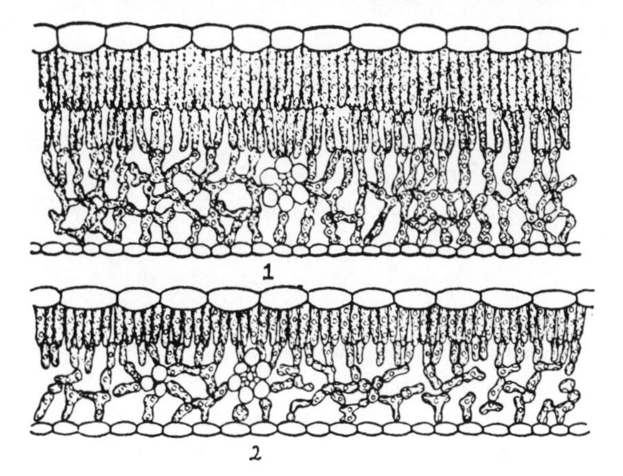

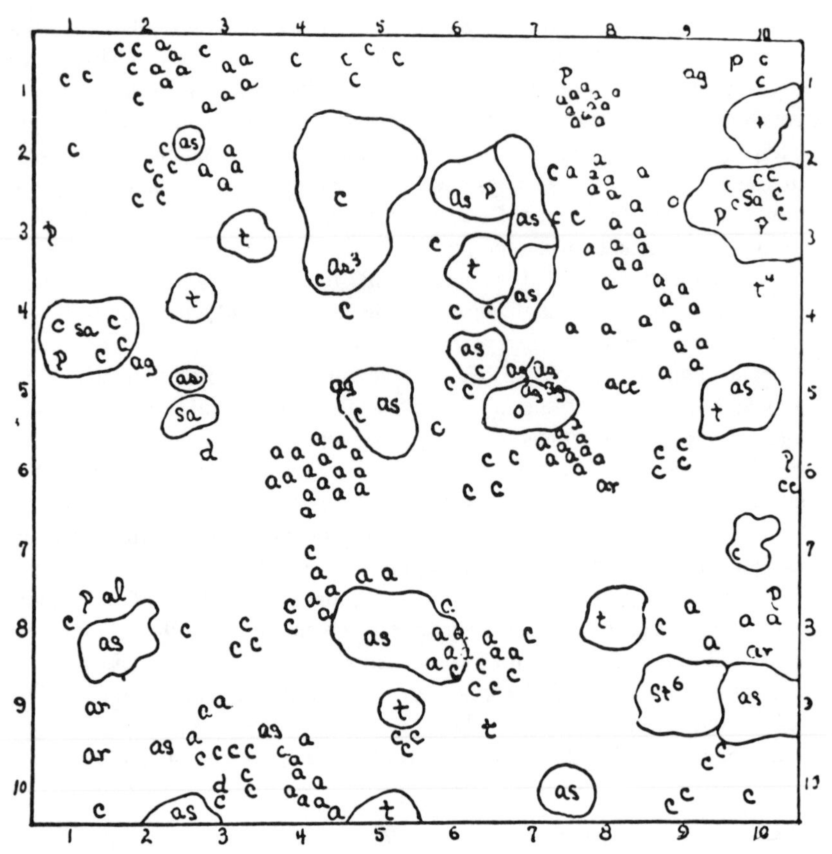

Figure 8.2. Chart of a quadrat. Letters represent individual plants of various species; outlines represent mats or rosettes. (From Frederic Edward Clements, *Research Methods in Ecology* [Lincoln, Neb.: University Publishing Company, 1905], p. 169.)

were, of course, important, but even these were discussed almost exclusively in physical terms. Plants competed primarily by shading one another or by removing water from the soil. In other words, competition was an indirect factor that could be reduced to changes in direct factors such as water content and light.[39]

For Clements, an ecological organism was a dynamic, spatiotemporal entity with relatively well-defined structures and functions. In the case of an individual plant these characteristics were fairly obvious. The structures of the individual plant included various organs, tissues, and cells. By 1900 the major functions of the individual plant, such as photosynthesis, respiration, and osmosis, had been fairly adequately explained in terms of chemistry and physics.[40]

In *Research Methods in Ecology,* Clements gave several examples of how both function and structure could be explained in terms of simple physical and chemical mechanisms. Relying largely on the doctoral research of his wife, Edith Schwartz Clements, he described what he believed to be the causal chain linking light intensity, photosynthesis, and the gross morphology of leaves.[41] Within broad limits, an increase in light intensity (a "stimulus") caused an increase in the rate of photosynthesis (a "response"). This increase in photosynthetic activity resulted in microscopic structural changes. For example, the size and number of starch grains increased in photosynthetic cells. More important, the number of chloroplasts would increase and arrange themselves to optimize the absorption of light.[42] These intracellular changes would cause alterations in the arrangement of cells and tissues, leading in some cases to gross morphological changes in the leaf. Although Clements admitted that conclusive proof of this simple causal chain was lacking, the desirability of explaining gross morphology in terms of physiological function, and ultimately in terms of physical causation, seemed obvious. According to Clements, "it seems a truism to say that the number and arrangement of the chloroplasts determine the form of the cell, the tissue, and the leaf, although it has not been possible to demonstrate this connection conclusively by means of experiment. In spite of this lack of experimental proof, this principle is by far the best guide through the subject of adaptation to light."[43] In support of this general causal scheme, Clements cited the well-known fact that shaded leaves and sun-exposed leaves on the same plant frequently exhibit distinct morphological features. Furthermore, he referred to numerous illustrations of the cross-sections of leaf tissue. These illustrations, demonstrating differences between the fine structure of leaves exposed to intense sunlight and those exposed to shade, suggested a cellular basis for gross morphology.

Like individual organisms, plant communities were characterized by definable structures and functions. Furthermore, plant communities exhibited more or less definite life cycles. Therefore, Clements argued that a plant community could be viewed as a complex organism.[44] He admitted that this concept might appear foreign to most botanists, and he cautioned that analogies between individual plants and plant communities should not be taken too literally. For example, communities did not have structures that corresponded exactly with individual organs such as leaves or roots. Nor could the processes in a community be directly compared with functions in the individual plant. The similarity was a much more general one. Both individual organisms and complex community organisms were spatiotemporal entities with definable structures, functions and life histories.

The structures of a typical plant community were less obvious than the organs, tissues, and cells of individual plants. These community structures included layers, zones and societies. In a forest, for example, the vegetation could be analyzed into fairly distinct vertical layers: the canopy formed by tall trees, one or more layers of shorter trees and shrubs, and an understory of herbs and mosses. Horizontally, the community could be dissected into vegetational zones and into more or less homogeneous societies of plants dominated by a single species.

These community structures were dictated by a number of processes or functions, perhaps the most important of which were invasion and reaction. As new plants invaded a community, the structure of the vegetation changed, but this was not a random process. Ultimately, invasion was controlled by discrete physical causes. For example, a particular species of tree might require intense sunlight for growth. Although this species might successfully invade a meadow, it would be unable to invade an established forest. Therefore, as for the individual plant, Clements argued that there was a simple causal chain linking physical factors, functions or processes, and ultimately the structure of the plant community.

Succession was, in one sense, simply a case of periodic invasions, a process controlled by the physical environment. However, the periodic nature of invasion was due to the fact that a community *reacted* with this physical habitat. "The initial cause of succession," Clements maintained, "must be sought in a physical change in the habitat; its continuance depends upon the reaction which each stage of vegetation exerts upon the physical factors which constitute the environment."[45] For example, as shrubs invaded a meadow they would cause a decrease in the amount of light reaching the soil. This change in a physical factor would trigger another wave of invasion. Therefore, succession could be envisioned as a series of actions and reactions culminating in a stable climax community in equilibrium with its physical environment (Figure 8.3).

Clements provided an elegant conceptual framework within which ecological phenomena could be explained. Both at the level of the individual and the community, causation followed a simple chain linking discrete physical factors, physiological functions, and ultimately biological structures.[46] Individual plants were parts of a community in much the same way that cells were parts of an individual plant. Although the cell was theoretically the structural and functional unit of the body, the laboratory botanist in 1900 generally studied physiological activities within the context of the individual organism. In the same way, the ecologist studied ecological processes, which involved individual plants, within the context of the complex community organism. According to Clements,

> it is evident that the causes or factors of the habitat act directly upon the plant as an individual, and at the same time upon plants as groups of indi-

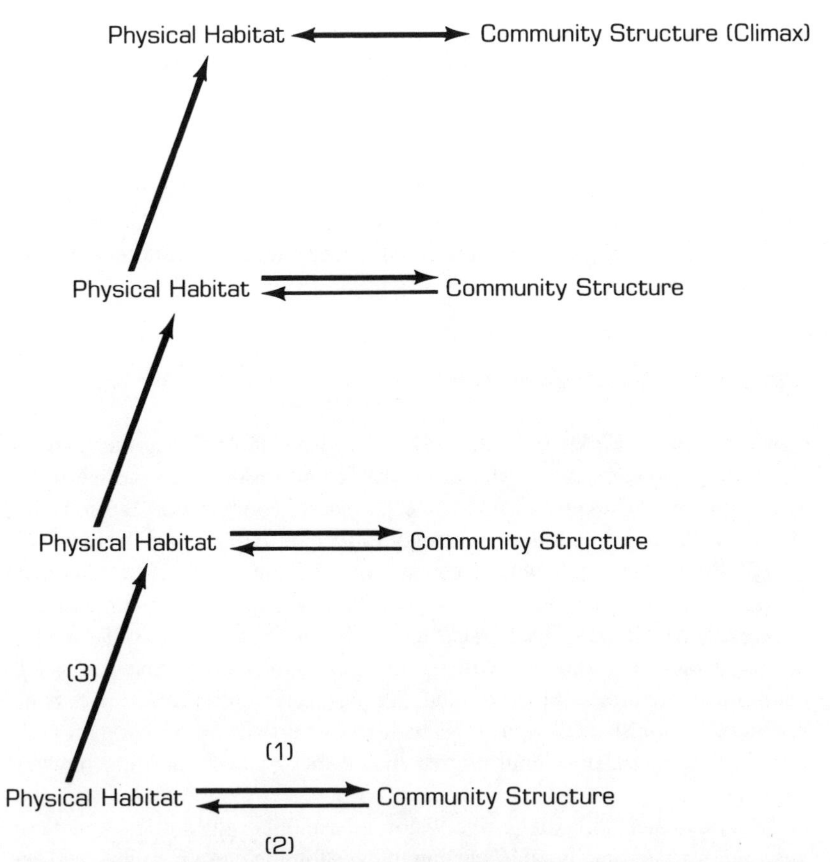

Figure 8.3. Clements's stimulus-response model of succession. A physical factor such as light intensity stimulates the process of (1) invasion, which causes a change in community structure. The community (2) reacts, causing (3) an alteration in the physical habitat. This periodic process continues until a climax community is established. The climax is in equilibrium with its physical environment. Consequently, no further major changes occur in the structure of the community.

viduals. The latter in no wise decreases the importance of the plant as the primary effect of the habitat, but it gives form to research by making it possible to consider two great natural groups of phenomena, each characterized by very different categories of effects. Ecology thus falls naturally into three great fundamental fields of inquiry: habitat, plant, and formation (or vegetation). To be sure, the last can be approached only through the plant, but as the latter is not an individual, but the unit of a complex

> from the formational standpoint, the formation itself may be regarded as a sort of multiple organism, which is in many ways at least a direct effect of the habitat.[47]

These direct effects of the habitat could be analyzed in much the same way that laboratory botanists studied physiological processes. The discrete physical forces of the environment could be accurately measured by instruments, and the structural changes within the community could be monitored by using permanent quadrats.

Clements's *Plant Succession*

Clements's *Research Methods in Ecology,* published in 1905, was the apotheosis of the physiological perspective in early plant ecology. A year before the book appeared, Ganong had called on ecologists to combine an environmental physics with an adaptational physiology to form a unified, experimental science.[48] This is precisely what Clements did. Clements provided ecologists with an elegant conceptual framework within which structure and function were explained in physicochemical terms. In contrast to descriptive natural history, ecologists could now study plant communities as organisms with recognizable structures, functions, and life histories. Furthermore, Clements presented ecologists with what appeared to be a rigorous methodology for discovering the causal mechanisms that linked the organism and the physical environment.

Paradoxically, this suggestive vision of a unified physiological ecology appeared just as some botanists began to question the close parallels between ecology and laboratory physiology. Indeed, Clements's book with its explicit physiological perspective focused attention on this problematic relationship. The reaction of botanists to the book was, therefore, ambivalent. On one hand, they were impressed by *Research Methods* because it seemed to place the new discipline of ecology on a solid scientific foundation. On the other hand, many botanists were either skeptical or hostile toward some of Clements's ideas.

This ambivalence is clearly illustrated by Cowles's review of Clements's book. Citing the many ingenious research techniques that Clements had developed, Cowles concluded: "One can scarcely praise this work too much; it is what is needed to prevent ecology from falling into swift and merited disfavor. If read and pondered, it will prevent the thoughtless from entering the ecological field, and it will serve the higher end of directing the thoughtful as to the methods of procedure."[49] While Cowles's review was generally quite favorable, he was critical of the implicit teleology in Clementsian ecology. Teleology, Cowles feared, implied an ultimate purpose in nature, a notion that would certainly bring ecology into disrepute. Cowles's choice of terms was unfortunate. The determinism so evident in Clements's explanations of individual adaptation and ecological succession was not the result of ultimate purpose; it resulted from simple physicochemical mechanisms that caused biological changes. Misunderstood as it was, this rigid determinism was one of the most controversial aspects of Clementsian ecology. It continues to draw fire from critics today.[50]

Quite a different form of criticism was leveled against Clements's *Research Methods* by the British ecologist A. G. Tansley and physiologist F. F. Blackman. Like Cowles, they were generally impressed by the book, judging it "the most ambitious and most important general work on Ecology" since the pioneering texts of Schimper and Wärming in the 1890s.[51] Although they acknowledged that Clements's organismal concept might be a useful way to think about communities, Blackman and Tansley criticized Clements for grossly oversimplifying the relationship between the physical environment and the organism. Simple stimulus-response mechanisms did not accurately reflect the true complexity of natural systems. According to the British critics, Clements was premature in embracing experimental physiology, and he ran the risk of misapplying crude physiological concepts in an inappropriate context. Tansley was not opposed to physiological ecology in principle, but he questioned the need for basing community ecology on an explicit physiological foundation.[52] Other ecologists also emphasized the problems of simply transferring physiological concepts and methods into ecology.[53]

If Blackman and Tansley were ambivalent toward Clementsian ecology, some physiologists were uncompromisingly hostile. For example, Charles Barnes, an experimental physiologist at the University of Chicago, was outspokenly critical of Clements's notion of simple causal mechanisms. According to Barnes, "there is a tendency to ascribe changes of function or form which have a very complex causation to a particular factor. . . . The offhand way in which some of the most elusive problems are settled by referring them to simple causes is appalling."[54] Barnes accused Clements of using "vague and inexact physiological cant" and proposing "'explanations' which are merely a form of words to cloak our ignorance."[55] Barnes's sarcasm was not directed specifically toward *Research Methods in Ecology,* but rather at Clements's less technical college textbook, *Plant Physiology and Ecology.*[56] However, by Clements's own admission the two books were based on the same ideas. Therefore, Barnes's critique could be seen as a scathing attack on Clementsian ecology in general. Most ecologists apparently saw it as such.[57]

Despite criticism of his ideas, Clements never totally abandoned the physiological perspective that characterized *Research Methods in Ecology.* The idea that communities are complex organisms and a functional approach to the study of ecology also characterized Clements's later works. In the years following 1905, however, Clements made subtle but significant changes in his ecological ideas. These changes undoubtedly reflected the criticism that his early book had received. To a certain extent, the changes also reflected Clements's maturation as an ecologist. While *Research Methods in Ecology* was the optimistic statement of a relative neophyte in a young and poorly defined discipline, Clements's most influential work, *Plant Succession* (1916) was based on extensive field research.

Certainly the most striking difference between *Research Methods* and *Plant Succession* is the lack of physiological rhetoric in the later work. In 1905 Clements had confidently claimed that ecology would become a truly experimental science—that it would, in effect, become a rational field physiology. In practice, however, the quantitative and experimental methods that Clements had outlined in his early books were expensive, time consuming, and in many cases impractical. Succession, for example, was a process that often required hundreds of years to complete. To study this process using permanent quadrats was not feasible, although Clements mused about the possibility of such studies being carried out by several generations of botanists.[58] In fact, Clements's actual field research on succession owed relatively little to the use of the quadrat. Although Clements and other early plant ecologists did use the quadrat to study succession, the method was generally employed to demonstrate conclusions already made rather than as a rigorous method to test alternative hypotheses. In short, in his own ecological research Clements owed more to the descriptive natural history that he had earlier criticized than to the experimental methods of the physiology laboratory. Therefore, although experimental physiology continued to serve as an inspiration for Clements's ecological theory, it did not serve as a practical foundation for most of his field research.

In *Plant Succession,* Clements also deemphasized simple mechanical causation. In his earlier book, Clements claimed that the environment could be analyzed into a few fundamental physical factors: light, humidity, and the water content of soil. Little consideration was given to interactions between physical factors. However, demonstrating the causal relationship between an isolated physical factor and a biological process was difficult to achieve in the field, and Clements was unable to do so convincingly. Not surprisingly, in *Plant Succession* and in his later writings Clements placed greater emphasis on more general environmental terms such as climate and factor-complex.

This shift away from discrete physical factors was a significant change in Clementsian ecology. Clements had been quite critical of nineteenth-century naturalists for failing to specify causal factors and for simply correlating differences in vegetation with general features of the environment. Ironically, Clements was forced to return to this more general notion of causality in *Plant Succession.* Equally significant was a recognition that the biotic environment often significantly influenced the course of succession. In *Research Methods in Ecology* Clements virtually ignored the ecological role of animals. However, in a study of succession in forests damaged by fire, Clements found that by eating seeds, squirrels could have a major impact on the regeneration of the forest.[59]

Finally, in contrast to his earlier views, Clements later stressed the complexity of ecological succession. In *Research Methods in Ecology* Clements portrayed succession as a relatively simple, mechanical process. According to this view, succession was a progressive process leading inevitably to the mature or climax community. Clements never completely rejected this view of succession, but his own field research demonstrated the complexities of succession. For example, in the same study of forests, Clements found that periodic forest fires in mountainous areas of Colorado sometimes prevented the theoretical climax community from developing. Lodgepole pines, which were particularly well adapted for rapid growth after forest fires, often dominated areas that theoretically ought to have developed into forests of Engelmann spruce and alpine fir.[60]

Conceptually, *Plant Succession* lacked the elegant simplicity of *Research Methods in Ecology.* Although the community was still conceived as a complex organism and succession as a developmental process analogous to ontogeny, Clements had been forced to hedge many of his conclusions. For example, he continued to argue that succession was inevitably progressive and that the climax community was in equilibrium with its physical environment; however, these laws of succession had numerous qualifications. Clements was forced to provide rather convoluted explanations for changes in vegetation that appeared to other ecologists to be retrogressive.[61] He had to admit that succession might be indefinitely stalled in a preclimax state, or conversely, that it might overshoot the theoretical climax and reach a postclimax stage.[62] Even under normal circumstances, Clements admitted, the climax was never in complete equilibrium with its environment. Indeed, minor structural changes would continue as the complex organism adjusted to fluctuations in climate.[63]

Qualifications notwithstanding, *Plant Succession* was an impressive book. Its five hundred pages were packed with voluminous information. Furthermore, the organismal metaphor continued to provide a provocative framework for interpreting data. Botanists immediately recognized Clements's monograph as a treatise to be reckoned with. Even Henry Allan Gleason, often portrayed as Clements's nemesis, was clearly impressed with the work. Although uncompromisingly critical of Clements's organismal ideas, Gleason was quick to admit the significance of *Plant Succession.*

> Ecological literature has recently been enriched by the publication of an exceedingly important book on the structure and development of vegetation. Not only does the book present a thorough and detailed analysis of vegetation, based on the researches of the author, but it also reviews the

> theories and summarizes the facts from a vast array of the ecological literature. For all its contents the working ecologist is grateful, although it is probable that some of the more radical ideas of the author may be accepted reluctantly and others may be rejected completely.[64]

Gleason's evaluation was quite perceptive. *Plant Succession* was exceedingly important; it has served as a focal point for ecological discussion throughout the twentieth century. Many of Clements's claims were controversial, and some of his ideas were rejected by most ecologists; yet the organismal concept has continued to attract a number of adherents.[65]

Conclusion

Ecology emerged at a time when many American biologists were shifting their attention away from description and classification, and toward a study of biological processes.[66] Indeed, many young biologists—zoologists and botanists—were impatient with what they considered to be traditional natural history. Instead, a number of these critics turned to experimental physiology for inspiration.[67] Some methods and concepts from physiology could be applied directly to ecological research. However, physiology also was attractive in a much more general way; by patterning ecology after the successful model of laboratory physiology, ecologists could legitimize their new area of biological research. According to Cittadino, "It was physiology that separated the biological sciences from natural history. The study of process and function, the use of the laboratory, the experimental method—these were held to be the ingredients of true science."[68]

Clementsian ecology can best be understood within this historical context. Frederic Clements shared this general enthusiasm for physiology. Like many of his contemporaries, he often used rather vague references to experimental physiology to distinguish his approach to ecology from what he considered to be a moribund, descriptive natural history tradition. Clements may have felt a particular need to legitimize ecology because his teacher, Charles Bessey, was openly skeptical of the new discipline, at times referring to it as a fad.[69] Complex instrumentation, quantitative methods, and experimentation, therefore, were desirable not only on purely scientific grounds; they also served another important purpose. Because these methods required "much patience and seriousness of purpose," their use would discourage the faddish amateur ecologists. "As a consequence," Clements concluded, "there will be a general exodus from ecology of those that have been attracted to it as the latest biological fad, and have done so much to bring it into disrepute."[70]

Clementsian ecology, however, aimed at being more than physiological rhetoric. The organism and its internal processes provided the subject matter for physiology. By comparing plant communities with individual organisms, Clements attempted to establish a unified physiological theory of ecology. Though very different, the complex community organism and the individual organism could be understood in terms of the same mechanical principles. In both cases, Clements claimed to trace a direct causal chain linking physical factors in the environment to biological functions, and ultimately to changes in biological structures. Whether the ecologist was studying adaptive changes in the individual plant or long-term successional changes in the plant community, the explanation could be reduced to simple stimulus–response relationships between the physical environment and the biological organism.

Clements's physiological ecology reflected a mechanistic bias shared by a number of early twentieth-century biologists.[71] Like other mechanistic schemes, Clements's vision of a unified physiological ecology was controversial. While critics acknowledged that his *Research Methods* was important because it attempted to define the new science of ecology and place it upon a rigorous foundation, they tended to be skeptical of Clements's mechanistic interpretations. For most ecologists, nature was too complex to be reduced to simple cause-and-effect relationships between individual physical factors and biological processes. Clements, himself, was forced to modify his mechanistic ideas, and he was never able to implement fully his physiological approach to ecology. Despite his early attacks on descriptive botany, Clements did not completely free himself from traditional natural history. In fact, a later generation of ecologists would criticize him for not advancing much beyond description and classification.[72] The criticism was unfair, but it underscores the difficulty of simply transferring physiological ideas and methods into the new discipline of ecology.

Historians have often drawn a rather sharp distinction between natural history and experimental biology.[73] However, a consideration of Frederic Clements's career illustrates how problematic such a dichotomy can be. Rhetorically, Clements himself distinguished between natural history and experimental biology, for he viewed these as conflicting scientific methodologies. He apparently considered this distinction necessary for defining and legitimizing the new discipline of ecology. But as a practicing scientist, Clements cannot be accurately categorized as either a naturalist or an experimentalist. Indeed, he was both. Clements employed some innovative experimental techniques and his vision of a rigorous experimental physiology inspired his research program. However, much of Clements's research on plant communities appears to be the work of a gifted and imaginative naturalist. This ambiguity was not completely idiosyncratic; it appears to have characterized twentieth-century ecology in general. Some ecologists have explicitly referred to their field as scientific natural history, while others have continued to argue for a more experimental and physiological approach to ecology.[74]

What was the lasting significance of Clements's claim that the plant community is an organism? In one of his last major works, Clements presented a rather grandiose assessment of the historical importance of this idea. According to Clements the organismal concept had been an " 'open sesame' to a whole new vista of scientific thought, a veritable *magna carta* for future progress."[75] Many ecologists today would dismiss this claim; for them Clements's organismal concept is a relic from the past, an erroneous notion that has been effectively refuted. Tobey draws a similar conclusion in his analysis of Clements's role as the leading figure in the development of a midwestern school of grassland ecologists. He notes that organismal ideas formed a part of the "Clementsian paradigm" that guided this school of scientists; however, both the school and its paradigm, after flourishing during the early decades of the twentieth century, declined rapidly during the 1940s.[76]

Tobey presents a compelling account of the intellectual and institutional development and decline of a research group. His conclusion about the fate of Clements's organismal concept is correct in the sense that few, if any, biologists today would accept the idea that communities *are* organisms. However, Clements was the most articulate early spokesman for a physiological perspective that continues to attract some ecologists. For these ecologists, the community may not *be* an organism, but in important ways it is similar to an individual organism. Not surprisingly, those ecologists who have been interested in the processes or "functions" of communities often continue to use organismal concepts.[77]

Although many of his ideas were later qualified or rejected, Clements's research had an enormous impact upon the early development of ecology. At the turn of the century many biologists complained that ecology was unfocused, or worse, that it was simply a passing scientific fad. Clements was one of a handful of creative botanists who provided intellectual direction to this fledgling discipline. More than any other early ecologist, Clements stressed the dynamic nature of communities and emphasized the fundamental importance of ecological processes. Together with Cowles, he established succession as the primary focus of early ecological research. This theory had a unifying effect on ecology, for it provided a set of problems, methods and ideas that zoologists, as well as botanists, considered significant. Victor Shelford, for example, who studied with Cowles at the University of Chicago and later worked with Clements, studied succession in a variety of animal communities. Later, Clements and Shelford coauthored one of the first ecological textbooks to address explicitly the connections between plant and animal communities.[78] Similarly, the British ecologist Charles Elton, citing the significance of Clements's early work on plant communities, used succession as an organizing principle in his influential book, *Animal Ecology*.[79] The ecologist Eugene P. Odum was not exaggerating when he claimed that the concept of succession was as significant for the early development of ecology as Mendel's laws were for genetics.[80]

Acknowledgments

I wish to thank Lindley Darden and Pamela Henson for critical comments on an early version of this article. Many helpful suggestions also were provided by the other contributors to this volume. I am particularly indebted to Keith Benson, Jane Maienschein, and Ronald Rainger for their helpful criticisms throughout this project.

Notes

1. Henry Chandler Cowles, "The Ecological Relations of the Vegetation of the Sand Dunes of Lake Michigan," *Botanical Gazette,* 1899, *27:* 95–117; 167–202; 281–308; 361–391.

2. A. G. Tansley, "The Use and Abuse of Vegetational Concepts and Terms," *Ecology,* 1935, *16:* 284–307; see also A. G. Tansley, "Henry Chandler Cowles," *Journal of Ecology,* 1940, *28:* 450–452.

3. Cynthia Eagle Russett, *The Concept of Equilibrium in American Social Thought* (New Haven: Yale University Press, 1966), pp. 65–66. Richard Hofstadter also cites numerous examples of the use of organismal metaphors in his *Social Darwinism in American Thought,* 2nd ed. (New York: George Braziller, 1955). Stephen J. Pyne notes the "uncanny resemblance" between the developmental analogies used in early discussions of ecological succession and the "frontier hypothesis" articulated in the 1890s by the historian Frederick Jackson Turner, *Fire in America: A Cultural History of Wildland and Rural Fire* (Princeton: Princeton University Press, 1982), p. 492.

4. Cowles, "Ecological Relations," p. 95.

5. Jonathan L. Richardson, "The Organismic Community: Resilience of an Embattled Concept," *Bioscience,* 1980, *30:* 465–471.

6. Eugene Cittadino, "Ecology and the Professionalization of Botany in America, 1890–1905," *Studies in History of Biology,* 1980, *4:* 171–198; H. A. Gleason, "Twenty-five Years of Ecology, 1910–1935," *Brooklyn Botanic Garden Memoirs,* 1936, *4:* 41–49; Robert P. McIntosh, *The Background of Ecology: Concept and Theory* (New York: Cambridge University Press, 1985), chap. 2; A. G. Tansley, "The Early History of Modern Plant Ecology in Britain," *J. Ecol.,* 1947, *35:* 130–137; Robert L. Burgess, "The Ecological Society of America: Historical Data and Some Preliminary Analysis," in Frank N. Egerton, ed., *History of American Ecology* (New York: Arno Press, 1977).

7. Ronald C. Tobey, *Saving the Prairies: The Life Cycle of the Founding School of American Plant Ecology, 1895–1955* (Berkeley: University of California Press, 1981); see also Paul B. Sears, "Plant Ecology," in Joseph Ewan, ed., *A Short History of Botany in the United States* (New York: Hafner, 1969).

8. Tobey, *Saving the Prairies.*

9. Burgess, "Ecological Society," p. 4.

10. Joseph Ewan uses this phrase to characterize Frederic Clements and H. L. Shantz in "Plant Geography," in Ewan, *A Short History of Botany,* p. 121.

11. Cittadino, "Ecology."

12. Ibid., p. 172.

13. W. C. Allee, Orlando Park, Alfred E. Emerson, Thomas Park, and Karl P. Schmidt, *Principles of Animal Ecology* (Philadelphia: W. B. Saunders, 1949), p. 44.

14. McIntosh, *Background*, p. 30.

15. Paul B. Sears, "Botanists and the Conservation of Natural Resources," in William Campbell Steere, ed., *Fifty Years of Botany: Golden Jubilee Volume of the Botanical Society of America* (New York: McGraw-Hill, 1958), p. 362.

16. W. F. Ganong, "The Cardinal Principles of Ecology," *Science,* 1904, *19:* 493. Ganong, a botanist at Smith College, had become interested in ecology while studying plant adaptation in Germany; see Cittadino, "Ecology."

17. Frederic Edward Clements, *Research Methods in Ecology* (Lincoln, Nebraska: University Publishing Co., 1905), p. 6; H. C. Cowles, "Research Methods in Ecology," *Bot. Gaz.,* 1905, *40:* 381–382.

18. Henry Chandler Cowles, "The Work of the Year 1903 in Ecology," *Sci.,* 1904, *19:* 879–885, p. 879. Cowles also discussed the chaotic state of ecology in "The Physiographic Ecology of Chicago and Vicinity; a Study of the Origin, Development, and Classification of Plant Societies," *Bot. Gaz.* 1901, *31:* 73–108, 73–76.

19. Cowles, "Work in Ecology," p. 880.

20. Clements, *Research Methods,* p. 12; V. M. Spalding, "The Rise and Progress of Ecology," *Sci.,* 1903, *17:* 201–210. Garland Allen has argued that many young biologists shared these views on experimentation and evolutionary theory, in *Life Sciences in the Twentieth Century* (New York: John Wiley and Sons, 1975), pp. 17–18.

21. I have discussed the development of this area of ecological research in "Ecologists and Taxonomists: Divergent Traditions in Twentieth-Century Plant Geography," *Journal of the History of Biology,* 1986, *19:* 197–214.

22. A. W. F. Schimper, *Pflanzengeographie auf Physiologischer Grundlage* (Jena: Gustav Fischer, 1898); idem, *Plant Geography upon a Physiological Basis,* trans. W. R. Fisher, ed. Percy Groom and I. B. Balfour (Oxford: Clarendon Press, 1903).

23. This enthusiasm for physiology seems to have been shared by many young biologists in several fields at the turn of the century; see Allen, *Life Sciences* pp. xv–xix, 81.

24. Clements, *Research Methods,* pp. 306–307.

25. Ibid., p. 13.

26. Ibid., p. 10.

27. Tobey, *Saving the Prairies,* pp. 35–47.

28. Ibid.; Hamilton Cravens, *The Triumph of Evolution: American Scientists and the Heredity-Environment Controversy, 1900–1941* (Philadelphia: University of Pennsylvania Press, 1978), chap. 1.

29. Roscoe Pound and Frederic E. Clements, *The Phytogeography of Nebraska,* 2nd ed., (Lincoln, Neb.: Botanical Seminar, 1900).

30. Clements, *Research Methods,* pp. 8–9.

31. Tobey, *Saving the Prairies,* pp. 84–87. A Spencerian influence has also been suggested by McIntosh, *Background of Ecology,* p. 43, and by Donald Worster, *Nature's Economy: The Roots of Ecology* (San Francisco: Sierra Club Books, 1977), pp. 212–215. Late in life, Clements suggested that Comte and Spencer had influenced other organismal biologists; however, he did not admit to such an influence on his own

work. See Frederic E. Clements and Victor E. Shelford, *Bio-Ecology* (New York: John Wiley and Sons, 1939), p. 24.

32. The idea of the community as organism first appeared in a collection of short articles: Frederic E. Clements, *The Development and Structure of Vegetation* (Lincoln, Neb.: Botanical Seminar, 1904). Cowles's study is cited in this volume.

33. Clements, *Research Methods,* p. 306.

34. The quadrat was originally used simply as a method for quantitative sampling of species for geographical surveys; see Tobey, *Saving the Prairies,* chap. 3; McIntosh, *Background of Ecology,* pp. 78–79.

35. David Hull, *Philosophy of Biological Science* (Englewood Cliffs, N.J.: Prentice-Hall, 1974), chap. 5; D. C. Phillips, "Organicism in the Late Nineteenth and Early Twentieth Centuries," *Journal of the History of Ideas,* 1970, *31:* 413–432. Tobey, *Saving the Prairies,* pp. 159–160, places Frederic Clements within an anti-mechanistic tradition in biology. In a contentious essay, Daniel Simberloff, "A Succession of Paradigms in Ecology: Essentialism to Materialism and Probabilism," *Synthese,* 1980, *43:* 3–39, identifies Clementsian ecology with Platonic idealism and essentialism.

36. Clements, *Research Methods,* p. 17.

37. Ibid., p. 5. For a brief discussion of Bonnier's experiments and their relationship to Clements's work, see Joel B. Hagen, "Experimentalists and Naturalists in Twentieth-Century Botany: Experimental Taxonomy, 1920–1950," *J. Hist. Biol.,* 1984, *17:* 249–270.

38. Clements, *Research Methods,* pp. 18–20.

39. Ibid., pp. 285–289.

40. J. Reynolds Green, *A History of Botany 1860–1900: Being a Continuation of Sach's "History of Botany 1530–1860"* (New York: Russell & Russell, 1909), Book 3.

41. Clements, *Research Methods,* pp. 129–144; Edith Schwartz Clements, "The Relation of Leaf Structure to Physical Factors," *Transactions of the American Microscopical Society,* 1905, *26:* 19–102.

42. Clements, *Research Methods,* p. 133.

43. Ibid., p. 138.

44. Ibid., p. 199.

45. Ibid., p. 256.

46. Ibid., p. 17.

47. Ibid.

48. Ganong, "Cardinal Principles," p. 494.

49. Cowles, "Research Methods," p. 382.

50. For example, see Simberloff, "Succession of Paradigms."

51. F. F. Blackman and A. G. Tansley, "Ecology in its Physiological and Phytogeographical Aspects," *The New Phytologist,* 1905, *4:* 199–203, 232–253, p. 203; Tansley, "Early History," p. 133.

52. A. G. Tansley, ed., *Types of British Vegetation* (Cambridge: Cambridge University Press, 1911), pp. 2–3.

53. Burton E. Livingston, "Present Problems of Physiological Plant Ecology," *American Naturalist,* 1909, *43:* 369–378.

54. C[harles] R. B[arnes], "Physiology and Ecology," *Bot. Gaz.,* 1907, *44:* 308.

55. Ibid., p. 307.

56. Frederic E. Clements, *Plant Physiology and Ecology* (New York: Henry Holt, 1907).

57. Paul B. Sears, "Clements, Frederic Edward," *Dictionary of Scientific Biography,* vol. 3 (New York: Charles Scribner's Sons, 1973), pp. 168–170.

58. Clements, *Research Methods,* p. 272.

59. F. E. Clements, "The Life History of Lodgepole Burn Forests," *Forest Service Bulletin,* 1910, *79:* 35–39.

60. Ibid., pp. 53–56.

61. Frederic E. Clements, *Plant Succession: An Analysis of the Development of Vegetation* (Washington, D.C.: Carnegie Institution of Washington, 1916), chap. 8.

62. Ibid., chap. 6.

63. Ibid., pp. 3, 99–100.

64. Henry Allan Gleason, "The Structure and Development of the Plant Association," *Bulletin of the Torrey Botanical Club,* 1917, *44:* 463–481, p. 463.

65. Richardson, "Organismic Community."

66. Cittadino, "Ecology."

67. Allen, *Life Science,* pp. xv–xix.

68. Cittadino, "Ecology," p. 179.

69. Ibid., p. 171.

70. Clements, *Research Methods,* p. 20.

71. Allen, *Life Science,* pp. 73–79.

72. G. E. Hutchinson, "*Bio-Ecology,*" *Ecology,* 1940, *21:* 267–268; see also H. L. Shantz, "Frederic Edward Clements," *Ecol.* 1945, *26:* 317–319.

73. Ernst Mayr, *The Growth of Biological Thought* (Cambridge, Mass.: Harvard University Press, 1982), pp. 540–550; Allen, *Life Sciences,* pp. 18–19.

74. Compare the discussion of scientific natural history in Charles Elton, *Animal Ecology* (London: Sidgwick and Jackson, 1927), p. 1, and Eugene P. Odum's discussion of functional ecology and the study of community metabolism in *Fundamentals of Ecology,* 2nd ed. (Philadelphia: W. B. Saunders, 1959), p. ix.

75. Clements and Shelford, *Bio-Ecology,* p. 24.

76. Tobey, *Saving the Prairies,* chap. 7.

77. Odum, *Fundamentals,* pp. ix, 25–27.

78. Clements and Shelford, *Bio-Ecology;* Keith Benson, "Experimental Ecology on the Pacific Coast: Victor Shelford and his Search for Appropriate Methods," unpublished manuscript.

79. Elton, *Animal Ecology,* chap. 3.

80. Odum, *Fundamentals,* p. 257.

Diane B. Paul and
Barbara A. Kimmelman

9 Mendel in America: Theory and Practice, 1900–1919

In September 1903, Willet Hays addressed the first meeting of the American Breeders Association (ABA). A founding member and guiding spirit of the association, Hays was then professor of agriculture at the University of Minnesota and director of the state experiment station; two years later he would be appointed Assistant Secretary of Agriculture. In his opening remarks, he lamented:

> Science has been content to remain at the task of proving for the ten thousandth time that Darwin's main contention is true but has allowed the great economic problems of evolution guided by man to remain almost a virgin field. Only recently have such men as Galton, Mendel, de Vries, Bateson and a few others, entered upon comprehensive lines of research and many of these have hardly grasped the vast economic interests which are at stake, nor have they seen the open doors of opportunity which might be entered by cooperation with the men who control the breeding herds and the plant-breeding nurseries.[1]

Hays's opinions were widely shared. Excited by developments in what would soon be called "genetics," most ABA members agreed with William Bateson that: "At this time we need no more *general* ideas about evolution. We need *particular* knowledge of the evolution of *particular* forms."[2] Francis Galton's law of ancestral heredity, Gregor Mendel's laws of dominance, segregation and independent assortment, and Hugo de Vries' mutation theory were all greeted enthusiastically. Because the ABA had a varied membership, there was more than one source of interest in the new discoveries. The chief factor, however, was a belief that these laws could readily be used to improve artificial selection. Some association members were particularly concerned with selection in humans. A greater emphasis on eugenics was promoted by Charles Davenport, who argued that "society must protect itself; as it claims the right to deprive the murderer of his life so also it may annihilate the hideous serpent of hopelessly vicious protoplasm."[3] The majority, however, were less concerned to improve humans than grapes, hogs, beans, or, especially, corn. Their primary concern, in Deborah Fitzgerald's phrase, was with "the business of breeding."[4]

Who were these early enthusiasts for Mendel? Why did they accord his work such a warm reception? How did their response compare with that of biologists whose interests were principally descriptive or theoretical, rather than applied? And how did their concerns with agricultural practice inform their program of research?

We find that particular economic pressures and practical demands on agricultural breeders, within the context of late nineteenth-century agricultural reform, encouraged actively interventionist, experimental techniques. These included hybridization and the crossbreeding of varieties.[5] In particular, a perceived crisis of wheat overproduction in the 1870s prompted the United States Department of Agriculture (USDA) to elaborate a policy of diversifying agricultural products. USDA officials saw creation of novelty as crucial to this program. More specifically, they aimed to increase variation and produce stable hybrids. In the 1880s and 1890s, the department expanded its own experimental breeding work along these lines and, significantly, promoted such work at state agricultural colleges and experiment stations.

Focusing on experimental creation of variation and the inheritance of particular characters, the work of scientists at agricultural institutions converged with that of an international group of botanists and hybridists interested in evolutionary problems (including de Vries and Bateson). After 1900, these contacts provided American agriculturalists with ready access to Mendel's work, while their technical and intellectual background prepared them to receive it enthusiastically. The strength of breeding and genetics programs at agricultural colleges and experiment stations in the decades that followed insured the pursuit of genetics at publicly supported institutions characterized by simultaneous commitments to the ideal of basic research and the practical demands of economic agriculture. The early work on hybrid corn, presented here as a case study, illustrates the importance of this context for the direction of genetics research.

Who Were the "Breeders"?

From 30 August to 2 September 1902, the International Conference on Plant Breeding and Hybridization met in New York City. Reviewing the conference for *Torreya,* Walter Cannon noted that, "generally speaking, the plant breeders had not taken advantage of the Mendelian theory in their work, and some of them did not know of Mendel or of his experiments before the Conference."[6] But they left it as converts to the new genetics. C. W. Ward, a carnation grower, was one of the commercial breeders in attendance. His remarks attest both to the degree and source of the excitement with which such breeders greeted Mendel's work: "I have known nothing of Mendel's theory or law until the day before yesterday," he said, "but what I have heard here regarding Mendel has awakened an increasing interest in the work of hybridizing and I shall secure his books and read them with the greatest interest, for if there is a fixed rule by which I can produce six inch carnations on four foot stems I certainly wish to learn that rule."[7]

One of the foreign visitors, speaking on "Practical Aspects of the New Discoveries in Heredity," was William Bateson.[8] Following the talk, Liberty Hyde Bailey of the state agricultural college at Cornell recommended Bateson's just-published book, *Mendel's Principles of Heredity: A Defence.*[9] "If you wish to follow this ['the Mendelian hypothesis'] with the greatest degree of accuracy, you should get Mr. Bateson's recent book," he urged, adding that: "I expect to use this book as a basis for all our work in plant breeding."[10] His advice was apparently heeded. On 3 October, Bateson wrote excitedly to his wife: "At the train yesterday, many of the party arrived with their 'Mendel's Principles' in their hands! It has been 'Mendel, Mendel all the way,' and I think a boom is beginning at last. There is talk of an International Assn. of Breeders of Plants and Animals and I am glad to be right in the swim."[11]

Commenting on this letter, Garland Allen suggests that: "It may perhaps seem curious that plant and animal breeders, with their primary concern for practical results, would have taken more readily as a group to Mendel's theoretical presentation than many of the more academic biologists."[12] He is certainly right to note that Mendelism made sense of breeders' results—both their successes and failures—and was thus greeted as a means to improve the efficiency of breeding practice.[13] The distinction between breeders and academic biologists is, however, somewhat misleading if the former are equated with farmers or "seedsmen."[14] To be sure, some "practical breeders" such as L. H. Kerrick and Eugene Funk were founding members of the ABA. This group also played an important role in garnering political support for the work of agriculturalists at state institutions, which was crucial to the establishment of genetics as an academic discipline.[15] But the enthusiasts referred to by Bateson were mostly scientists, not seedsmen. Of course they were scientists of a particular kind, both institutionally and in respect to their aims. Employed at agricultural colleges and agricultural experiment stations, rather than at arts and sciences institutions, they were concerned with the implications of Mendelism for practice as well as theory.

The dominant force at the 1902 New York Conference was Bateson; his lead paper combined a straightforward account of Mendel's laws with a discussion of their applied, and especially commercial, importance. To the breeders he argued:

> Now when we come to the question of the significance of these things to the breeder and to the hybridist, it will be found that the significance is exceedingly great. I am afraid of saying that we have reached a point when the practical man who is doing these things with a definite, economic object or commercial object in view can take the facts and use them for his definite advantage. But we do for the first time get a clear sight of some of the fundamentals on which he will in future work, and it cannot be now very many years, if the investigations go on at the present rate, before the breeder will be in a position not so very different from that in which the chemist is: —when he will be able to do what he wants to do, instead of merely what happens to turn up.[16]

The following two papers, by C. C. Hurst (Bateson's close friend and colleague) and Hugo de Vries, focused on Mendel as well.[17] In all, ten participants either presented papers or made extended remarks promoting Mendelism. Seven were Americans: Willet Hays of the Minnesota Agricultural Experiment Station, Liberty Hyde Bailey of Cornell, S. A. Beach of the New York State Experiment Station, Walter Austin Cannon of Columbia and the New York Botanical Garden, and O. F. Cook and W. J. Spillman, both of the USDA. If members of this group are to be characterized as "breeders," it follows that many breeders active in promoting Mendelism were academic biologists. Of course, they were academic biologists of a particular kind, both institutionally and in respect to their goals. These biologists were generally affiliated with the USDA or state agricultural colleges and experiment stations and they aimed to combine practical public interests with theoretical science.

Our paper details the crucial, yet historically neglected, role of this group in introducing and popularizing Mendel's work. But we should note that biologists with a primarily theoretical orientation were also generally receptive to Mendelism (in sharp contrast with naturalists, who were decidedly cool).[18] T. H. Morgan was a severe critic, but his views on Mendel were atypical. While some biologists who employed the method of experimental breeding (either to improve selection or to answer questions regarding the physical basis of heredity or its relation to evolution) were indifferent to the new genetics, few were actively hostile. Indeed, most were enthusiastic, irrespective of whether they were employed at state agricultural institutions or elite colleges and laboratories. Walter Sutton, Nettie Stevens, E. B. Wilson, E. M. East, William E. Castle, Charles Davenport, and George H. Shull, for example, were all avid Mendelians, whose principal concerns were theoretical. The last four also belonged to the ABA.

If Allen asserts too sharp a separation between breeders and academic biologists, Jan Sapp rejects the distinction altogether. In his view, "breeder" is practically synonymous with "geneticist" until at least 1915, because in the early years "Mendelian investigators recognized no distinction between pure science and applied science."[19] But although many academic biologists, particularly those employed at state agricultural institutions, pursued both pure and applied research, they certainly recognized "pure" and "applied" as categories; indeed, the proper balance between them was a matter of intense concern. Moreover, as we have seen, there were breeders whose interests were purely applied and geneticists whose interests were largely (in a few cases, entirely) theoretical. Eugene Funk was not a "geneticist," nor George Shull a "breeder." The relationship between practical breeders and geneticists was addressed by Hays in his opening speech at the first meeting of the ABA:

> The producers of new values through breeding are brought together as an appreciative constituency of their servants, the scientists. They are ready to aid in securing all needed means for scientific research in problems relating to heredity, provided the scientists can develop methods of research which will aid the breeders in more rapidly improving the plants and animals. The scientists, on the other hand, are ready to emerge from the cloister of species and genus grinding in the study of historic evolution, and cooperate with practical breeders in the study of breed and variety formation and improvement. . . . No less of an incentive, at least to the scientists, is the possible solution of some of the intricate problems of development in plants, in the lower animals, and in man.[20]

What united virtually all ABA members—whether commercial seedsmen, editors of farm journals, USDA officials, or researchers at state agricultural colleges or elite arts and sciences institutions—was their enthusiasm for the technique of experimental breeding. Some hoped to answer theoretical questions, others to improve agricultural practice; many aimed to do both. But however disparate their motivations, those who used experimental methods generally expressed a keen interest in Mendel's work.

The USDA and the Reception of Mendelism

No organization played a more important role in the dissemination of Mendelism than the USDA. At the 1902 New York Conference, eleven of the seventy-five participants were employed directly by the USDA; many more were affiliated with state agricultural colleges and experiment stations. At the first two meetings of the ABA (1903 and 1905), seventeen of forty-five papers were presented by USDA officials; these included the majority of papers dealing primarily with Mendelism, such as Spillman's "Mendel's Law in Relation to Animal Breeding" and Webber's "Explanation of Mendel's Law of Hybrids."[21]

As early as 1901, the USDA's *Experiment Station Record,* which functioned as an information clearinghouse for the experiment stations, published a detailed synopsis of Mendel's work based on Bateson's communication to the Royal Horticultural Society, "G. Mendel, Experiments in Plant Hybridization."[22] The *Record* summarized Mendel's concepts of dominance and independent assortment of different pairs of characters, and discussed the significance of his statistical ratios. It also quoted Bateson's claim that Mendel's laws were "worthy to rank with those that laid the foundation of the atomic laws of chemistry."[23]

An abstract of Tschermak's article on the inheritance of characters when crossing peas and beans immediately followed this report. According to the *Record,* Tschermak's work tested Mendel's predictions and, while losing something of the generality of his results, nevertheless underscored its "importance for theoretical and plant breeding purposes."[24] A few issues later, the *Record* provided a synopsis of Bateson's earlier communication to the Royal Horticultural Society on "Problems of heredity as a subject for horticultural investigation," which was "largely a review of the work of Mendel and de Vries."[25]

By choosing to abstract this work in detail, and through its editorial remarks, the *Experiment Station Record* conveyed to its readers a profound appreciation of the scientific value of Mendelism and its implications for horticultural investigation. In the years immediately following, the *Record* reported both American and European work, noting obvious limitations as well as evidence of its potential applicability to practical problems.[26] Thus, news of Mendel's work and the experiments and controversies it prompted was available to experiment-station personnel in every state.

The USDA also helped popularize Mendelism through its Graduate School of Agriculture, inaugurated in July 1902 (only two months before the New York Conference). Seventy-five students attended, of whom twenty-seven were faculty at agricultural colleges and thirty-one assistants in the agricultural colleges and experiment stations. According to Liberty Hyde Bailey: "Perhaps the two agencies most responsible for the dissemination of the Mendelian ideas in America were the instruction given by Webber and others in the Graduate School of Agriculture at Columbus last summer, and the prolonged discussion before the International Conference on Plant-Breeding at New York last September."[27]

The USDA was founded in 1862, the same year as passage of the Morrill Land-Grant Act, which established most of the country's agricultural colleges. In the 1870s and 1880s, the federal government began to increase markedly its commitment to agricultural research. As Margaret Rossiter has noted, the degree of federal support is striking, given the depressed condition of contemporary agricultural science: there had been no intellectual successes since the development of agricultural chemistry in the 1840s and 1850s and low enrollments in agricultural subjects persisted at state colleges.[28] Nevertheless, the Hatch Act, funding the experiment stations, was passed in 1887 and the Morrill Act, primarily for establishing black colleges of agriculture, in 1890. During the same period, appropriations for the USDA itself increased dramatically, as did the number of its employees.

These agricultural appropriation acts, reflecting greater federal intervention in agriculture, were contemporaneous with others that signaled an end to the laissez-faire ideology characteristic of nineteenth-century America. Both the Interstate Commerce Act and Hatch Act were passed in the same year; so were the Sherman Anti-Trust Act and the second Morrill law. As the United States entered international commerce following the Civil War, both industrial and agricultural production were thought too important to be left to chance. The government, with the aid of various interest groups, was now prepared to intervene in the interest of the national economy.[29]

In response to the social and economic crises that marked this period, federal administrators and researchers advanced the cause of science-based agriculture. They held that agricultural problems could be addressed most effectively through the application of work in the natural and social sciences, pursued by experts at centralized institutions and disseminated to farmers. Their program was a response, and potential antidote, to the campaign for structural changes advocated by populists. It also helped decide a longstanding debate among agricultural educators in favor of those who believed that state colleges and research institutions should pursue basic research (not just vocational training). The expansion of science-based agricultural curricula, the federal funding of state experiment stations, and the development of extension services were thus elements in a linked program of scientific, social, and professional reform. In this context, manipulative experimental techniques such as hybridization and crossbreeding served a dual role. They helped define agricultural expertise in terms of schooled scientific skill rather than day-to-day practice; and they promoted active intervention to achieve practical goals.

These techniques were central to the USDA's response to a crisis resulting from the overproduction of wheat, with the consequent glutting of world markets and decline in the value of wheat and wheatstuffs. The great expansion of productivity in American agriculture that created this surplus preceded by more than a decade the official promotion of hybridization by the USDA. The purpose of intensive and specialized hybridization work was therefore not a generalized increase in productivity, which had been achieved by other means. It was part of the refinement of the system—a means to cope with a specific crisis of agricultural production.

The massive increase in U.S. production of wheat in the 1870s, according to USDA statistician J. R. Dodge, was attributable to three factors: the ready availability of fresh agricultural land; the penetration of the railroads into areas previously inaccessible to markets; and, most important, an extraordinarily inflated demand resulting from several years of European crop failures.[30] The problem, apparent by 1880, had intensified by the middle of the decade. With the wheat crop of 1884 five times the size of the 1830 crop, and with prices falling in the world market as European production recovered, the problem of American overproduction began to concern USDA officials.[31]

In 1885, Dodge suggested a solution ultimately adopted by the department. He directly linked the overproduction of wheat to the underproduction of other valuable commodities that the nation now imported.[32] The solution required that farmers decrease the acreage devoted to wheat production and extend cultivation of other crops.[33]

With this goal, the USDA developed a three-part corrective program aimed at diversification of the nation's agricultural products: decreasing the acreage of wheat in the production; decreasing U.S. reliance on particular import items; and increasing exports of specialty items of high quality, for which there was strong market demand.[34] Generally, this solution involved an intensification of economic and scientific research, centrally directed by the USDA, and expert analysis of the general problems of national agricultural production. In 1887, Commissioner of Agriculture Norman J. Colman explicitly linked efforts to diversify with the search for new products.

> It is an important question, in view of the rapid increase of available rural labor, tending to overproduction of the fruits of the soil and the cheapening of their value, what can be done to give greater variety to the products of agriculture? What can this Department do towards the introduction of new plants and development of new rural industries?[35]

In his statistician's report for 1889, Dodge stressed that solving the problem of overproduction "required the fullest and promptest information concerning new fruits, fibers, or products of economic plants."[36] Rather than depressing production, it should be encouraged in new directions, particularly toward the cultivation of products hitherto imported.[37] Throughout the 1890s, the secretaries of agriculture echoed these sentiments; the American farmers should pursue a favorable balance of trade through the "substitution in our own markets of home-grown for foreign-grown products."[38]

As diversification, quality improvement, and increased self-sufficiency became central to the USDA's response, various branches of economic botany assumed greater importance within the department. Seed introduction, plant exploration, botany, and horticulture gained in status and acquired new institutional structures. Botanical specialties such as plant pathology, pomology, and agrostology achieved independent divisional status for the first time.[39] Scientists working in these divisions found hybridization and cross-fertilization valuable for both rhetorical and practical purposes.

In 1885, William M. King, chief of the Seed Division, drew an explicit connection between crop improvement, hybridization, and reformist conceptions of social and scientific progress. In pursuit of the division's purpose, to "promote the interests of all classes, in whatever industrial pursuits . . . by an increased improvement in both quantity and quality of agricultural products," it should strengthen ties with foreign governments that would promote the exchange of plants and seeds.[40] King also had specific ideas concerning the fate of these imports. Hybridization was the chief means to seed improvement, according to King, and he provided a two-page summary of hybridization techniques, citing no less an authority than Charles Darwin in support of the contention that cross-fertilization was generally beneficial and self-fertilization injurious.[41] His discussion placed in the fore the improvement of wheat and creation of new varieties, not surprising given the specter of overproduction and the importance of grain for the export market. King reproduced a letter from A. E. Blount, of the State Agricultural College of Colorado, whose hybridization work in wheat involved "over 300 varieties of seed obtained from almost every wheat-producing country in the world."[42]

In 1886, the report of H. E. Van Deman of the new Division of Pomology directly addressed a crucial aspect of the department's diversification program—namely, the production of fruits for export. The apple and citrus fruits were seen as particularly significant; Van Deman's staff was seeking unusual varieties, both domestic and foreign, of these and other fruits.[43] The following year, as the acquisition of plants and seeds from foreign countries continued, the pomologist emphasized the importance of the "science of breeding" in the creation of artificial hybrids of economic importance.[44]

Hybridization was thought to apply to "all cultivated plants," thus cutting across boundaries traditional in agricultural and horticultural research.[45] This aspect of hybridization was of immense practical and institutional significance. Numerous specialty divisions of the USDA (as well as other agricultural research organizations) could offer support. Furthermore, its universality implied that the technique was based on fundamental natural laws. Hybridization thus validated agriculture as a biological science, a feature of special importance to agricultural scientists working in an academic context. Experiment station researchers and academic reformers ultimately capitalized on both the technical and institutional implications of hybridization in their promotion of experimental breeding techniques.

Within the USDA, the laboratories of the Division of Vegetable Pathology became the locus of hybridization and breeding investigations. Division chief Beverly T. Galloway insisted on the crucial relationship between vegetable pathology and vegetable physiology, pointing to the "urgent necessity of a thorough study of the normal physiology of a plant, as a groundwork for pathological investigations."[46] Scientific practice within this conception of the study of plant diseases required "the aid of many branches of science,"[47] plant breeding among them. Indeed, Galloway argued that the problems of plant breeding were inseparable from work in plant pathology and physiology, because the conditions of development dictated by the inheritance of the organism were as crucial to a successful crop as were the conditions of the environment in which the crop was grown. The goal was to uncover principles that "will enable the grower to not only modify his conditions to suit the plants, but to modify the plants to suit the conditions."[48]

By the close of the century, studies of inheritance in plants and experimental improvement through breeding and selection were secure elements in the research of the division, which had been appropriately renamed the Division of Vegetable Physiology and Pathology. Division researchers had undertaken crossbreeding with grapes, oranges, pineapples, pears, and wheat and were soon to work with cotton, in all cases seeking to combine excellent quality of the fruit or grain with hardiness in the face of climatic severities or with resistance to specific plant diseases. Discussing the work of the division in 1898, A. F. Woods reported more than 20,000 crosses of raisin grapes, 116 crosses of pear varieties, the production and propagation of hundreds of hybrid pineapples, oranges, and other citrus fruits, and the breeding of wheat for disease resistance and yield.[49] Several important plant breeders began their professional careers within this division, including Herbert J. Webber, who would be an important advocate of Mendelism within agricultural institutions.[50]

During the 1880s and 1890s, then, efforts of USDA administrators and scientists to generate new products promoted hybridization and varietal crossing. The cultivation of specialty items to replace some imports and provide new products for export was central to their program of economic reform. Moreover, the use of hybridization, presented as a technique demanding specialized skills, also played an important role in scientific reform, which penetrated not only the USDA but the nation's agricultural colleges and experiment stations.

Research and Reform: The Place of Hybridization

By the 1880s, USDA officials had developed a strong research orientation. The scientific work conducted under their auspices was accompanied by a marked appreciation for the power of science.[51] At the same time, with passage of the Hatch Act ensuring that agricultural experimentation would receive significant public support, USDA administrators were determined to exert more control over the state experiment stations.[52] The scientific research ethos permeating the department inevitably played a role in management of the stations, as shown by the annual reports filed by the secretary of agriculture. Each year the secretary complained that farmers and state legislators who demanded practical help misunderstood the purpose of the experiment stations, whose mission was original investigation. James Wilson's report for 1898 is typical. After noting that experiment stations were not the only means for educating the farmer (who could make use of agricultural colleges, farmers' institutes, and boards of agriculture), he argued:

> It is the business of the experiment station, on the other hand, to advance knowledge of the facts and principles underlying successful agriculture and to teach the farmer new truths made known by their investigations. The act of Congress creating the Stations clearly defines their functions to be the making and publishing of original investigations. Whenever a station has neglected this and merely endeavored to educate the farmer, we find a weak station, and whenever a station has earnestly devoted itself to original investigations, we find a strong station.[53]

The Office of Experiment Stations (OES) was created by the commissioner of agriculture, under the authority of the Hatch Act, to oversee the work of the stations and provide a central clearinghouse for station research.[54] Its commitment to station research increased under Alfred C. True, who became director in 1893.[55] At the same time, a number of botanists and horticulturalists at the agricultural colleges and stations, some with training or research experience at elite American and European institutions, were happy to comply. In fact, they had pioneered experimental work at their home institutions and introduced it in their teaching.[56] Plant breeding, already recognized as an important experimental technique with agricultural applications, assumed its place in the armamentarium of reform. Scientific reformers like True, endeavoring to achieve their goals while maintaining a commitment to practical applications, promoted breeding work as part of the movement to transform the agricultural experiment stations into scientifically oriented research centers.

The leadership of the OES at this time was important because, despite the enthusiasm of like-minded investigators at some institutions, horticultural work at most stations consisted chiefly of variety testing, some plant pathology, and experimentation with culture methods.[57] The horticultural work of many stations was weakly developed, and in 1889 four stations specified that no varietal improvement would be attempted in vegetable crops.[58] But the same economic forces that prompted development of the USDA's diversification and crop improvement program operated in the states with even greater immediacy, and in the years following passage of the Hatch Act, many stations adopted programs of variety improvement, to proceed chiefly by selection techniques. By 1889, twenty-three stations reported active or planned programs of variety improvement, and eight specified that hybridization would be part of these efforts.[59]

True and his staff were not content merely to document a growing interest in hybridization. In 1890, the OES published results of a questionnaire that it had presented to the stations, probing the nature and extent of botanical work by station researchers. Most questions pertained exclusively to applied research. Two questions, however, specified research areas with no explicit mention of practical applications. One concerned plant physiology. The other asked pointedly: "Are you experimenting in cross-fertilization and hybridization in the hope of obtaining better knowledge of the laws that underlie these processes?"[60]

The research areas singled out were unambiguously stamped with OES approval, and the questionnaire therefore served a propagandistic, as well as informational, purpose. The responses from station botanists reveal that they were prepared to be encouraged. Of thirty-eight stations responding, twenty indicated that they were already engaged in cross-fertilization and hybridization work. Seven of the twenty specified that their work in this area was limited, but six indicated that crossbreeding work was expected to be a major aspect, in some cases a specialty, of their research. Three others that had not yet begun such work reported plans to initiate it in the near future.[61]

As important as the actual extent of hybridization work is the evidence of growing interest in the subject among station researchers. Since the 1888 station reports to the OES, the number of institutions undertaking crop improvement through hybridization as well as selection had increased from seven to twenty. The particular economic pressures within the various states ensured the application of the technique to virtually all horticultural and field crops. At the Florida station, investigators hybridized peaches and oranges; at Arkansas, strawberries; at Michigan, wheat; at Massachusetts and Indiana, fruit trees; and at Iowa, corn. Other stations quickly followed suit.[62]

Significantly, this late nineteenth-century work in hybridization and crossbreeding was undertaken by personnel at publicly supported research and teaching institutions, unlike such work earlier in the century, which was pursued chiefly by private individuals or commercial concerns. The public institutions, numerous and geographically dispersed, were charged with education and technical training: researchers, and more importantly students, gained experience in growing, propagating, manipulating, and hybridizing plant materials—skills crucial for research in plant inheritance. The public institutions thus provided a new context for hybridization studies, a formally structured academic context, where workers undertook research beneath a standard proclaiming commitment to science as the basis for practical advancement. Although improved varieties of agricultural crops were undoubtedly the ultimate goal of the USDA or station-sponsored work, the wording of the OES's 1890 questionnaire explicitly presented the aim of such work as "better knowledge of the laws that underlie these processes." At the stations, as at the USDA, the significance and institutional success of crossbreeding lay in its dual implications for practical applications and scientific theory. The technique was particularly appropriate for the agricultural institutions, struggling to combine their economic and social service role with allegiance to the values of academic science.

The rediscovery of Mendel's work in 1900 thus occurred at a time when hybridization was of unprecedented importance in American agricultural research. The scientific reform of agricultural research and education ensured that many breeders would be college professors or college-trained researchers. For these men, breeding work had a scientific goal, the investigation of universal laws of inheritance, as well as a practical one. Station investigators were thus sensitive to problems uniting intellectual and commercial concerns—and hence particularly concerned with the nature of variation—before 1900. Station botanists sought to understand the relationship between certain types of crosses and the appearance of sterility in offspring, a result obviously to be avoided in efforts to produce self-sustaining lines of improved varieties. Several researchers studied the influence of crossbreeding and hybridization on the expression of heritable characteristics in search of regularities in their transmission.[63] Workers with corn at the Illinois station considered commercial features such as size of the ear as well as patterns in the inheritance of kernel color, and then examined the greater or lesser stability (or constancy) of hybrid effects in subsequent generations.[64] In short, at many of the agricultural experiment stations researchers were engaged in the study of variation—its appearance, alteration, and constancy through several generations.

Mendel, Bateson, and American Agriculturists

In this same period, variation also assumed a central place in biological investigations. Darwinian evolutionary theory, whether accepted or rejected, defined the problems of biological research in the second half of the nineteenth century. Chief among the serious objections to Darwin's theory were perceived inadequacies in his treatment of the origin and transmission of variation. For scientists concerned with evolutionary issues, the production of new varieties through cross-fertilization and hybridization represented a valuable experimental method for investigating these problems. William Bateson published his *Materials for the Study of Variation* in 1894, providing scientists with a handbook of "experimental evolution," and Hugo de Vries's concern with establishing a truly experimental study of evolution prompted his search for both hybrid constancy and spontaneous appearances of saltatory variation.[65]

The conjunction of research interests among students of evolutionary theory and practical breeders proved crucial for the rapid success of Mendelism in the United States. The 1899 Conference on Hybridization and Crossbreeding, convened in London by the Royal Horticultural Society, brought these groups together. In attendance were British, American, and continental scientists with wide-ranging theoretical, practical, and commercial interests in botanical hybrids.

The participants included William Bateson, Hugo de Vries, and C. C. Hurst. Also present, by invitation, were American agricultural scientists Herbert J. Webber, David Fairchild, and Walter Swingle of the USDA and Willet M. Hays of the Minnesota experiment station. Liberty Hyde Bailey, also invited but unable to attend, sent a paper.[66] The conference proceedings reveal that theoretical investigators, practical agriculturalists, and commercial breeders had common interests in hybridization.

William Bateson was a masterful presence; his conference paper, focusing on transmission of discrete characters in hybrid crosses between closely related individuals, emphasized the value of experimental hybridization for evolutionary theory. He also discussed the problem of swamping, raised regularly in critiques of Darwinism.[67] De Vries explained his most recent breeding experiments, interpreting the creation of apparently stable hybrid crosses as a possibly crucial mechanism of evolution.[68] Hurst discussed his cross-breeding experiments at length, and elaborated a law of "partial prepotency."[69] While both Bateson and de Vries addressed problems of concern to practical investigators, Hurst was the most thorough and explicit in linking the work of practical breeders and students of evolution. He insisted that the results of crossbreeding experiments seemed to "bear directly upon the problems of inheritance and variation," and pointed to the problem of hybrid constancy as central to breeding practice.[70] He thus provided a powerful justification for pursuing basic research on hybridity as the basis for further practical achievements.

The Americans spoke and wrote almost exclusively of practical advances in plant breeding in the United States. Thus Webber lauded the successes of the USDA's hybridization work in oranges, pineapples, pears, apples, wheat, corn, and cotton.[71] Hays's contribution indicates that, in general, the American focus was on broad characteristics of economic significance, such as vigor, hardiness, and size.[72] But despite some differences in orientation, the American agricultural scientists were obviously excited by the spirit of scientific cooperation, and seemed impressed by the British and European studies. They also made a strong impression on their British audience, some of whom expressed envy of the institutional support available for hybridization work in the United States.[73]

Between the 1899 conference in London and the 1902 conference in New York, Mendel's work had been rediscovered, and Bateson had embarked on a campaign to promote it. Because American agriculturists were already familiar with Bateson's views and because the American agricultural research apparatus was attuned to the relevant issues, Bateson's success at the 1902 conference is understandable. The editorialist for the *Experiment Station Record* crowed that "there was an almost universal acceptance of Mendel's law regarding the appearance of dominant and recessive hybrids."[74] Although not every participant was persuaded either of the generality of Mendel's laws or their practical importance, the enthusiasm was widespread—among seedsmen as well as scientists. We have tried to explain Mendel's appeal to agricultural scientists. But what was his appeal to seedsmen?

Marketing Mendel

The answer is partly that Mendelism offered a plausible explanation for the extreme difficulty in obtaining varieties that would "breed true." Specific results that had long puzzled practical breeders included the "reversions on crossing" discussed by Darwin, the greater variability of new types, and the problem of fixing hybrids. It was doubtless interesting to know why some varieties could apparently not be fixed, despite repeated selection, and why success was so long in coming with others. However, these breeders were also practical, interested in knowledge as a means to power, and power as a means to profit, not as an end in itself. The laws of heredity were sold to breeders as a set of rules for efficient selection, worth "a total of hundreds of millions of dollars' worth of added annual income with but little added expenditure."[75] The laws of dominance and segregation were unabashedly advertised as a means to make money; Spillman could assert that "if Mendel's law is true, it is worth millions of dollars to the breeders of plants in this country."[76]

Among the scientists, only Liberty Hyde Bailey was publicly doubtful. "The wildest prophecies have been made in respect to the application of Mendel's law to the practice of plant breeding," he wrote in 1903.[77] Bailey's caution reflected his doubt about the generality of Mendel's results. But it also reflected his realization that even if Mendel's laws were both true and universal, they would have no immediate dramatic effect on the practice of plant breeding. Before Mendel, breeders selected; after Mendel, they would do the same. They might be moved to keep better records, select individuals, and select from a larger number of plants. The main difference, however, was that they could now provide plausible explanations for their successes and failures. They would learn that the "rogue characters" they each year hoed out resulted from "the fortuitous union of recessive germs."[78] But they could not eliminate these rogues except by the methods they always used. Their traditional maxims had been: "Avoid breeding for antagonistic characters," "Breed for one thing at a time," "Know what you want," "Have a definite ideal," and "Keep the variety up to standard."[79] They now knew that these maxims made sense. But Mendelism did not, could not (indeed, cannot) offer a fixed rule by which to "produce six inch carnations on four foot stems." It could not, at this point, do much for commercial breeders at all. That situation would change dramatically with two linked developments: a Mendelian interpretation of the effects of inbreeding (and crossbreeding) and invention of the double-cross method of breeding—work that made possible the development of hybrid corn.

The Invention of Hybrid Corn

Hybrid corn has been repeatedly characterized as "the greatest success story of genetics."[80] The belief that hybrids are responsible for vast increases in yield has, however, recently been challenged by Jean-Pierre Berlan and Richard Lewontin.[81] In their view, the story of hybrid corn illustrates the success of seed companies, not science. Traditionally, farmers harvested their next year's seed from their own plants. But one cannot use seed obtained from hybrids without suffering substantial declines in yield. Thus farmers must buy their seed anew each year. This feature of hybrids—and not any intrinsic superiority in respect to yield, disease resistance, or other important traits—explains seed companies' huge investment in their development. Conventional comparisons of hybrids with open-pollinated varieties are therefore beside the point. Hybrids, of course, do better. They have been intensively improved for the last sixty years. Had mass selection of open-pollinates been pursued with equal zeal, they should now out-perform hybrids.[82] But seed companies have no incentive to improve a product that anyone can reproduce.

However, hybrid corn was initially developed by scientists, not seedsmen. Why should they have cared if breeders were commercially successful? To answer this question, it is necessary to sketch briefly some developments in maize genetics between 1905 and 1919.

Hybrid corn developed out of the work of three geneticists: George H. Shull, then at the Carnegie Station at Cold Spring Harbor, Edward M. East, then at the Connecticut Agricultural Experiment Station, and East's student, Donald F. Jones, also at Connecticut. East and Shull were particularly concerned with the analysis of quantitative characters. All worked with corn, an ideal subject for such study. Naturally open-pollinated, each kernel may be fertilized with pollen from a different plant. In a single ear of corn, therefore, the researcher can obtain a large and highly variable population. Moreover, that variability is reflected in such easily measurable characteristics as the size, shape, number, and color of different kernels. It was also a crop of great, and steadily increasing, economic importance. Between 1866 and 1900 the total corn acreage tripled while production quadrupled; by the turn of the century, twice as much corn was produced as wheat, the second most valuable crop.

Thus theoretical and practical interests combined, in the early years of American genetics, to focus attention on corn. The careers of East and Shull nicely illustrate this point. East began as a chemist at Illinois, working with C. G. Hopkins to develop strains of corn with high oil and low protein, and low oil and high protein content (to improve its value as livestock feed). In 1905, he began experiments to examine the effects of inbreeding, work he expanded after moving to Connecticut later that year. Shull, on the other hand, initially used corn to test a criticism of de Vries's *Oenothera* studies (that his mutations were artifacts of selfing a species that was naturally cross-fertilizing). This work spurred an interest in testing the effects of selfing and crossing on the expression of a purely quantitative character.[83] Because it was such an easy trait to measure, Shull chose the number of kernel rows in an ear of corn.

Using hand pollination, Shull inbred a number of lines (thereby reducing their variability). As inbreeding progressed, the plants declined in respect to such desirable traits as size and strength of the stalks, number of ears, and resistance to disease; the decline in "vigor" corresponded with the increase in homozygosity and ultimately leveled off.[84] Shull assumed that his inbreeding had resulted in the isolation of "pure lines" or "biotypes" similar to those described by Wilhelm Johannsen in beans. When he then crossed these lines, the offspring were not only superior to their parents in size and general vigor; they sometimes surpassed the original open-pollinated corn plants. (Working independently, East had observed similar effects of inbreeding and crossing, but did not connect them to Johannsen's pure lines.)

East and Shull were hardly the first to note that deterioration often accompanies inbreeding, and an increase in general luxuriance or vigor the crossing of closely related strains. However, the cause of "inbreeding depression" remained obscure, as did its relation to hybrid vigor. The former was generally assumed to result from an accumulation of injurious individual variations, which in turn produced "unbalanced constitutions." Inbreeding was thus viewed as a process of continual degeneration.[85] Shull argued that deterioration did not result from self-fertilization per se. In his view, it was an indirect effect of the isolation of distinct biotypes (or pure lines). Hybrid vigor resulted from their mixture, and was therefore simply the converse of inbreeding depression.

Shull also recognized that it was almost impossible for a corn plant to self-fertilize, given the lightness of the pollen shed by the male flowers and the location of the female flowers halfway down the stem. He therefore concluded that virtually every plant in a field of corn is naturally a hybrid—although one resulting from a "promiscuous" process of fertilization. To exploit fully the benefits of hybrid vigor, he proposed to substitute a process that was completely controlled. His "pure line method of corn breeding" would maintain otherwise useless inbred lines of corn solely for the purpose of utilizing the vigor obtained from their crossing. Shull argued that a policy of simple selection for the best individuals would not be effective, because the value of the resulting strains differed not only in their pure state but also in their various hybrid combinations. Ordinary methods of selection could take only the former into account. However, it was not possible to predict the relative vigor of hybrids simply from the value of the pure lines that produced them. "The object of the corn-breeder" Shull wrote, "should not be to find the best pure-line, but to find and maintain the best hybrid combination."[86] But the technique he suggested produced seed corn too expensive for commercial use.

The ultimate source of hybrid vigor is moreover as obscure in Shull's account as it was in pre-Mendelian works, such as Darwin's.[87] According to Shull, vigor results from crossing because crossing greatly increases the mixture of biotypes in the new hybrid strains. But *why* should mixing biotypes be beneficial? On this point, Shull is silent. In early 1909 however, he read a paper by East, also opposing the view that inbreeding per se was deleterious, in which East did advance an explanation of inbreeding depression and hybrid vigor.[88] More accurately, he proposed two.

Following Davenport, East first considered a simple Mendelian account of the effects of inbreeding—that it uncovers deleterious recessives (whose effects are masked by crossing). However, he believed this hypothesis inadequate for it failed to account for both developmental and genetic effects. East assumed that sexual reproduction has two functions: to recombine hereditary characters and to stimulate development. He suggested that this beneficial stimulation would be increased by the crossing of "two strains differing in gametic structure."[89] (That fertilization serves to "rejuvenate the egg" as well as create new genetic combinations was then a common belief.[90]) Hybrid vigor is thus largely attributable to "the physiological stimulation of heterozygosis," a phrase that Shull shortened to "heterosis" in 1914.[91]

From this standpoint, hybrid vigor results from the beneficial effects of hybridity per se. "In other words," Shull wrote, "hybridity itself,—the union of unlike elements, the state of being heterozygous,—has, according to my view, a stimulating effect upon the physiological activities of the organism."[92] As a corollary, it was not possible to produce pure lines as productive as hybrids; and because the quality of hybrids deteriorates after the first generation, the only way to obtain maximum yields was to return each year to the original combination.[93] Farmers could not effectively reuse their seed. "Like the mule, CROSSED CORN has the advantage of hybrid vigor," asserted an early advertisement; like the mule, it is also (effectively) sterile. As Shull approvingly noted: "When the farmer wants to duplicate the splendid results he has had one year with hybrid corn, his only recourse is to return to the same hybridizer from whom he secured his seed the previous year and obtain again the same hybrid combination."[94]

If vigor is a function of the degree of heterozygosis, hybrids are grounded in an unfortunate (from the farmers' perspective) fact of nature. However, the view that heterozygotes were inherently superior had become a minority position long before the commercial development of hybrids. From the beginning, it had a rival—a truly Mendelian interpretation of inbreeding. Even in the nineteenth century, a few breeders held that in crosses parents usually possess different defects that tend to cancel out in their progeny.[95] After 1900, this insight was easily rephrased in Mendelian terms—that is, in the course of selection for various traits, breeders create strains that are homozygous for deleterious genes elsewhere in the genome. A hybrid formed between two inbred strains would acquire normal, dominant alleles at most of these loci.

The dominance interpretation of hybrid vigor was not immediately compelling because it predicted results that did not completely accord with observation (such as a skew distribution in the second generation). These anomalies disappeared, however, when linkage or the involvement of at least twenty genes was assumed. Moreover, the concept of physiological stimulation arising in some unknown way from heterozygosity was both distressingly vague and unsupported by any evidence.[96] Even East was later to admit that it was "an assumption for which there was no proof, and which was not illuminating as a dynamic interpretation."[97]

By 1919, when East and his student Jones published their influential book *Inbreeding and Outbreeding*, the "heterosis concept" was already in retreat.[98] (Jones invented the double-cross method of breeding, which made the production of hybrid seed commercially viable.[99]) They themselves concluded that the dominance interpretation of hybrid vigor was correct—hence, that pure lines were in theory more desirable than hybrids.[100]

In the 1920s and 1930s, the dominance explanation was generally accepted. In the 1940s, it would be challenged by Fred Hull, who argued that intrinsic heterozygote superiority (which he termed "overdominance") provided a partial explanation of hybrid vigor in corn.[101] This view was popularized by Jay Lush, author of the leading text on animal breeding.[102] Thus the East/Shull thesis of vigor resulting from the physiological stimulation produced by unlike gametes would eventually reappear in new, Mendelian garb. By then, however, virtually the entire corn belt had been planted in hybrids.[103] If the dominance explanation was widely accepted in the 1930s, how were hybrids justified? In part, as an expedient.

East and Jones believed that mass selection of open-pollinated varieties would ultimately produce lines as good or better than hybrids. In *Inbreeding and Outbreeding*, the success of the dominance interpretation of hybrid vigor is characterized as a "happy result." Why? Because the physiological stimulation hypothesis "locked the door on any hope of originating pure strains having as much vigor as first generation hybrids."[104] Hybrids produced a uniform field and (under some circumstances) a rapid boost in yield.[105] But selection could ultimately do the same, and more. For East and Jones understood that dominance was rarely complete. "Perfect dominance, except in more or less superficial characters, rarely occurs, and even when it does occur, it may be merely an appearance rather than a reality," they wrote. The consensus was "that there is no such thing as perfect dominance, that the heterozygote merely approaches the condition of one or the other parent more closely."[106]

The conclusion is obvious. If dominant genes do not completely mask the effects of deleterious recessives, selection should eventually produce pure lines superior to hybrids. In their words: "if dominance is but partial, this [homozygous] individual, through the very fact of its homozygous condition, will be even more vigorous than those of the first hybrid generation."[107] Thus, they predicted that hybrids would ultimately be replaced by pure lines.[108] Yet when they considered the mechanics of plant and animal improvement, the only method described was hybridization, and hybrids, in fact, remained the only approach to improvement of corn. By the 1930s, mass selection of open-pollinated varieties was no longer discussed.

East and Jones also provide a clue to the reason why. After explaining that Jones's double-cross method might appear complex, they write:

> It is not a method that will interest most farmers, but it is something that may easily be taken up by seedsmen; in fact, it is the first time in agricultural history that a seedsman is enabled to gain the full benefit from a desirable origination of his own or something that he has purchased. The man who originates devices to open our boxes of shoe polish or to autograph our camera negatives, is able to patent his product and gain the full reward for his inventiveness. The man who originates a new plant which may be of incalculable benefit to the whole country gets nothing—not even fame—for his pains, as the plants can be propagated by anyone. There is correspondingly less incentive for the production of improved types. The utilization of first generation hybrids enables the originator to keep the parental types and give out only the crossed seeds, which are less valuable for continued propagation.[109]

East and Jones believed that commercial breeders would not have sufficient incentive to improve plants until they could prevent farmers from using their own crops to propagate the next generation. Hybrids in effect conferred the equivalent of a patent right on new varieties of seed. East and Jones knew that, in theory, hybrids were not the only, or even best route, to improvement of corn. They themselves identified the alternative. But that method, as a breeder associated with Jones wrote, would probably "spoil the prospects of any one thinking of producing the seed commercially."[110] Without the commercial incentive provided by hybrids, they believed that corn would not be improved. With the breeders, East and Jones also thought it fundamentally unjust that the creators of new plants and animals should fail to profit by their inventions.[111]

They thus faced a dilemma. East and Jones held a scientific theory according to which pure lines should produce maximum increases in yield. But they also held a social theory (reflecting the facts of their actual world), according to which improvement of corn required an incentive for commercial producers that only hybrids could offer. Their prediction that pure lines would one day replace hybrids was therefore naive. Who would improve the open-pollinated varieties? The answer might seem obvious: state universities and their affiliated experiment stations. After all, a commitment to the welfare of the farmer and the general public was at the heart of experiment station ideology. However, public institutions proved unable to resist the clamor for hybrids, on the part of both farmers (excited by seed company advertising and focused on short-term gains) and large seed producers (whose representatives dominated university crop advisory committees).[112] Thus the interests of those who aimed to make seed a commodity ultimately prevailed over those who opposed it. But that is another story.[113]

Conclusion

The development of hybrid corn was no simple matter of the transfer of theoretical science from an elite academic to an applied commercial context. Much of the theoretical work that made hybrids possible was pursued at institutions concerned with improving the efficiency and productivity of agriculture. In the 1880s, agricultural administrators began to promote hybridization as part of an effort to produce commercially viable new varieties. Ongoing interest in hybridization, in turn, underlay an enthusiastic response to Mendelism among researchers at the USDA and at agricultural colleges and experiment stations. However, agricultural leaders and researchers were also committed to basic research and scientific reform. The agricultural disciplines and institutions were themselves flourishing hybrids.

Mendelism between 1900 and 1910 was thus an applied science, in the literal sense of both; it was surely applied, and it was certainly science. The rapid development of genetics within an agricultural context, where breeding, selection techniques, hybridization, and even evolutionary issues had been addressed in the late nineteenth century, endowed Mendelism in the United States with a strongly practical and popular aspect. It also ensured that fundamental problems in genetics would be addressed within institutions oriented to practical ends—and that the subsequent development of genetic research would often reflect dominant social and economic interests in American agriculture.

The case study of hybrid corn illustrates these points. George Shull was not trained at an agricultural institution nor did he ever work at one; but his early experience of Mendelism brought him close to horticultural and agricultural researchers and to involvement with practical issues. Edward East was trained and did his early work at agricultural colleges and experiment stations before moving to Harvard's Bussey Institution, and Donald Jones worked at the Connecticut station throughout his career. That their research ultimately provided an important breakthrough for commercial agriculture is comprehensible within the complex constraints and constituencies of these scientists' sponsoring institutions.

Numerous centennial and celebratory volumes document the achievements of the USDA, state experiment stations, and agricultural colleges. However, these institutions have generally been given short shrift by historians of biology. Their attention has focused on the elite colleges and laboratories dedicated to the ideal of pure research.

From the perspective of researchers at Johns Hopkins, Columbia, or Woods Hole, agricultural colleges and experiment stations were doubtless at the periphery of the new biology. However, we see no reason to privilege their standpoint. It certainly does not accord with the self-conception of those employed at agricultural institutions. They were generally proud of their mandate: to serve the public interest. As we have seen, that goal did not exclude a commitment to basic research. On the contrary, much fundamental work in biology, especially genetics, emerged from these institutions. For this reason alone, they deserve greater attention from historians of biology. Of further interest is the fact that their research agendas reflected, in a particularly blunt way, political and economic interests. To consider these institutions is thus to broaden our understanding both of American culture and American science.

Acknowledgments

We would like to thank several individuals. Barbara Kimmelman gratefully acknowledges the support of her doctoral advisor, Robert Kohler. Diane Paul is indebted to Richard Lewontin and to two other visitors in his population genetics laboratory during 1986–87. Michel Veuille persisted in asking questions about the American reception of Mendelism that inspired this research. The case study that concludes this essay draws heavily on work, some of it unpublished, by Jean-Pierre Berlan and Lewontin. We are also grateful to Deborah Fitzgerald for sharing her research on the development of hybrid corn in Illinois and for many helpful comments on successive drafts of this essay.

Notes

1. Willet M. Hays, "Address by Chairman of Organization Committee," *Proceedings of the American Breeders Association,* 1905, *1:* 9–15.
2. William Bateson, "Hybridisation and Cross-breeding as a Method of Scientific Investigation," *Journal of the Royal Horticultural Society,* 1900, *24:* 59–66.

3. Charles B. Davenport, "Report of the Committee on Eugenics," *The American Breeders Magazine,* 1910, *1:* 126–129. See also Barbara A. Kimmelman, "The American Breeders Association: Genetics and Eugenics in an Agricultural Context," *Social Studies of Science,* 1983, *13:* 163–204.

4. Deborah K. Fitzgerald, "The Business of Breeding: Public and Private Development of Hybrid Corn in Illinois, 1890–1940" (Ph.D. dissertation, University of Pennsylvania, 1985).

5. During this period, investigators generally used the terms hybridization, crossbreeding, and cross-fertilization to mean, respectively, interspecific crosses, intervarietal crosses, and intraspecific crosses between male and female of different plants. Darwin based his crucial analogy between interspecific hybrids and intervarietal mongrels on such distinctions. In practice, the analogy proved more powerful than the distinction, and many investigators used the terms interchangeably. We will see that, after 1910, hybrids came to acquire a more precise, technical meaning.

6. Walter A. Cannon, "Review of Proceedings: International Conference on Plant Breeding and Hybridization," *Torreya,* 1905, *1:* 12–14.

7. C. W. Ward, "Improvement of Carnations," *Memoirs of the Horticultural Society of New York,* 1904, *1:* 151–155.

8. William Bateson, "Practical Aspects of the New Discoveries in Heredity," *Mems. Hort. Soc. N.Y.,* 1904, *1:* 1–8.

9. William Bateson, *Mendel's Principles of Heredity: A Defence* (Cambridge: Cambridge University Press; New York: Macmillan Co.; 1902). Alan Cock notes that the book was published on 3 or 4 June 1902 in Britain and must have appeared almost simultaneously in the United States (private communication).

10. Liberty Hyde Bailey, "Comments," *Mems. Hort. Soc. N.Y.,* 1904, *1:* 8.

11. William Bateson to Beatrice Bateson, 3 October 1902, William Bateson papers, University of Cambridge Library, letter G-3D-05; typed transcript at G8G01D. We gratefully acknowledge the Library's permission to quote from Bateson's letter. The originals of the collection prefaced G have recently been transferred to the Library from the John Innes Institute, which now has photocopies. Mrs. Rosemary Harvey, archivist at the John Innes, kindly supplied copies of Bateson's 1902 letters from America.

12. Garland Allen, *Life Science in the Twentieth Century* (New York: John Wiley and Sons, 1975), p. 52.

13. Ibid.

14. "Seedsmen," or "practical breeders," were primarily farmers who also bred particular varieties of seed/grain for purpose of sale. A few of these farmers (such as the Funks) engaged in breeding for sale as a major enterprise but for most it was a minor activity. Some seedsmen (mostly urban horticulturalists) were businessmen *simpliciter.*

15. Barbara A. Kimmelman examines the founding and early years of research departments in genetics within the agricultural colleges at Ithaca, N.Y., Madison, Wis., and Berkeley, Calif., in "A Progressive Era Discipline: Genetics at American Agricultural Colleges and Experiment Stations, 1890–1920" (Ph.D. dissertation, University of Pennsylvania, 1987).

16. Bateson, "Practical Aspects," p. 3.

17. C. C. Hurst, "Notes on Mendel's Methods of Cross Breeding," *Mems. Hort.*

Soc. N.Y., 1904, *1:* 11–16; Hugo de Vries, "On Artificial Atavism," *Mems. Hort. Soc. N.Y.*, 1904, *1:* 16–23.

18. This point is illustrated by the diverse character of articles on Mendelism published in American journals between 1901 and 1903. The first to appear was Charles Davenport's "Mendel's Law of Dichotomy," in the *Biological Bulletin,* 1901, *2:* 307–310. It was quickly followed by E. B. Wilson, "Mendel's Principles of Heredity and the Maturation of the Germ-cells," *Science,* 1902, *16:* 991–992; Walter Sutton, "On the Morphology of the Chromosome Group in Brachystola magna, *Biol. Bull.*, 1902, *3*: 24–39; W. J. Spillman, "Exceptions to Mendel's Law," *Sci.*, 1902, 16: 709–710 and 784–796; R. A. Emerson, "Preliminary Account of Variation in Bean Hybrids," *15th Annual Report of the Nebraska Experiment Station,* 1902; and Walter A. Cannon, "A Cytological Basis for the Mendelian Cases," *Bulletin of the Torrey Botanical Club,* 1902. Other early accounts include Liberty Hyde Bailey, "A Discussion of Mendel's Law and its Bearings," Address before the Society for Plant Morphology and Physiology, Washington, D.C., 29 Dec. 1902, published as "Some Recent Ideas on the Evolution of Plants," *Sci.*, 1903, *17:* 441–454; Walter Sutton, "The Chromosomes in Heredity," *Biol. Bull.*, 1902, *4:* 231–251; and William Castle, "The Laws of Heredity of Galton and Mendel and some Laws Governing Race Improvement by Selection," *Proceedings of the American Academy of Arts and Sciences,* 1903, *38:* 535–548; reprinted as "Mendel's Law of Heredity," in *Sci.*, 1903, *18:* 396–406. Compare with the lack of interest expressed by the *Botanical Gazette.* The first mention of Mendel is a dismissive comment by the editor, John Merle Coulter, in a review of the third edition of Liberty Hyde Bailey's *Plant Breeding* (*Botanical Gazette,* 1904, *37:* 471–472). The *American Naturalist* was also unimpressed. Other than a passing reference in a Botanical Note of 1902, there is no mention of Mendelism until 1904, and then only in Charles Davenport's book reviews. Editorial notes and articles first appear in 1907.

19. Jan Sapp, "The Struggle for Authority in the Field of Heredity, 1900–1932: New Perspectives on the Rise of Genetics," *Journal of the History of Biology,* 1983, *16:* 311–342.

20. Willet M. Hays, "Address by Chairman of Organization Committee," p. 10.

21. W. J. Spillman, "Mendel's Law in Relation to Animal Breeding," *Proc. ABA,* 1905, *1:* 171–176; H. J. Webber, "Explanation of Mendel's Law of Hybrids," *Proc. ABA,* 1905, *1:* 138–143.

22. William Bateson, "G. Mendel: Experiments in Plant Hybridisation," *J. Royal Hort. Soc.*, 1901, *26:* 1–32.

23. U.S. Department of Agriculture, Office of Experiment Stations, Abstract of G. Mendel, "Experiments in Plant Hybridization," *Experiment Station Record, 1901–1902,* vol. 13 (Washington, D.C.: Government Printing Office, 1902), p. 744.

24. Ibid., p. 745.

25. Ibid., p. 1004.

26. See, for example, U.S. Department of Agriculture, Office of Experiment Stations, Abstracts of W. F. R. Weldon, "Mendel's Law of Alternative Inheritance in Peas," C. C. Hurst, "Mendel's Law Applied to Orchid Hybrids," and Carl Correns, "Apparent Exceptions to Mendel's Law of Dissociation in Hybrids," *Experiment Station Record, 1902–1903,* vol. 14 (Washington, D.C.: Government Printing Office, 1903), pp. 466–467, 569. Also see U.S. Department of Agriculture, Office of Ex-

periment Stations, Abstracts of A. D. Darbishire, "On the Bearing of Mendelian Principles of Heredity on Current Theories of the Origin of Species," Erich von Tschermak, "Further Studies in Crossing Peas, Stocks, and Beans," and David Starr Jordan, "Some Experiments of Luther Burbank," *Experiment Station Record, 1904–1905*, vol. 16 (Washington, D.C.: Government Printing Office, 1905), pp. 232, 263, 773–774. Jordan's study indicated that Burbank's results did not generally conform to Mendel's laws.

27. Bailey, "A Discussion of Mendel's Law," pp. 445–446.

28. Margaret Rossiter, "The Organization of the Agricultural Sciences," in Alexandra Oleson and John Voss, eds., *The Organization of Knowledge in Modern America, 1860–1920* (Baltimore: Johns Hopkins University Press, 1979), pp. 211–248.

29. Our interpretation of the Progressive Era accords with the work of Sidney Fine, *Laissez Faire and the General Welfare State: A Study of Conflict in American Thought 1865–1901* (Ann Arbor: University of Michigan Press, 1956); Samuel P. Hays, *The Response to Industrialism 1885–1914* (Chicago: University of Chicago Press, 1947); Harold U. Faulkner, *Politics, Reform, and Expansion 1890–1900* (New York: Harper and Row, 1959); Robert H. Wiebe, *The Search for Order 1877–1920* (New York: Hill and Wang, 1967); and James Weinstein, *The Corporate Ideal in the Liberal State 1890–1918* (Boston: Beacon Press, 1968). See also Samuel P. Hays, "Introduction: The New Organizational Society," in Jerry Israel, ed., *Building the Organizational Society* (New York: The Free Press, 1972), pp. 1–15.

30. U.S. Department of Agriculture, *Report of the Commissioner of Agriculture for the year 1885*, "Report of the Statistician," by J. R. Dodge (Washington, D.C.: Government Printing Office, 1885), p. 372.

31. For example, see U.S. Department of Agriculture, *Report of the Commissioner of Agriculture for the year 1880*, "Report of the Statistician," by Charles Worthington (Washington, D.C.: Government Printing Office, 1880), p. 210; Dodge, "Report of the Statistician," pp. 372–376; and U.S. Department of Agriculture, *First Report of the Secretary of Agriculture, 1889*, "Report of the Secretary of Agriculture" by J. M. Rusk (Washington, D.C.: Government Printing Office, 1889), p. 14.

32. Dodge, "Report of the Statistician," pp. 379–380.

33. Ibid., p. 373.

34. A valuable discussion of the growing importance of U.S. international markets (although, like its subjects, it disparages the significance of the agricultural sector) is Walter LeFeber, *The New Empire: An Interpretation of American Expansion 1860–1898* (Ithaca: Cornell University Press, 1967). Works addressing this theme in agriculture include Fred A. Shannon, *The Farmer's Last Frontier: Agriculture, 1860–1897* (New York: Farrar and Rinehart, 1945); Alan I. Marcus, *Agricultural Science and the Quest for Legitimacy: Farmers, Agricultural Colleges, and Experiment Stations, 1870–1890* (Ames: Iowa State University Press, 1985); and David Danbom, *The Resisted Revolution: Urban America and the Industrialization of Agriculture, 1900–1930* (Ames: Iowa State University Press, 1979), although it deals with a slightly later period.

35. U.S. Department of Agriculture, *Report of the Commissioner of Agriculture for the year 1887*, "Report of the Commissioner of Agriculture," by Norman J. Colman (Washington, D.C.: Government Printing Office, 1888), p. 8.

36. U.S. Department of Agriculture, *Report of the Secretary of Agriculture for*

the year 1888, "Report of the Statistician," by J. R. Dodge (Washington, D.C.: Government Printing Office, 1889), p. 201.

37. Ibid., pp. 201–202. See also U.S. Department of Agriculture, *Report of the Secretary of Agriculture for the year 1891,* "Report of the Statistician," by J. R. Dodge (Washington, D.C.: Government Printing Office, 1892), pp. 301–306.

38. U.S. Department of Agriculture, *Report of the Secretary of Agriculture for the year 1892,* "Report of the Secretary of Agriculture," by J. M. Rusk (Washington, D.C.: Government Printing Office, 1893), p. 10; also U.S. Department of Agriculture, *Report of the Secretary of Agriculture for the year 1893,* "Report of the Secretary of Agriculture," by J. Sterling Morton (Washington, D.C.: Government Printing Office), p. 16.

39. The agricultural appropriation act of 1881 statutorily established a divisional organization for the USDA, at which time the divisions of Seed, Gardens and Grounds and of Botany were founded. The Division of Pomology was established in 1886; the Division of Vegetable Pathology achieved independence from the Division of Botany in 1891, and became the Division of Vegetable Physiology and Pathology in 1895; in that year, the Division of Agrostology was also founded. The Section Foreign Seed and Plant Introduction was established in 1897. See U.S. Department of Agriculture, Division of Publications, *Historical Sketch of the U.S. Department of Agriculture: Its Objects and Present Organization,* by Charles H. Greathouse, Bulletin no. 3 (Washington, D.C.: Government Printing Office, 1898), pp. 33–40; Fred Wilbur Powell, *The Bureau of Plant Industry, Its History, Activities and Organizations,* Institute for Government Research, Service Monographs of the United States Government no. 47 (Baltimore: Johns Hopkins University Press, 1927), pp. 1–9; and David Fairchild, *The World Was My Garden, Travels of a Plant Explorer* (New York: Charles Scribner's Sons, 1939), pp. 105–107.

40. U.S. Department of Agriculture, *Report of the Commissioner of Agriculture for the year 1885,* "Report of the Chief of the Seed Division," by William M. King (Washington, D.C.: Government Printing Office, 1885), p. 47.

41. Ibid., p. 51.

42. Ibid., p. 53.

43. U.S. Department of Agriculture, *Report of the Commissioner of Agriculture for the year 1886,* "Report of the Pomologist," by H. E. Van Deman (Washington, D.C.: Government Printing Office, 1886), p. 260.

44. U.S. Department of Agriculture, *Report of the Commissioner of Agriculture for the year 1887,* "Report of the Pomologist," by H. E. Van Deman (Washington, D.C.: Government Printing Office, 1888), pp. 627–628.

45. U.S. Department of Agriculture, *Report of the Secretary of Agriculture for the year 1892,* "Report of the Superintendent of Gardens and Grounds," by William Saunders (Washington, D.C.: Government Printing Office, 1893).

46. U.S. Department of Agriculture, *Report of the Secretary of Agriculture for the year 1892,* "Report of the Chief of the Division of Vegetable Pathology," by B. T. Galloway (Washington, D.C.: Government Printing Office, 1893), p. 246.

47. B. T. Galloway, "Division of Vegetable Physiology and Pathology," *Yearbook U.S.D.A. 1897* (Washington, D.C.: Government Printing Office, 1898), p. 104.

48. Ibid., p. 106.

49. Albert F. Woods, "Work in Vegetable Physiology and Pathology," *Yearbook*

U.S.D.A. 1898 (Washington, D.C.: Government Printing Office, 1899), pp. 264–266. See also U.S. Department of Agriculture, *Report of the Secretary of Agriculture for the year 1899,* "Report of the Secretary of Agriculture," by James Wilson (Washington, D.C.: Government Printing Office, 1899), p. 15.

50. As illustrative of the division's work during this period and of the intellectual problems addressed by its researchers, see Walter T. Swingle and Herbert J. Webber, "Hybrids and Their Utilization in Plant Breeding," *Yearbook U.S.D.A. 1897,* pp. 383–421.

51. For example, see U.S. Department of Agriculture, *Report of the Secretary of Agriculture for the year 1889,* "Report of the Botanist," by George Vasey (Washington, D.C.: Government Printing Office, 1889), pp. 377–381; and U.S. Department of Agriculture, *Report of the Secretary of Agriculture for the year 1890,* "Special Report of the Assistant Secretary: the Scientific Work of the Department in its Relations to Practical Agriculture," by Edwin Willits (Washington, D.C.: Government Printing Office, 1890), pp. 59–73.

52. Rossiter, "The Organization of the Agricultural Sciences," pp. 213–215.

53. U.S. Department of Agriculture, *Report of the Secretary of Agriculture for the year 1898,* "Report of the Secretary of Agriculture," by James Wilson (Washington, D.C.: Government Printing Office 1898), p. 47.

54. U.S. Department of Agriculture, Office of Experiment Stations, *Organization of the Agricultural Experiment Stations of the United States,* Norman J. Colman, Bulletin no. 1 (Washington, D.C.: Government Printing Office, 1884).

55. See Charles E. Rosenberg, "The Adams Act: Politics and the Cause of Scientific Research," in Charles E. Rosenberg, *No Other Gods: On Science and American Social Thought* (Baltimore: Johns Hopkins University Press, 1976), pp. 174–175.

56. This group included William James Beal, Charles E. Bessey, and Liberty Hyde Bailey. On their efforts, see Andrew Denny Rodgers, *John Merle Coulter, Missionary in Science* (Princeton: Princeton University Press, 1944), esp. pp. 52–64, and idem, *Liberty Hyde Bailey, A Story of American Plant Sciences* (Princeton: Princeton University Press, 1949), esp. pp. 22–181, 242–247. See also Richard Overfield, "Charles E. Bessey: The Impact of the 'New Botany' in American Agriculture, 1880–1910," *Technology and Culture,* 1975, *16:* 162–181.

57. See U.S. Department of Agriculture, Office of Experiment Stations, *Digest of the Annual Reports of the Agricultural Experiment Stations in the United States for 1888,* Bulletin no. 2, pt. 1 (Washington, D.C.: Government Printing Office, 1884). For a discussion of constraints on scientific work at the stations during this period, see Rosenberg, "Science, Technology, and Economic Growth: The Case of the Agricultural Experiment Station Scientist, 1875–1914," in Rosenberg, *No Other Gods,* pp. 153–172.

58. The stations were South Dakota, Maryland, Massachusetts, and Mississippi. See U.S. Department of Agriculture, Office of Experiment Stations, *List of Horticulturalists of the Agricultural Experiment Stations in the United States, with an Outline of Work in Horticulture at the Several Stations,* by W. B. Alwood, Bulletin no. 4 (Washington, D.C.: Government Printing Office, 1889).

59. Ibid. The stations were Colorado, Florida, Kansas, Michigan, New York (Geneva), Missouri, New Jersey, and Ohio.

60. U.S. Department of Agriculture, Office of Experiment Stations, *List of Bota-*

nists of the Agricultural Experiment Stations in the United States, with an Outline of the Work in Botany at the Several Stations, Bulletin no. 6 (Washington, D.C.: Government Printing Office, 1890), p. 6.

61. Ibid., pp. 7–23, for responses from the stations in alphabetical order by state. The states with strong programs planned in crossbreeding were Illinois, Indiana, Kansas, Michigan, New Jersey, and Ohio.

62. See H. J. Webber and E. A. Bessey, "The Progress of Plant Breeding in the United States," *Yearbook U.S.D.A. 1899* (Washington, D.C.: Government Printing Office, 1899), esp. pp. 478, 480–481, 487–488.

63. Office of Experiment Stations, *List of Botanists,* p. 6.

64. The Illinois and Kansas stations' corn work is summarized in the U.S. Department of Agriculture, Office of Experiment Stations, *Handbook of Experiment Station Work: A Popular Digest of the Agricultural Experiment Stations in the United States,* Bulletin no. 15 (Washington, D.C.: Government Printing Office, 1893), pp. 81–89 (entry under "Corn, crossing").

65. William Bateson, *Materials for the Study of Variation* (London: 1894); the culmination of de Vries' late-nineteenth-century work was his *Die Mutationstheorie,* vol. 1 (Leipzig: 1901) and vol. 2 (Leipzig: 1903).

66. Wilfred Mark Webb, "The International Conference on Hybridisation and Cross-breeding," *Nature,* 1899, *60:* 305–307. Others invited but not present (in addition to Bailey) were David Fairchild of the USDA, Luther Burbank, and J. M. MacFarlane of the University of Pennsylvania.

67. William Bateson, "Hybridisation and Cross-breeding as a Method of Scientific Investigation."

68. Hugo de Vries, "Hybridisation as a Means of Pangenic Infection," *J. Royal Hort. Soc.,* 1900, *24:* 69–75.

69. C. C. Hurst, "Experiments in Hybridisation and Cross-breeding," *J. Royal. Hort. Soc.,* 1900, *24:* 90–127. "Prepotency" suggests a sire's ability to impress his traits on his progeny. It generally referred to an overall "type," rather than single characters.

70. Ibid., p. 90.

71. Herbert J. Webber, "Work of the United States Department of Agriculture in Plant Hybridisation," *J. Royal Hort. Soc.,* 1900, *24:* 128–145; see also L. H. Bailey, "Progress of Hybridisation in the United States of America," *J. Royal Hort. Soc.,* 1900, *24:* 209–213.

72. Willet M. Hays, "Breeding Staple Food Plants," *J. Royal. Hort. Soc.,* 1900, *24:* 257–265.

73. See Webb, "International Conference on Hybridisation and Cross-breeding," p. 307.

74. Office of Experiment Stations, *Experiment Station Record* (1902), p. 205.

75. W. M. Hays, "Address by Chairman of Organization Committee."

76. W. J. Spillman, *Mems. N.Y. Hort. Soc.,* 1904, *1:* 155.

77. Bailey, "A Discussion of Mendel's Law," p. 450.

78. Bateson, "Practical Aspects," p. 2.

79. Bailey, "A Discussion of Mendel's Law," p. 450.

80. L. C. Dunn, *A Short History of Genetics* (New York: McGraw-Hill, 1965), p. 125. The following list provides a few representative comments: Paul Mangelsdorf

writes that the increased yields resulting from hybrids "contributed not only to this country's war effort but also the rehabilitation of Europe after the war," in "George Harrison Shull," *Genetics,* 1955, *60:* 2–3. L. J. Stadler claims that "It is . . . no exaggeration to say, speaking in terms of the over-all national economy, that the dividend on our research investment in hybrid corn, during the war years alone, was enough to pay the money cost of the development of the atomic bomb," quoted in G. Shull, "Hybrid Seed Corn," *Sci.,* 1946, *103:* 547–550. Bentley Glass states that the increase in yield was "enough to pay for the Manhattan Project" and even "most significantly, increased production [which] permitted the United States to ship vast quantities of food abroad after the war, thus preventing famine and pestilence," in "Shull, George Harrison," *Dictionary of American Biography,* suppl. 5 (New York: Scribners, 1977), p. 629. Richard Crabb writes that hybrid corn helped "tip the scales in our favor in a war to the finish," in Richard Crabb, *The Hybrid-Corn Makers: Prophets of Plenty* (New Brunswick, N.J.: Rutgers University Press, 1948), p. 13.

81. Jean-Pierre Berlan and Richard Lewontin, "The Political Economy of Hybrid Corn," *Monthly Review,* 1986, *38:* 35–47; and Berlan and Lewontin, "Breeders' Rights and Patenting Life Forms," *Nat.,* 1986, *322:* 785–788, esp. pp. 787–788.

82. They should do better, rather than equally well, because of the ubiquity of partial dominance (discussed below).

83. G. H. Shull, "Beginnings of the Heterosis Concept," in John Gowan, ed., *Heterosis* (Ames: Iowa State College Press, 1952), pp. 14–48.

84. Shull, "A Pure-line Method of Corn Breeding," *Proc. ABA,* 1909, *5:* 51–59.

85. Edward M. East and Donald F. Jones, *Inbreeding and Outbreeding: Their Genetic and Sociological Significance* (Philadelphia: J. B. Lippincott, 1919), p. 168.

86. Shull, "A Pure-line Method of Corn Breeding," p. 52.

87. Charles Darwin, *The Effects of Cross and Self-fertilization in the Vegetable Kingdom* (London: John Murray, 1876).

88. E. M. East, "The Distinction between Development and Heredity in Inbreeding," *Am. Nat.,* 1909, *43:* 173–181.

89. Ibid., p. 177.

90. John Farley, *Gametes and Spores* (Baltimore: Johns Hopkins University Press, 1982), pp. 203–208.

91. G. H. Shull, "Duplicate Genes for Capsule Form in *Bursa bursa-pastoris,*" *Zeitschrift für Induktive Abstammungs- und Vererbungslehre,* 1914, *12:* 97–149, on p. 127.

92. Ibid., p. 126; also quoted in G. H. Shull, "What is Heterosis?" *Genetics,* 1948, *33:* 439–446, on p. 440.

93. G. H. Shull, "The Composition of a Field of Maize," *Proc. ABA,* 1908, *4:* 296–301, on p. 301.

94. Shull, "Hybrid Seed Corn," p. 549.

95. Conway Zirkle, "Early Ideas of Inbreeding and Crossbreeding," in Gowen, *Heterosis,* pp. 1–13.

96. See Frederick D. Richey, "Hybrid Vigor and Corn Breeding," *Journal of the American Society of Agronomists,* 1946, *38:* 833–841, and G. F. Sprague, "Heterosis in Maize: Theory and Practice," in R. Frankel, ed., *Heterosis: Reappraisal of Theory and Practice,* (New York: Springer-Verlag, 1982), p. 50.

97. E. M. East, "Heterosis," *Genet.,* 1936, *21:* 375–397, on p. 375.

98. E. M. East and D. F. Jones, *Inbreeding and Outbreeding*.

99. The inbred plants used to produce hybrids were so depressed that they generated little (hence expensive) seed. In the double-cross, the plants that produce the seed are themselves hybrids.

100. "Homozygosity, when obtained with the combination of all the most favorable characters, is the most effective condition for the purpose of growth and reproduction." East and Jones, *Inbreeding and Outbreeding,* p. 187.

101. Fred Hull, "Recurrent Selection for Specific Combining Ability in Corn," *J. Am. Soc. Agron.*, 1945, *37:* 134–145.

102. J. L. Lush, *Animal Breeding Plans,* 3rd ed. (Ames, Iowa: The Collegiate Press, 1945). Lush's influence is noted in James F. Crow, "Muller, Dobzhansky, and Overdominance," *J. Hist. Biol.*, 1987, *20:* 351–380.

103. By 1945, 99.9 percent of corn acreage in Iowa and 98.1 percent in Illinois and Indiana had been planted in hybrids. USDA, *Agricultural Statistics, 1947* (Washington, D.C.: Government Printing Office, 1949), table 48, p. 43.

104. East and Jones, *Inbreeding and Outbreeding,* p. 182.

105. Such a response occurs when there are large numbers of *already existing* inbred lines, maximizing the chance of finding a favorable cross.

106. East and Jones, *Inbreeding and Outbreeding,* pp. 177–178.

107. Ibid., p. 182.

108. Ibid., pp. 169, 181. Jones characterizes hybrids as a "makeshift measure" in "Selection in Self-Fertilized Lines as the Basis of Corn Improvement," *J. Am. Soc. Agron.*, 1920, *12:* 77–100, on p. 95.

109. East and Jones, *Inbreeding and Outbreeding,* p. 224.

110. George S. Carter to Henry A. Wallace, 12 May 1925, quoted in Jack Kloppenburg, *First the Seed: The Political Economy of Plant Biotechnology* (Cambridge: Cambridge University Press, 1988). The Wallace papers are deposited at the University of Iowa Library and are available on microfilm.

111. Compare Jones, "Selection in Self-fertilized Lines," p. 87 and Willet Hays, "Distributing Valuable New Varieties and Breeds," *Proc. ABA*, 1905, *1:* 58–65.

112. See Fitzgerald, "The Business of Breeding," esp. chap. 5 and conclusion.

113. See Kloppenburg, *First the Seed,* and Fitzgerald, "The Business of Breeding."

Scott F. Gilbert

10 Cellular Politics: Ernest Everett Just, Richard B. Goldschmidt, and the Attempt to Reconcile Embryology and Genetics

Reflecting on embryology in the 1930s, Johannes Holtfreter stated:

> We managed more or less successfully to keep our work undisturbed by humanity's strife and struggle around us and proceeded to study the plants and animals, and particularly, the secrets of amphibian development. Here, at least, in the realm of undespoiled Nature, everything seemed peaceful and in perfect order. It was from our growing intimacy with the inner harmony, the meaningfulness, the integration, and the interdependence of the structures and functions as we observed them in dumb creatures that we derived our own philosophy of life. It has served us well in this continuously troublesome world.[1]

The attempts to reintegrate embryology and genetics during the last years of the 1930s represent the last chapter in the emergence of American biology. When had American biology finished "emerging"? I suspect that stage was reached when it had successfully resisted the last attempts to reintegrate it into European-dominated traditions of inquiry. For genetics, this occurred in the late 1930s when Richard B. Goldschmidt and Ernest Everett Just separately countered the American school of genetics with European alternatives. Goldschmidt and Just both attempted to place genetics into a physiological framework. Goldschmidt was the director of the genetics section of the Kaiser Wilhelm Institute before fleeing the Nazis and coming to America in 1936. For Goldschmidt, the "static genetics" of T. H. Morgan, centered on individual particulate genes, was to be replaced by "physiological genetics" wherein the gene did not exist as an individual unit, and its activity, not its location, was the focus of research.

E. E. Just was a black American embryologist who had left America in 1931 to work in Europe. His emphasis on the importance of cytoplasmic factors in heredity was well within the European framework of Carl F. Correns, Fritz von Wettstein, and Alfred Kühn. For both Goldschmidt and Just, the relationship between nucleus and cytoplasm became a key issue. For American geneticists, Morgan had used the nuclear envelope as a conceptual and disciplinary barrier. Geneticists study the transmission of genetic traits within the nucleus, embryologists study the expression of those traits in the cytoplasm. This division was to allow each discipline to proceed separately. Because European geneticists did not recognize that boundary, the separation of genetics from embryology did not occur in Europe, where the dominant perspective of biology came from physiology. Just and Goldschmidt, two "American" biologists with European affinities, each tried to return American genetics to the physiological traditions. As Just pointed out in 1936, only when the genes are placed "within the domain of physiology" could "genetics become a branch of biology."

A House Divided

Thomas Hunt Morgan was an embryologist who inadvertently founded the gene theory in 1911.[2] While the Mendelian geneticists had been analyzing the segregation of characters from one generation to another, Morgan investigated whether changes in the nuclear composition of an organism affected its development. He began asking this question of ctenophore eggs and sea urchin embryos, and the results convinced him that it was the cytoplasm that controlled development and inheritance. Through 1910, Morgan remained the major American critic of the Sutton-Boveri synthesis of Mendelism and cytology. Only when his *Drosophila* studies demonstrated that factors for eye color, body color, wing shape, and sex all segregated with the X-chromosome did Morgan reluctantly propose the physical linkage of these genetic traits.

The years 1911 to 1915 saw the emergence of a new discipline—genetics. Although genetics would eventually come to influence all areas of biological study, the first to feel its effects was its parent discipline, embryology, for the experimental embryology pioneered by Roux and Weismann in the 1880s saw the problems of inheritance and development as the same. Even as late as 1910, embryologist Morgan stated, "We have come to look upon the problem of heredity as identical to the problem of development."[3] However, in the years following 1910, Morgan drove a wedge into embryology, splitting it into two divisions comprising the embryologists and the new geneticists.

Geneticists could not develop their own discipline without constructing a research program separate from that of the rest of the embryologists. To this end, Morgan employed Wilhelm Johannsen's distinction between genotype and phenotype. Johannsen had argued that heredity should only be considered as the transmission of genetic traits from one generation to another. The emergence of the phenotype was of secondary importance and belonged to the realm of embryology. Sapp and Allen have shown that Johannsen's distinctions allowed Morgan to shift his attention from the cytoplasmic realm of the phenotype to that of the nuclear genotype.[4]

In the mature formulation of his genetics, *The Theory of the Gene,* Morgan stated that much unwarranted criticism of genetics had come "from confusing the problems of genetics with those of development." This separation was extremely important for the emergence of genetics as a new discipline, and he argued that "the sorting out of characters in successive generations can be explained at present without reference to the way in which the gene affects the developmental process." Thus, Morgan had separated the transmission of hereditary traits (genetics) from the expression of those traits (embryology).[5]

Yet Morgan himself never completely abandoned his primary devotion to embryology, and he returned to active embryological research after he left Columbia University. Already one year after *The Theory of the Gene* was printed, Morgan published *Experimental Embryology.* These two excellent textbooks demonstrated Morgan's continued knowledge and expertise in both areas. Therefore, when his *Embryology and Genetics* appeared in 1934, those scientists who desired the resynthesis of the two disciplines had high hopes. Yet, although this volume provided a survey of both genetics and embryology, it did not attempt to integrate them. Boris Ephrussi, who was later to play a major role in such reunifying efforts, recalled his own response.

> I said I found the book very interesting, but I thought that the title was misleading because he did not try to bridge the gap between embryology and genetics as he had promised in the title. Morgan looked at me with a smile and said, 'You think the title is misleading! What is the title?' 'Embryology and Genetics,' I said. 'Well,' he asked, 'is not there some embryology and some genetics?' This shows how polarized I was on the gap between embryology and genetics, and how anxiously I was waiting for somebody to bridge it.[6]

Many biologists wished to reconcile the two fields. The small community of mechanistic biologists felt embarrassed by the widening gap between two of its most successful disciplines. Speaking of the separate courses taken by genetics and developmental physiology, F. R. Lillie had remarked that: "There can be no doubt, I think, that the majority of geneticists, and many [developmental] physiologists certainly, hope for and expect a reunion. The spectacle of the biological sciences divided permanently into two camps is evidently for them too serious a one to be regarded with satisfaction."[7]

However, after mentioning various attempts to reconcile the two fields, Lillie pessimistically concluded that developmental physiology and genetics must remain separate. "Those who desire to make genetics the basis of physiology of development will have to explain how an unchanging complex can direct the course of an ordered developmental stream. . . . The dilemma at which we have arrived appears to be irresolvable at present."

Like many embryologists, Lillie had good reason for being suspicious of genetics as an explanation for development. First, as the quotation illustrates, the chromosomal repertoire was believed by geneticists to be constant in every cell. Yet development was defined by cellular change. "It would, therefore, appear to be self-contradictory to attempt to explain embryonic segregation by behavior of genes which are *ex-hype* the same in every cell." Second, heredity was perceived to be controlled by nuclear chromosomes. Development on the other hand was manifest in the cytoplasm. As Lillie stated, "The germ exhibits the duality of nucleus and cytoplasm; the geneticist has taken the former for his field, the embryologist the latter."[8]

Differentiation was seen as caused by intercellular relationships (a cell becoming a different structure when placed in a different part of an embryo), and such relationships were "mediated through the cytoplasm, not through the nucleus." Lillie thought that the genes constituted the basis of a physiological reaction system to the environment but were not responsible for specifying particular developmental characteristics.

These were the points that Morgan failed to address in his 1934 book. Indeed, the synthesis had not progressed very far in the seven years since Lillie's essay. Morgan noted that "the interlocking of these two has become a subject of absorbing interest" and that his book would attempt to "point out in a simple way their interrelations." However, Morgan's goals were actually more superficial. He was not so much interested in discussing "interrelations" as he was "points of contact." The latter was Morgan's own metaphor and aptly described his views.[9] Throughout this book, Morgan portrayed the two disciplines as exclusive spheres touching at a common point. This "common meeting place of embryology and genetics was found in the relationship between the hereditary units in the chromosomes, the genes, and the protoplasm of the cell where the influence of the genes comes to visible expression." This relationship was expressed simply: "The initial differences in the protoplasmic regions may be supposed to affect the activity of the genes. The genes will then in turn affect the protoplasm, which will start a new series of reciprocal reactions. In this way, we can picture to ourselves the gradual elaboration and differentiation of the various regions of the embryo."[10]

This framework of nucleocytoplasmic interaction was not new but was an updating of Hans Driesch's classic embryological statement from 1894.[11] Furthermore, Morgan could offer no direct experimental evidence in favor of this hypothesis, and he returned to explanations used at the turn of the century by E. B. Wilson to explain the nuclear role in programming the cytoplasm. Thus,

Morgan's analysis of sinistral coiling of the snail's shell—the only developmental mutant discussed in his book—is almost identical to Wilson's analysis of molluscan development in 1894 and 1904. Some material is postulated to pass from the nucleus to become active in the cytoplasm. In the case of snail coiling, the gene is active in the production of the oocyte cytoplasm itself.

Although Morgan's book is generally a review of the two divergent disciplines, he did present a new and very liberal interpretation of the gene. The two fundamental properties of such a unit, he wrote, were its power to grow and divide and its power to cause changes in the chemical and physical nature of the cytoplasm. The evidence for the former rested on the cytological demonstration of chromosomal replication, whereas the evidence for the latter, admittedly circumstantial, was that changes in hereditary characteristics can be traced back to particular loci on particular chromosomes. He did not hold to the view of "genic balance"—that is, that all the genes are active in every cell and that the phenotype is the summed product of all the individual influences. In that view, a mutation or environmental assault would perturb this equilibrium in one way or another. This concept had the advantage of fitting into current models of general physiology and homeostasis, and many embryologists (such as Lillie) tended to assume it. Morgan, however, thought that this view was "quite inadequate to explain the sequence of changes through which the embryo passes." Morgan also did not believe that "different batteries of genes come into action as development proceeds," for this was inconsistent with Driesch's data wherein nuclei given different positions in the embryo directed the differentiation in accord with their new locations. So Morgan did not propose any mechanism for differentiation that would work according to his model of nucleocytoplasmic interaction. However, the last page of this book suggests that the nuclear genes may not be the unchangeable entities that geneticists had (and until very recently still have) assumed. "It is, however, conceivable that the genes also are building up more and more, or are changing in some way, as development proceeds in response to that part of the protoplasm in which they come to lie, and that these changes have a reciprocal influence on the protoplasm." [12] Morgan's refusal to integrate genetics and embryology and his extremely flexible, even epigenetic, view of the gene opened the way for others to attempt the synthesis.

The Rival Professions

Although embryologists in the 1920s generally hoped that genetics would return to the fold, by the mid 1930s many embryologists were reacting against the new genetics. The geneticists had become too successful. Jan Sapp, and Diane Paul and Barbara Kimmelman (this volume) show that with the interest in applied breeding techniques (in both animal husbandry, plant breeding, and eugenics), genetics quickly asserted itself as the premier biological science in America. The years from 1915 to 1932 were characterized by "the establishment of university chairs of genetics; by the founding of an academic journal, *Genetics* . . . and by the emergence of a purely academic genetics society, quite separate from the American Genetics Association. The Genetics Society of America was founded in 1932." [13]

The geneticists believed they had mastered the mechanisms of chromosome transmission and were now looking in the direction of gene expression.[14] Expression, of course, is epigenetic development and had been left to the embryologists to unravel. But the embryologists had their own set of problems and interests, and entering the nuclear realm to identify the products of these so-called genes was not one of them. The organizer experiments from Hans Spemann's laboratory, the intercellular gradient theories of C. M. Child, the intracellular gradients discovered by J. Runnstrom and Sven Hörstadius, the limb development fields of R. G. Harrison, and the pluripotency of neural crest cells shown by Benjamin Willier and Mary Rawles were fascinating phenomena worthy of any embryologist's attention. Thus, a few geneticists started to venture into the realm of gene expression. Most notably, Ephrussi and G. W. Beadle began their analysis of the genetic control of the development of eye pigment.[15]

In 1936, two years after Morgan's book, a joint session of the American Society of Zoologists, the American Society of Naturalists, and the Genetic Society of America was conducted on "Genetics and Development." Like Morgan's book before, the discussions had some genetics and some development, but there was little "crossover." The discussants were E. E. Just, E. W. Sinnott, G. W. Beadle, and V. C. Twitty. Just began his lecture by limiting his discussion to the embryological events of fertilization and cleavage as if those were the only embryonic stages where such a discussion was possible. "In discussing the phenomena of the process of animal embryogenesis . . . from which we may attempt to derive a theory of development and heredity, I must obviously limit myself to those changes that take place before the embryo is delineated." [16] He relegated Mendelian nuclear characters to secondary status as those finishing touches occurring after the cytoplasm had built the basis of the embryo, and he argued that neither nucleus nor cytoplasm, alone, is a functioning biological entity. Sinnott noted that it was impossible to know *how* a gene controlled something until one had learned *what* a gene controlled. His lecture, on the genetic control of shape in gourds and melons, concentrated on the problem of allometric growth. Beadle reviewed the inheritance of eye color in *Drosophila,* showing that diffusible substances involved in pigment production are deficient in certain mutants. Last, Twitty summarized the genetics of pigment pattern in salamanders. Here, interspecific hybrids and grafts showed that both the egg and sperm influence pigment development and distribution.

Not only was there no agreement among the speakers, but the embryologists were hostile to the notion that genetics and embryology might be two approaches to study the same phenomenon. Harrison, in his review of the session, noted that "the embryologist . . . is more concerned with the larger changes in the whole organism and its primitive systems of organs than with the lesser qualities associated with gene action." Just put it more succinctly, stating that he was more interested in how the embryo made a back than in the formation of the bristles on the back and more interested in the developmental construction of the eye than in the synthesis of eye pigments.[17]

Immediately after this session, Harrison, as outgoing chairman of the section on zoological sciences of the American Association for the Advancement of Science, presented a lecture on "Embryology and Its Relations." It is obvious that he did not think that the realms of embryology and genetics were coextensive, and he wanted to keep the geneticists on their own turf.

> Now that the necessity of relating the data of genetics to embryology is generally recognized and the "Wanderlust" of geneticists is beginning to urge them in our direction, it may not be inappropriate to point out a danger in this threatened invasion.
>
> The prestige of success enjoyed by the gene theory might easily become a hindrance to the understanding of development by directing our attention solely to the genom, whereas cell movements, differentiation, and in fact all developmental processes are actually effected by the cytoplasm.[18]

Harrison was not alone in his fears for embryology. N. J. Berrill, one of the founders of the Growth Symposium that later became the Developmental Biology Society, recently characterized the geneticists of the 1930s as "marauding intruders."[19] He likened their behavior to that of a corporation that aggressively subsumed other companies. "The geneticists," he said, "felt that they had all the answers, and all my life, they've been pushing." Berrill should know, because he had to defend embryology at McGill University against the encroachments by geneticist C. L. Huskins. Huskins wanted to unite the zoology, botany, and genetics departments with genetics on top.

This fight was synecdochical for the larger battle, for it was largely one of methodological orthodoxy versus methodological pluralism. The geneticists claimed that development could be approached as an epiphenomenon of genetic control and therefore that genetics could best obtain the answers to developmental questions. In fact, all biology was seen as epiphenomenal of the genetic processes, so it was natural for them to assert that genetics should be primary. For embryologists, however, developmental biology was a collection of problems. Genetics was only one approach of many. When one looks at the roster of speakers at the first Growth Society meetings in 1939, one is impressed by the different perspectives represented.[20] Warren Lewis discussed tissue culture techniques, J. Needham spoke on the biochemistry of the organizer, Oscar Schotte's talk dealt with regeneration, and E. W. Sinnott presented material on plant morphogenesis. Papers by P. W. Gregory and Otto Glaser concerned growth and size relationships, while papers by Curt Stern and C. H. Waddington separately discussed the role of genes in development. There was even a philosopher, J. H. Woodger, to close the session.

The embryologists celebrated this heterogeneity. Berrill was delighted that "representatives of the field[s] of agriculture, bacteriology, biochemistry, biophysics, botany, cytology, embryology, endocrinology, genetics, histology, mathematics, pathology, philosophy, physiology, and zoology concentrated on a single issue, and considerable correlation and conceptual integration was accomplished."[21] The single issue was, of course, development, and the listing of subjects was pluralistically alphabetical.

The American embryologists saw embryology and genetics as two intersecting spheres, one representing embryology, gene expression, phenotype, and the cytoplasm, and the other representing genetics, gene transmission, genotype, and the nucleus. However, the geneticists (who had originally established those boundaries) were beginning to see the spheres as enclosing the same domains, and genetics and embryology as simply two approaches to the same subject. Moreover, they believed that the genetic approach was superior to any other. As long as the geneticists confined their activities to those within the nucleus, the embryologists felt secure. It was only when the geneticists sought to cross the nuclear envelope into the cytoplasm that the embryologists became worried.

In Europe, the boundary of the nuclear membrane had never been formalized. The German school of genetics, never greatly interested in the gene localization program of Morgan's school, had been focusing since the early 1920s on the physiology of gene expression. However, as Harwood has shown, the German geneticists were split between those who accorded the nuclear genome absolute authority over cellular functions (a theory called *Kernmonopol* by its detractors) and those scientists who saw the cytoplasm as having an equal, if complementary, role in directing development. The advocates of *Kernmonopol* referred to the supremacy (*Überlegenheit*) of the nucleus and its dominating role (*dominierende Rolle*) in development. As one such advocate warned, if the cytoplasm contained hereditary determinants, "then the gene would be dethroned (*beherrschenden Platz entthront*) from its position of controlling development and evolution and would be forced to assume a secondary role."[22]

During the 1920s and 1930s, however, many geneticists, such as Correns, von Wettstein, and Kühn, criticized this notion, claiming that the structure of the cytoplasm carried genetic potentials as well. In an analogy that would be extended by both von Wettstein and Just, the geneticist H. Nachtsheim noted that "the plasma is the building material (*Baumaterial*) for the chromosomes" and that the type of chromosome could be influenced by the type of cytoplasm contained in the cell.[23] Although the center stage of genetics and development was occupied by English-speaking scientists, the German controversies remained very important, for both Just and Goldschmidt grounded their respective theories in this soil. As in America, there appeared to be little discussion in Germany between the embryologists and the geneticists.[24] However, in Europe, the embryologists were paramount, and the geneticists were the ones struggling to make inroads.

In America, genetics had been remarkably successful and was starting to enter where the Europeans had long been theorizing, into the realm of gene expression. Here, the boundaries had been firmly demarcated. "The cytoplasm may be ignored genetically," Morgan had declared in 1926; but just as Morgan had chased his embryological problems into the nucleus, so the geneticists were chasing their problems right back into the cytoplasm.[25] In so doing, the geneticists laid claim to embryology.

In any union or reunion of disciplines, the problem of professional hierarchy becomes acute. If the geneticists were content to study the stable nuclear genotype and the embryologists were satisfied to study the emergence of the changing cytoplasmic phenotype, all would be well. Each field could develop (or evolve) on its own. However, if the subject matter of embryology and genetics was actually the same (as in a resynthesis of the two fields), who was best suited to study such a united field, the geneticists or the embryologists? The relationship between the nucleus and the cytoplasm became critical in these discussions because of an implicit analogy: *Genetics is to embryology as the nucleus is to the cytoplasm.* If the nucleus were seen to control the cytoplasmic phenotype, then the geneticists would have the right to guide the field. Conversely, if the potentials for development were cytoplasmically located, the nucleus (and the geneticists) would play a subservient role. Given the boldly assertive nature of the newly organized geneticists, the embryologists, not surprisingly, tried to show that the fields were not coextensive, whereas the geneticists pushed for a synthesis. The implicit professional analogy between nucleus and cytoplasm should be remembered whenever such syntheses are being proposed, for the nucleus and cytoplasm became code words for genetics and embryology, respectively.

By 1938, genetics and embryology remained separate disciplines. While the geneticists were formulating a genetic approach to development, embryologists persisted in ignoring new ideas in genetics. Thus, Spemann's enormously influential book *Embryonic Development and Induction* (1938) completely ignored all of genetics. His only acknowledgment of the significance of the nucleus was his notion that nuclear transplantation experiments might show whether Weismann's view of nuclear determination was correct.[26]

Similarly, Paul Weiss's important embryology textbook *Principles of Development* relegated genetics to a single footnote saying that Curt Stern had seen "striking correlations between chromosomal aberrations and the morphology of the mutant cells [which] indicate a nuclear foundation of differentiation potencies."[27]

The remainder of this essay will study two syntheses of embryology and genetics that were published in 1938 and 1940. These attempts at reunion, published respectively by geneticist Richard Benedict Goldschmidt (1878–1958) and embryologist Ernest Everett Just (1883–1941) were also attempts to reunite (or subjugate) American genetics into the matrix of German biology. They represented the two poles of continental thinking on genetics and embryology. In rejecting this reintegration, American biology demonstrated its independence from Europe.

The Outsiders: R. B. Goldschmidt and E. E. Just

In America in 1938, the breach between embryology and genetics had not been healed, and not everybody sought such a reunion. Yet in the next two years, four major volumes attempted a reunion of the two disciplines. Two books, authored by E. E. Just and Richard B. Goldschmidt, were considered as mature statements of their respective authors. Goldschmidt's *Physiological Genetics* sought to subsume development into a large framework of genetics. Indeed, development was seen as the epiphenomenon of activities directed by nuclear genes. Just's *The Biology of the Cell Surface* belittled the role of the genes, giving them minor roles to play in the essentially cytoplasmic process of development.

In 1938 both Just and Goldschmidt were in similar positions. Both were cultured, sophisticated, and arrogantly proud men who had been exiled from their homelands and from their disciplines. Goldschmidt was a German citizen of Jewish descent working in America because of the genocidal policies of the Third Reich. Just, conversely, was a black American who felt forced to work in Europe because of racial discrimination in American universities.

By 1938, Goldschmidt had already alienated himself from the majority of geneticists with a series of increasingly serious breaks with the genetic "orthodoxy" of the Morgan School. First, Goldschmidt had disagreed with the simple chromosomal genetics of sex determination espoused by Morgan, Bridges, and Sturtevant, preferring instead the physiological approach of the German school that he had helped lead. Reflecting on his lectures at Woods Hole in 1915, Goldschmidt wrote that only Jacques Loeb, a physiologist, "understood the significance of my work in trying to bring dynamic viewpoints into genetics."[28]

Second, Goldschmidt denigrated the basis of the grand synthesis of genetics and evolution into neo-Darwinism. (He later claimed that he was a neo-Darwinian before neo-Darwinism and saw its flaws as the others were adopting it.) Crucial to this synthesis was the belief that the gradual accumulation of small mutations led to distinct species. There was no qualitative difference in the genetic mechanisms that produce races, species, or higher taxa. However, by the mid-1930s, Goldschmidt was claiming that microevolution (the evolutionary changes within species) and macroevolution (the origin of divergent species and higher taxa) were caused by different mechanisms. The synthesis of Mendelism and Darwinism could explain microevolution, but the current genetic theories could not explain the creation of new species. He now suggested that species differences might arise either by "macromutations" involving chromosome structure or by regulatory mutations early in development—processes very different from the structural gene mutations then known by geneticists. To make matters more difficult for himself, he grouped natural selection and special creation together as "extreme suppositions," neither of which contains the whole truth.[29]

Third, Goldschmidt disagreed with the very corpuscular nature of the gene itself. He interpreted the recent papers of J. Schultz, B. Glass, and N. P. Dubinin as indicating that the presence of a portion of the chromosome did not determine whether or not it is optimally active. Rather its position within the chromosomal complex determined its activity. This position-effect phenomenon could not be explained by classical genetics and allowed Goldschmidt to postulate "a theory of the germ plasm in which the individual genes as separate units will no longer exist."[30] Moreover, whereas most of his colleagues were interested in the transmission of hereditary factors, Goldschmidt concentrated on their expression.

Goldschmidt had indeed "struck a hornets' nest" in the United States. The Neo-Darwinians' counterattack succeeded in burying his work for nearly fifty years. Gould has reported on the depth of this neglect, stating that Goldschmidt "suffered the worst fate of all: to be ridiculed and unread."[31] Elsewhere, Gould has compared Goldschmidt to Orwell's Goldstein, the object of the daily two-minute hate sessions in *1984*.[32] Certainly by 1938, Goldschmidt was an outsider both to his field and to his country.

Goldschmidt was, in one important sense, much more fortunate than Just. Neither of them could aspire to the heights of the career they had entered. Goldschmidt made most of his reputation while working as Richard Hertwig's assistant, and he could never hope to get a tenured position in a university even though he essentially ran Hertwig's laboratory and taught Hertwig's courses. "A number of times my name had been in the running, but in every case academic anti-Semitism had decided against me. Thus I longed to get away from the university into a pure research position. But there was none in Germany, and I had to resign myself."[33] Goldschmidt was not wrong in his assessment, as T. J. Horder and P. J. Weindling have shown.[34] In 1914

Theodor Boveri, the premier cytologist in Europe, wrote to his former student, Hans Spemann, telling him he disliked Goldschmidt's face and he did not want yet another Jew to become a director of a Kaiser-Wilhelm Institute. However, when the Kaiser-Wilhelm Institute for Biology was founded in Berlin, the president of the Kaiser-Wilhelm Society, theologian Adolf Harnack, made it clear that religion would not play a role in the selection of the department directors. Thus, Goldschmidt escaped the Prussian university system to become a division director of the Kaiser-Wilhelm Institute for Biology in 1913.

Just's appointment to Howard University was not an escape from the throes of American racial antagonisms. As Kenneth Manning has shown in his excellent biography of Just, Howard University abused their star biologist.[35] Just was graduated from Dartmouth College with honors in both biology and English. Howard was the only place in the country where a black scientist had a chance to rise to any position of responsibility and power, and Just reached this "height" as soon as he left college. After that, the administration of Howard withheld funds that Just had raised, gave him enormous teaching assignments, and made certain that he could not go anywhere else. By working at Woods Hole and University of Chicago, Just received his doctorate under Frank Lillie; but still, there was no other place in America where Just could pursue his work. Earlier, Just had gone to Dohrn's Stazione Zoologica in Naples, and in 1930 he decided to leave America permanently. His work on fertilization and parthenogenesis in marine organisms was appreciated more in Europe than in America, and he was unable to get funds to continue his work in his native land. Unfortunately, Just's timing could not have been worse, for Mussolini was in the process of nationalizing the scientific enterprises in Italy, and Just was interned briefly as an enemy alien. *The Biology of the Cell Surface,* the culmination of Just's biological theories, was written in Paris, away from both homes, America and Italy. Just's work, like that of Goldschmidt's, met with polite neglect. Although his 1931 paper in *Naturwissenschaften*[36] had provided the first evidence for functional changes in the cell surface during development, it was all but ignored, and even when cited (as in Heilbrun's books), it was not discussed. When research on the cell surface began again after World War II, Just's work was quickly forgotten.[37]

Manning's biography of Just chronicles the hardships that a competent black American scientist met in securing employment, respect, and funds during the first half of this century. Yet one of the best expressions of the outsider's education comes from Goldschmidt, himself, and probably holds equally true for the sensitive, self-confident black American scientist.

> I think, actually, that nobody has a better chance to see the ugly side of human nature than an intellectual Jew who has succeeded in life. Thinking of the innumerable instances when I was stabbed in the back by those who

> breathed deference in my face, fellows intellectually and morally below me, I am surprised that I am not a pessimist. No doubt, this thorn in my flesh has had an immense influence in shaping my character. It has made me cautious and remote, unwilling to show the warmth of my nature unless I know the other man thoroughly. It has taught me to look through people and to analyze them. It has forced me to learn to control my temper, to appear quiet on the surface when I am burning, to appear distant when I long for friendship, to develop self-observation and self-control to a perfection—all of which is frequently misinterpreted as coolness or haughtiness. It has also produced an unnecessarily deep contempt for the second-rater, the go-getter, the social peacock, and clubman. And it certainly killed a number of qualities that I otherwise would have developed, qualities I consciously forced into the background because I knew that their development would expose me to slights and hatred that my soul was not sufficiently robust to drop off lightly. What is the use of aspiring to leadership of men if every half-wit, scoundrel, or Philistine can knock you out with the single word 'Jew'?[38]

Just was never fully accepted by the American embryological community (and, as Pauly shows in this volume, at Woods Hole, it was indeed a community), and he burned his bridges to America when, in 1930, he walked out of the Marine Biological Laboratory's (MBL) tribute to Lillie, saying that he had known more collegiality during one year in Europe than during all his time at the MBL.[39]

But Just was an outsider to embryology for other reasons as well. His contemporaries did not think his research was important, and experimental embryology had left Just behind. The heated debates between Loeb and Lillie over fertilization had all but been forgotten in the 1930s. Textbooks paid no account to what had been a dramatic struggle twenty years before. There was enormous excitement in embryology, and this excitement had moved from fertilization to embryonic determination. This was the era of the great transplantations. By reciprocal transplantation Spemann had demonstrated the importance of gastrulation in determining embryonic cell fate, and he and Hilde Mangold capped those experiments by showing the existence of the "primary embryonic organizer." Hörstadius recombined different tiers of sea urchin blastomeres to discover gradients of preformed substances that informed those cells how to develop, and Harrison transplanted salamander limbs in different positions to discover the laws by which organisms retained their polarity. Niu, Twitty, and Willier (the last mentioned having begun his graduate work under Lillie in the same year Just received his doctorate) discovered the ways in which neural crest cells migrate and differentiate, and Weiss investigated the ways in which neurons migrate to their target tissue. A new research program had been established, one that Joseph Needham has christened "*Gestaltungsgesetze,* the rules of morphological order."[40]

Whereas neither Lillie nor Loeb was working on fertilization after 1920, Just continued studying these same problems of fertilization and was still using the older, less invasive techniques. His work and his methods, though still scientifically valid, had "gone out of fashion." Why was this the case?

Just's retention of old methods and problems has many explanations. One explanation focuses on his professional responsibilities.[41] Other investigators at Woods Hole often had two research interests, one that they pursued during the summer (when the embryos abounded at the MBL) and another that they pursued during the school year while landlocked and cold. Due to his teaching responsibilities, Just could pursue research only in the summer and was denied the chances to work on other lines of inquiry. Similarly, other investigators had a regular influx of graduate students, who could stimulate new research ideas and keep their advisors up to date. Just had neither collaborators nor a flow of new graduate students to spur him into newer areas. Another explanation is that Just's research plan was still a viable part of European, if not American, biology. Starting in 1930, many of Just's papers were sent to German journals, such as *Protoplasma,* that were more sympathetic to the role of cytoplasmic factors in development.

There are other explanations, too. Just was a perfectionist who did not like to leave a problem unsolved. The importance of the cortical cytoplasm during fertilization had been suggested by Lillie and documented by Just. Just did not want to leave the field until he had established incontrovertible proof that this was the case. Also, Just felt a respect and sense of loyalty to Lillie. He had cast himself as Lillie's "bulldog" against Loeb, and he was fighting his mentor's cause. There may still be another reason. As we shall see, Just "identified" himself with the cortical cytoplasm. Most biologists conceived of the cytoplasm as being dominated by the nucleus and merely responsive to its demands. The cortical cytoplasm was ignored by almost every biologist, some of whom thought that it was not even a living part of the cell. It was indeed ignored by embryologists and spurned by geneticists. To Just, the cortical cytoplasm was the most exciting part of the cell, guiding all intercellular communication, controlling cell functions, regulating early development, and serving as the vanguard of animal evolution. The nervous system of humanity, he claimed, was derived from the cortical cytoplasm. As I will try to show hereafter, Just was to fight the benign neglect of the cell periphery. Just identified with the cause of the cortical cytoplasm and in fighting its cause was fighting his own. Just's use of the cell as a model of society will become explicit in this essay.

Thus, these two syntheses of embryology and genetics, published in the late 1930s, were both written at a time when their respective authors were living in exile from their homeland and scientific communities. Yet two books having the same synthetic goal could hardly be more different. There is but one reference (Lillie's 1927 paper wherein he despairs of synthesis) in common between the two books. Just's book concentrates on those embryological phenomena manifested by cytoplasmic changes during the early development of marine invertebrates, whereas Goldschmidt's volume focuses on those later stages of insect development that he can prove to be under genetic control. As Just had said earlier, he was himself more interested in the back than the bristles on the back. Goldschmidt, however, delighted in bristles and wing hairs, and felt that they were just as much a product of development as any other structure.[42]

The Biology of the Cell Surface and *Physiological Genetics* are intensely personal books, each reflecting the idiosyncratic positions of its respective author.[43] Just and Goldschmidt used their books to promote a view of development at odds with those of their peers, and in doing so, each man quoted nearly his entire scientific corpus. Both represent attempts of two highly original scientists to integrate what they considered to be the important data on the awesome problem of how an organism is constructed from a fertilized egg. Both seek to place American genetics into the German type of developmental physiology. I propose to discuss these two books on three levels: first, as straight scientific texts (as their respective authors no doubt intended them to be read); second, as professional texts involving partisan claim-staking by two rival professional groups both seeking to study the same phenomenon; and third, as political texts, inasmuch as there is an implicit metaphor between the proper nuclear regulation during development and the proper action of a central governing body in a society. The relationship of the nucleus to the cytoplasm established in the scientific text becomes the way in which the relationship of genetics to embryology is seen in the professional text and the relationship that a central government bears to its people in the political text.

The Cellular Federalism of E. E. Just

The Biology of the Cell Surface was an attack on the mechanistic and reductionist view of development promulgated by the geneticists on the one hand and the biochemists on the other. Just's work attempted to accomplish two tasks thought to be mutually exclusive. First, it sought to counter the genetic mechanistic view with a cellular holism. To this end, Just redefined the scientific vocabulary used to describe developmental phenomena and elevated the cytoplasm at the expense of the nucleus. Second, it tried to integrate genetics and embryology, as both the nucleus and the cytoplasm played necessary roles in cell differentiation. In this synthesis, Just posited that all the potentials for development were present in the cytoplasm, and gave the nucleus a necessary, but secondary function.

Just began his book with a seventy-four page defense of cellular holism against the reductionist research programs of the geneticists and biochemists of the 1930s. His view of the cell as the irreducible unit of life was expressed on the title page in a verse from Goethe containing a remarkably apt pun.

> Natur hat weder Kern
> Noch Schale,
> Alles ist sie mit einemmale.[44]

The noun *Kern,* translated here as kernel (inasmuch as Goethe's analogy is that of a fruit), also possesses the cytological meaning of cell nucleus. This unity of the cell was to be the hallmark of Just's book. Therefore, he attacked the radical separation of nucleus from cytoplasm and the supposed hegemony of the nucleus over the other regions of the cell.

> In general, the organization of living matter, that is, of protoplasm, appears as consisting of two components, a nuclear and a cytoplasmic. Although most often these are set off as two distinct regions, as a sphere (nucleus) within a sphere (cytoplasm), this sharp differentiation is not invariable. For several reasons, as will be shown beyond, much of modern biological investigation has centered upon the nuclear component as though it were indeed the kernel of life. Not only has the cytoplasmic component been relatively neglected but also have those protoplasmic systems which lack sharply defined and set-off nuclei received scant attention. . . . Because of the rapid rise of genetics, hegemony in the protoplasmic organization has been ascribed to the chromosomal structure of the nucleus and the cytoplasm has been subordinated as though it be a mere protective and nutritive shell. It is no part of the purpose of this book to minimize the achievements of genetics and the investigations on chromosome-structure, all outgrowths of descriptive studies on protoplasmic organization. Instead, inasmuch as life, as we know it so far, resides in the whole system, the pages which follow aim to show how far life-processes are related to the dual and reciprocal components, nuclear and cytoplasmic structure.[45]

Just's critique of genetics was similar to that of other embryologists such as Harrison and Lillie. He was willing to accept the chromosomal theory of inheritance but could not see the unchanging chromosomes directing embryogenesis. First, he did not see how a chemical gene could persist intact throughout numerous cell divisions. Having no concept of molecular replication in his mind, he supposed that if a gene were a molecule controlling development, it must become half a molecule after the first cell division and a quarter-molecule after the second cleavage. Second, he followed Lillie and Harrison in asking: "How could genes be responsible for differentiation, if they are the same in every cell?" Not averse to using ridicule, Just chided geneticists such as Demerec who had recently spoken on the embryologist's home turf, Woods Hole: "Untutored savage man made his god as big as possible because his god could do everything. It remained for the geneticists to make one of molecular size, the gene. Here obviously infinite minuteness means infinite capacity. According to one geneticist Demerec, the gene has almost magic power. Here is physico-chemical biology with a vengeance."[46]

Having disposed of the genetic theory of development, Just elevated the roles of the cytoplasm. He did this in two approaches. First, he redefined the terms of cell biology, which he thought had been warped by genetics. Second, he summarized the scientific evidence for the importance of the cytoplasm for developmental regulation. Just was extremely concerned with the precision of the scientific language, and of the sociological function of language. He alluded to the relationship between language and professionalization, saying that the last thirty years had seen genetics "develop almost to the proportions of a separate science—at least it has a very rich vocabulary of its own." In his next paragraph, Just would use his own rhetorical devices to exclude geneticists not only from embryology, but from the entire science. Here, he contrasted geneticists who accepted the gene theory of development with "biologists, on the other hand," who harbored doubts.[47]

The Biology of the Cell Surface is full of carefully phrased redefinitions. Just redefined such terms as cytoplasm, cell membrane, life, fertilization, and cell division in a context of embryological holism and against genetic reductionism. For instance, he described in detail the events of fertilization in four species of marine invertebrates to demonstrate that "the fertilization-process in these four examples resolves itself into two phases—an external, that concerns the ectoplasm, and an internal, that concerns the nuclei." After demonstrating that the binding of sperm and egg is a complete, well-orchestrated phenomenon involving adjacent cell surfaces and that in some experimental (parthenogenesis) and natural (*Rhabditis*) instances, fertilization occurs without nuclear fusion, Just concluded: "To retain the old definition of fertilization as the union of egg and sperm-nucleus is to violate both fact and logic." Similarly, Just found that "common usage has been loose in giving the term, cell division, the meaning of the division of the nucleus." After demonstrating numerous cases where cells divide without mitotic figures and where nuclei divide without cell division (as in early insect eggs), he stated, "It becomes obvious in the light of what has been said that nuclear and cytoplasmic division are separate phenomena. . . . Cell division is to be defined as the division of the cell body. . . . Finding it impossible to relate division of the cytoplasmic mass to the nucleus, we turn to the cytoplasm itself."[48]

As firm as Just was against nuclear "hegemony," he was equally hard on those who he thought misrepresented the cytoplasm. Foremost among these malefactors were the biochemists who sought to explain embryogenesis by breaking embryos apart and measuring chemical reactions. Needham's epic *Biochemical Embryology* had just been published in 1931, complete with a magisterial historical prologue giving the new biochemical methodology a classical pedigree. Just belittled those activities and, in a metaphor linking biochemistry with colonialism, depicted biochemists as conducting "punitive expeditions against the egg." Moreover, Just was adamant against using life as a means and not an end in itself. "The cell is never a tool. . . . Living matter is never an excuse and living phenomena never an opportunity for the display of the investigator's physico-chemical knowledge."[49]

Just sought the mystery of life, not its mastery.[50] This attitude was more akin to the naturalist than to the experimental embryologist of the 1930s. Indeed, Just's approach was very much that of the naturalist, making detailed observations without perturbing his organisms experimentally. Throughout *The Biology of the Cell Surface,* he criticized those modern researchers who experimented on organisms that are already damaged and who manipulated them so harshly that no value could be attached to their results. This lack of care, he believed, was due to a lack of respect for living phenomena. "Scientists degrade eggs by calling them 'material.' They do not respect their specimen, nor do they respect the integrity of life." Rather, "those experiments which alter a normal process the least have today especially great value in the study of the egg and its development. . . . By experiment we here slightly exaggerate, there lightly fret the tones out of which the harmony of the living state arises."[51] This approach is a far cry from the major embryological research program of the 1930s, which was characterized as dissecting the whole into smaller and smaller parts "quite heedless as to how far analysis into the nonvital may be possible."[52]

This brings us to the main thrust of Just's book, his experiments showing the importance of the cell cortex in development. Just began his analysis of the cell surface with a review of fertilization. What interested Just was not the movements of the nuclei or even the rearrangements of the cytoplasm that are so evident during fertilization, but the immediate effects brought about by the attachment of sperm and egg. To Just, all the other effects are secondary to the real drama that was occurring at the egg cell surface. To observe these events, some of which take place in a matter of seconds, requires persistent and careful observation of the most perfectly normal eggs. If the eggs were damaged, any conclusions that might be drawn from them were useless, and an observer who was not careful and persistent would likely dismiss the small transient events as meaningless.

Just's first paper, in 1912, established his reputation as a meticulous observer of natural phenomena. Here, Just demonstrated that the plane of the first cleavage of *Nereis* eggs is determined by the point of sperm entry.[53] This observation implied that the particular point of the cell surface that bound the sperm played a decisive role in the future development of the organism. For the next eight years, Just focused his research on detailed observations of the fertilization reaction in marine organisms. In 1915, he published another paper on the fertilization reaction of *Nereis,* which supported the work of his mentor, Frank Lillie, against that of Loeb. The debate between Lillie and Loeb over the nature of fertilization had been heated for several years, and Just's research added more strength to Lillie's model. What is striking in the rhetoric of this 1915 report is that Just supported his case by using Loeb's own words against their original author.[54] Just's literary technique is in marked contrast to the style of Lillie's own paper (published next to Just's in the same volume), which is a detailed rebuttal of Loeb couched in the most polite and carefully phrased language. It is no mystery why Loeb (and his students) developed an antagonism to this brash newcomer to their field.[55]

Unlike most researchers of his era, Just held that "fertilization is essentially a process of the egg." The sperm was secondary and, in some species, eliminated. Just's major support for this view came in 1919 when he published two papers on sea urchin fertilization. In the first paper, he dissected the initial minute of fertilization into a series of reactions on the egg surface. First, the sperm did not bore its way into the egg; rather the egg pulled it in. About ten seconds thereafter, a blister formed at the point of sperm entry. Droplets dispersed from this point, and the membrane that had been glued to the egg surface began to peel off. Moreover, "before the actual elevation of the membrane, some cortical change beginning at the point of sperm entry sweeps over the egg immunizing it to other sperms; the direct opposite pole of the site of sperm entry is the last point affected." This cortical change precedes the actual beginning of membrane lifting, because "before the membrane begins lifting at the site of sperm entry, sperm can no longer enter at any point of the egg." Only afterwards does one see the formation of the fertilization membrane, starting at the point of sperm entry. Just had observed what are now referred to as the fast and slow blocks to polyspermy, and he interpreted them as such. His interpretation is particularly significant in that he repeated the assertion that "the membrane is merely the sign and consequence of more profound cortical changes."[56]

In the second paper of 1919, Just gave support to Lillie's fertilizin hypothesis by showing that the ability of sea urchin eggs in the water to agglutinate sperm (presumably by fertilizin) correlated with the fertilizability of the eggs. This fertilizin theory was important to Just because one of the bases of Lillie's model was that "fertilizin is located in the cortex of the egg."[57] These two papers of 1919 give the impression that Just championed the cortex even more than he championed fertilizin. The fertilizin model was worthwhile solely because it showed the importance of the cortical cytoplasm, and not the converse.

From this point until 1931, Just played many variations on this same theme, stressing the responsiveness of the cortical cytoplasm. He could readily turn a research report into a polemic, and did so in 1929, in reply to a series of investigations by R. Chambers. After questioning the validity of Chambers's data by showing how he had used improperly prepared eggs, Just gave his own description of the minute, delicate filaments in the cortical cytoplasm of echinoderm eggs. Just then passed from data to propaganda.

> The reactivity of the cell as a whole—its individual and peculiar response to stimulation with attendant measurable physical and chemical changes—is largely, if indeed not wholly, a cortical (ectoplasmic) phenomenon. Cortical changes in ova are, therefore, no mere epiphenomena: they constitute the sine qua non of cellular life. In responding to and propagating the effects of the initial event in the fertilization-reaction, the attachment of the spermatozoon, the egg exhibits cortical changes which eventually modify the whole protoplasm and direct the course of ontogeny.[58]

In *The Biology of the Cell Surface,* Just referred to these strands in an important statement where various metaphors of the cell cortex (ectoplasm) are contrasted.

> All these considerations and data indicate that the surface-cytoplasm cannot be thought of as inert or apart from the living cell-substance. The ectoplasm is more than a barrier to stem the rising tide within the active cell-substance; it is more than a dam against the outside world. It is a living mobile part of the cell. It reacts upon and with the inner substance and in turn the inner substance reacts upon and with it. It is not only a series of mouths, gateways. The waves of protoplasmic activity rise to heights and shape the surface anew. Without, the environment plays upon the ectoplasm and its delicate filaments as a player upon the strings of a harp, giving them new forms and calling forth new melodies. But these are too nice for the undiscriminating ear of man.[59]

For Just, the sperm triggered the egg to develop but did not play a major role afterwards. It was merely the finger that plucked a well-tuned string to call forth a resonance. This idea is also expressed in his analysis of Kruger's observation that in the *Rhabditis* egg the sperm activated the egg to develop but remained inert afterwards, never uniting with the egg nucleus. That sperm was not essential for development was further shown by parthenogenesis, in which development is initiated by artificial means. The analysis of artificial parthenogenesis was a major thrust of Just's research program from 1919 to 1930, as it enabled him to look specifically at the cortical cytoplasm reaction system that began development. Just saw a "rhythmical movement of water" causing regional dehydration in portions of the egg and concluded that the sperm or parthenogenetic agents first caused a dehydration of the cortex, which subsequently dehydrated the cytoplasmic ground substance, which thereby brought about mitosis. It is this reallocation of cortical water that he believed brought about development.

When parthenogenetic agents displaced water from the cortex, development ensued. Just discovered that one of the agents capable of causing this response was ultraviolet irradiation. As early as 1926, Just had observed that *Nereis* eggs exposed to ultraviolet light developed abnormally. Moreover, many of these abnormal embryos showed a localized defect, "which is traced back to the site of cortical injury by radiation."[60] It is not surprising, then, that as late as 1932 Just thought he had turned the tables on the geneticists, dismissing radiation-induced mutations as mere epiphenomena of the real site of injury—the cell cortex. "Normal chromosome distribution and combinations depend upon the integrity of the cortex; their aberrant behavior is the effect of the loss of this integrity. . . . This would mean, therefore, that chromosome-behavior is not a primary one, but rather the expression of the ectoplasmic reactions."[61]

Thus, Just contended that mutations were the result of cortical disruption and not of direct injury to chromatin. This concept is extremely important, for it shows that for Just, genetics was subservient to development (or similarly, that the nucleus is subservient to the cytoplasm). A mutation is not a defect in a gene, but a defect in the cortical cytoplasm that directs development. Starting at this point, Just could speculate as to how genetics and embryology are related. His first such speculations are in this article analyzing mutation. He returned to Boveri's original experiments on dispermic eggs, "as a possible starting-point from which we may begin an attempt at the union, nowadays seemingly hopeless, of genetics and the physiology of development." Just interpreted Boveri's data to show that the aberrant chromosomal arrangements were possible only because of the "weakened conditions in the cytoplasm which make dispermy possible." After this, Just put forth his view (hearkening back to Driesch) that the cell is a system wherein nucleus and cytoplasm reciprocally interact with one another."[62]

Just believed that the nucleus and its chromosomes are normal cellular structures that are constructed de novo from cytoplasmic stuff after each mitosis. But Just claimed that some relationship clearly had to exist between genetics and embryology "since heredity is expressed during the process of development." Because the geneticists had failed to unite the two, a new theory was required. Just's theory is an inverted Weismannism wherein all the hereditary potentials for development exist in the cytoplasm of the fertilized egg. However, these potencies are all in an inhibited state. The nucleus exists to absorb these inhibitors from the cytoplasm differently in each cell. "Genetic restriction then depends upon the removal by the nucleus of certain materials from the cytoplasm, leaving others free. The free materials determine the character of the cell. . . . With each cleavage each nucleus fixes all material other than that which makes the blastomere what it is. The potencies for embryo-formation are all present in the uncleaved egg."[63]

There is a genetically based progressive restriction in potency as the nuclear chromosomes absorb the various agents of differentiation. "Thus, finally, every cell in the most complex organism has in its nucleus all the potencies except that one in the cytoplasm that makes the cell specific." In the germ cells, these potencies would be released by the chromosomes. In an extremely important statement, Just declared, "Every cell in an organism becomes what it is because its cytoplasm has free its particular potencies whilst the nucleus binds all others. These latter would, if left unbound in the cytoplasm, act as obstacles to the display of special potencies."[64]

The chromosomes are constructed of the unused substances that would have caused the cell to differentiate in another fashion. For Just, the nucleus was not the throne room of the cell; it was its refuse dump, a necessary, but far from noble, position.

Science is a creative human product, and scientific writings, like music, art, and literature, are historical artifacts. Looking at Just's scientific work in its historical context allows us to see two hidden agenda contained therein. The first concerns the relationship between embryology and genetics. Just viewed "differentiation and heredity as merely two expressions of development."[65] Genetics was thought to be subservient to embryology, just as the nucleus was seen to be formed from and be subservient to the potencies of the cytoplasm. In Just's model, the cytoplasm is where the answer to heredity and differentiation is to be found, not in the nucleus. In this newly synthesized field, the embryologists travel the straight and noble path.

Another hidden agendum concerns government. The developing organism is a polity of interacting cells, and each cell contains a nucleus and cytoplasm. The way that Just expressed the genetic regulation between these cellular elements is fascinating. In effect, Just postulated a noble cytoplasmic populace that contained all the potencies needed for the body or body politic. The nucleus acted to withdraw certain "obstacles" from the cytoplasm such that a specific potential could be expressed. In another cell, this obstacle is itself a specific potential. Thus, the nucleus, the central government of the cell, allows the expression of cytoplasmic potential by suppressing other possible potentials. This view reflects a specific solution to a problem that was being debated at that time by black scholars. The 1930s was the era of black migration from the South to the North and of the ensuing ethnic battles in the cities they entered. (In 1935, the *Herald Tribune* claimed that a second Italo-Ethiopian War was being fought in the streets of New York City.) It was the era of Langston Hughes and Richard Wright. Blacks had their potentials, too, but could only express them if other potentials/obstacles were removed. The Harlem Renaissance before the Depression showed how great those potentials could be when realized. The use by Just of the word "obstacle" instead of the more technical word "inhibitor" is instructive. The embryo—that ideal of organization—is modeled like an ideal society, a society that allows the optimal expression of its constituents' potentials. Just never enjoyed political debates; they bored him. Perhaps he already had an ideal society in mind and was trying desperately to bring that idea before the public. This ideal of government was his cellular republic, and Just explicitly viewed the egg cell as a microcosm. Just knew that people tend to extrapolate from nature to politics, and he felt that Kropotkin's mutual-aid hypothesis of evolution was a valid extrapolation of nature into human behavior. Moreover, "the means of cooperation and adjustment is the ectoplasm," because the ectoplasmic cortex of the egg is incorporated into the nervous system of adult vertebrates.[66] For Just, the agent of evolution was not the nucleus, but the cortical cytoplasm, the marginalized populace. Just viewed the cell and the embryo as perfectly balanced societies. Within an embryo, each cell had a defined role. The type of cell it became was determined by the cytoplasmic potentials remaining after the others were locked away in the nucleus. All potentials were needed in a balanced social or physical organism.

Just's federalism of the cell was a rebellion against contemporary views of the *Zellenstaat,* which were more authoritarian, and was the cytological analogue of the "physiological democracy" of his friend W. C. Allee.[67] In viewing the embryo as a society where each cell is allowed to develop a particular potential, Just saw a model of a society where each group of people could express its potential once certain "obstacles" were removed. It was a federalism that allowed the minorities to express themselves in a local fashion and, at the same time, contribute to the general welfare of the society. This was precisely the urgent political question that was being addressed by the leading black sociologists of Just's time. W. E. B. DuBois called it the problem of the majority. "Granted that government should be based on the consent of the governed, does the consent of a majority at any particular time adequately express the consent of all? Has the minority, even though a small and unpopular and unfashionable minority, no right to respectful consideration?"[68]

Indeed, DuBois inadvertently used an embryological metaphor when he demanded that the majority must not crush the self-development of a minority population. The goal of the American Negro, said DuBois, is "to be a coworker in the kingdom of culture, to escape both death and isolation, to husband and use his best powers and latent genius."[69] This was precisely the dialectic that the embryo had solved, groups of cells achieving self-development, expressing their unique potential to the betterment of the organism. But there was another, deeper parallel; for the cell itself had a government. The embryo was a federal republic of its constituent cells, and each of the cells harbored the potentials (as embryologists knew from Driesch's work) of every other type of cell. According to Just, each embryonic cell was directed to create a nucleus from its cytoplasmic material. These materials were the agents that would express certain potentials, and they were inhibited from so doing by being kept in the nucleus. Only one set of potentials could be expressed in any cell. The nucleus did not give any orders to the cytoplasm. Rather, the order was coming from the entire embryo. As Driesch had shown, the community of cells together determined the fate of each individual cell. The cell was not ruled by the nucleus; for that matter, it was not ruled by the cytoplasm either. The cytoplasm was far more important than the nucleus, to be sure, as it contained the developmental potentials and reacted with other cells and with the external environment; but the total community of embryonic cells is what determined the fate of a particular cell.

The development of an organism is a history of its cellular interaction. Therefore, to Just, development was a property of the cytoplasm; for only the cytoplasm could respond. The most responsive part of the cytoplasm—as Just showed from fertilization studies—is the cell cortex, the outermost cytoplasmic rim. It is this peripheral rim of material that Just championed as the prime mover of development, evolution, and intelligence. Whereas the mainstream of cellular biologists believed and assumed that the cytoplasm took its instructions from the central nucleus, Just believed in the primacy of the cortex. I believe further that he made some psychological self-identification with the object of his meticulous observations. He thought the peripheral rim of cytoplasm beautiful, sensitive, creative, powerful, important, and disregarded. In short, his view of the cortex mirrored his view of the black in American society. As we will see later, the heroes of his book are both the cell surface and himself. By championing the cause of the cortex, he subtly championed his own.[70]

There is excellent scientific argument in the works of E. E. Just. But there is more. There is a professional polemic designed to prevent his profession from falling into the hands of geneticists, and there is a political model of a federal republic that would recognize the potentials of his race. The three agenda are never far apart, and each one informs the other. What is particularly fascinating is that the same mixture of science, professionalization, and cytopolitics is found in the work of Just's geneticist contemporary, Richard Goldschmidt.

The Nuclear Aristocracy of Richard Goldschmidt

Physiological Genetics (1938) was Goldschmidt's attempt to reduce embryology to a subset of genetics. On the very first page of this volume, he redefined heredity to include "the mechanism of heredity," which he called "static genetics," and the "problem of development" which he called "dynamic genetics." Proclaiming that "development is to be linked specifically with the function and action of the gene," he preferred to call the latter study "physiological genetics." This part of genetics, wrote Goldschmidt, "was practically banned from advanced treatises and textbooks of genetics, and the opinion has developed and has even been voiced that it is not worth while to mention a field in which nothing is known with certainty." Goldschmidt admitted that "we know next to nothing of the action of the heredity material in controlling development," but he continued that he would now "present the entire material available."[71]

Because Goldschmidt saw development as an epiphenomenon of gene activity, even the most complex patterns of embryonic ontogeny were seen as driven by genes. "Development is, of course, the orderly production of pattern, and therefore, after all, genes must control pattern."[72] Goldschmidt assumed that if a problem is developmental, then it is in essence a genetic problem.

Goldschmidt then looked at the mechanisms by which the nucleus controlled development and offered two propositions—the first on the role of timing in gene activation and the second on the developmental inefficacy of the cytoplasm. The temporal basis of differentiation was one of the ideas that Goldschmidt tried in vain to get other geneticists to accept. Embryologists asked, "if all the genes in all the cell nuclei were the same, how can the genes control development?" Goldschmidt answered that the timing of gene activity was crucial. "The genes controlling pattern act by producing definite reactions of definite velocity."[73] Goldschmidt had analyzed a series of bizarre mutations called homoeotic mutants. Here, the development of a particular embryonic structure follows the development of another particular structure. For example, *Drosophila* having the dominant mutation *Aristapedia* have legs developing where their antennae should be. In some alleles of this locus, part of the antenna is converted into a leg structure (such as a tarsus or trochanter), whereas other alleles produce the entire transformation of antenna to leg. Goldschmidt analyzed these mutants and concluded that the aberrant development was not due to the elaboration of different materials by different genes, but to the aberrant timing of gene activity. "Here, then, a mutant gene changes an embryological process by shifting its initiation to a different point in time." Small changes in timing could create complex morphological changes.[74]

Goldschmidt looked at all genetic variations as changes in development. To him, the study of wing patterns or eye pigmentation was just as much a developmental problem as the development of the wing or eye themselves—and more readily analyzed. Like Just, Goldschmidt wanted to study these processes with the least possible perturbation. Goldschmidt saw the analysis of mutations as being superior in this respect to surgical manipulation.

> But certain processes [of development] may be changed without deleterious consequences; and if this is done by genetic change, we call it a mutation. Such considerations, obvious as they may seem to be, make us expect that the action of the mutated genes upon development cannot be of a different type from any other changes of development induced by experimental agencies; in both cases, something changes the detailed course of some developmental processes.

As evidence of this, Goldschmidt brought forth his data on phenocopies, organisms where experimentally produced abnormalities mimic certain mutants of the untreated organism. He provided numerous examples of temperature altering the pigment pattern of one race of butterflies in such a way as to make it resemble that of another. By relating the timing of heat shocks to the developmental pattern and by correlating the timing of eye color formation in various mutations with the time at which that color first develops, Goldschmidt concluded that "the mutant gene produces its effect, the difference from the wild-type, by changing the rates of the partial processes of development."[75] He was to use this principle of timing not only to explain mutations, homoeosis, and phenocopies, but also morphogenetic patterns and embryonic induction. In all these instances, development is reduced to being an epiphenomenon of genetics.

But there was other evidence as well. In 1934, J. Hämmerling had published his remarkable observations upon the development of *Acetabularia*.[76] These experiments (which Just ignored in his book) were critical in convincing biologists of the developmental importance of the nucleus. Hämmerling had shown that the nucleus of this unicellular protist controls the morphogenesis of its complex cap. Moreover, if the nucleus of one species were transplanted into a decapitated stem of another species, that stalk would regenerate a cap characteristic of the nuclear donor. To Goldschmidt, "this shows that actually the genes within the nucleus control the production of specific formative stuffs (not unspecific as in the hormonic type) which diffuses through the cytoplasm to the place of its form-controlling function." He saw this as "forging an interesting link between genetics and experimental embryology" and observed that it demonstrates "that such processes of distribution and arrangement of cytoplasmic components occur under control of genes."[77]

Goldschmidt thus saw genes as controlling the production and distribution of cytoplasmic materials, but he did not see the cytoplasm as controlling the nucleus. That reciprocal relationship, alluded to in the earlier discussion of the views of Morgan and Driesch, had been abandoned by Goldschmidt and Just. Whereas Just made the relationship vectoral, with the cytoplasm influencing the nucleus only, for Goldschmidt the nucleus influenced the cytoplasm, and the cytoplasm did as the nucleus demanded.

In Goldschmidt's model of the embryonic cell, the cytoplasm is the substratum upon which the genes act. It is not, itself, active. In all cases, the cytoplasm was considered to be the substratum on which the genes work, although the processes within the cytoplasm may take this independent course once started by the action of the genes. The word substratum is well chosen, for it not only denotes the background on which genes act, but it also connotes an enzymatic property as well. Genes act on the substratum-cytoplasm as enzymes act on substrates. The activation of the gene, then, is merely the preparation of the substrate. In the dividing egg, "one substratum is transformed into two or more different ones, which now provide the proper substratum for the activation of new genes." This process does not necessarily mean that some new cytoplasmic stuff feeds back into the nucleus to activate a specific gene. Rather, Goldschmidt's cytoplasmic "preparations" for gene activity include alterations of pH, temperature, and the presence of correct cofactors—precisely those elements that allow enzymes to react on substrates. The cytoplasm, then, exists to be modified by the action of the nucleus. It is a series of substrates to be acted upon by the gene-enzyme agents derived from the nucleus. "Thus we conclude that the cytoplasm is mainly the substratum for genic action, in which all those decisive processes take place which constitute development and which are steered by the genes."[78]

In Goldschmidt's model, the cytoplasm carries no potentials. It is impotent and subservient, a far cry from the potent, active cytoplasm proposed by Just; and its passivity is equally political. Goldschmidt's book can also be read as science, as professional polemic, and as utopian fantasy. Goldschmidt was the head of a genetics department and a firm believer in the reducibility of all developmental problems to genetic ones. Were genetics and embryology to be reunited, the geneticists would control the field. The relationship between nucleus and cytoplasm parallels that of geneticist and embryologist.

There is a political element here, as well. Goldschmidt viewed the cell as a monarchy run by the nucleus. This analogy was not peculiar to Goldschmidt. One finds it in other German biologists, and it is explicit in the work of Goldschmidt's advisor, Richard Hertwig. Writing on "Die Protozoen und die Zelltheorie" in the first article of the *Archiv für Protistenkunde,* Hertwig states that "I would like to compare the one-nucleus cell to an absolute monarchy; the achievements of such a political system result from the mass of people, the directives from monarch."[79] Hertwig compared multinucleated cells with oligarchies, emphasizing that the nucleus always gives directives such that "nothing would change in the unitedness of a political system, even if the 'oligarchs' allowed a division of labor to enter the leading roles."

This view of the nobility of the nucleus was agreeable to Goldschmidt's view of both himself and science. Richard Goldschmidt strove for a noble life. He considered himself an aristocrat, a self-aware king in the scientific world of interbellum Germany. Goldschmidt felt that artists and scientists were the truly free individuals, persons whose creative talents were responsible to no one. Indeed, the scientist and the artist were merged in Goldschmidt's personality. When Goldschmidt met Segovia in Japan, the latter gave him a personal concert. "Only kings used to be able to have such an experience." This ability to "create without outside interference or control" gave the scientist the royal duties of noblesse oblige.[80]

Goldschmidt as head of the genetics division of the Kaiser-Wilhelm Institute, a world-famous lecturer, art collector and critic, music connoisseur, and founder of journals, prized excellence. In his autobiography, his favorite most highly prized adjective is "noble." This word referred to spiritual rather than to worldly excellence, and he used it to describe individuals such as his father, nursemaid, or best childhood friend who had lived lives of hardship.[81] This nobility of spirit is stressed by those individuals who remember Goldschmidt well.

Goldschmidt shocked his contemporaries by attributing this nobility to his Jewish background, and he began his autobiography with a mythic pedigree that emphasized his Jewishness.

> I come from an old German-Jewish family. This fact may convey little meaning to most people, for it is perhaps not generally known that the German Jews are a group of people who can trace their origin, at least in a general way, farther back into gray antiquity than the oldest known family, the K'ung (Confucius) in China. The reason this can be done is that many German-Jewish families, like my own, belong to the caste of the Levites, the literary and teaching caste since Moses' time some three thousand years ago. The Levites kept to themselves through the centuries except for intermarriage with the priestly caste, the Cohens, and thus the members of the Levite caste are the product of an age-long selection of intellectual performance. . . . When the Romans had conquered Germany and erected the Limes Germanicus, the fortified frontier against the barbarians, Roman Jews of Levite families were settled along the Limes to teach the savage Teutons the amenities of Mediterranean agriculture.[82]

This is a truly royal pedigree wherein Goldschmidt claimed that he was not merely the fortuitous product of German high culture; rather, he was one whose ancestors created it, instructing the barbaric tribes in the ways of civilization. Moreover, Goldschmidt saw himself as the culmination of centuries of selective breeding for high intelligence. He returned to this idea later when he reflected upon a poster that the Nazis had circulated which showed all the positions of authority occupied by the Goldschmidt family. "I think that the Nazi poster could well be used as a chart demonstrating the effect of long selection of favorable hereditary traits upon the improvement of human families."[83]

Kingship metaphors abound in Goldschmidt's writing. One of the most revealing is Goldschmidt's rationale for expanding genetics into embryology, for it depicts an active genetics and a passive embryology. The latter was represented as tillable land: "geneticists will continue to worry about the problem of genetic action and take the risk of climbing over the fence erected by some jealous embryologists, who, while claiming the kingdom for themselves, do not set out to till its soil."[84]

Goldschmidt considered himself an aristocrat of the spirit. At the same time, he had good reason to fear the proletariat, for he had been a victim of two instances of mass hysteria. In 1918, while at Yale, Goldschmidt was arrested as "an extremely dangerous German" and was sent, under armed guard, to an internment camp in Georgia. Although he blamed the war-fever on the Justice Department and popular press, he found that the American public was only too willing to become a hateful mob. His supporters had to fear mob violence and personal attacks, said Goldschmidt, and "their noble actions brought many unpleasant experiences upon them during the following year."[85] He also recounted the servility and readiness with which the German people accepted Hitler, and related how nobody protested the mass murders and how one widow quietly accepted her husband's death. Decent people, he reflected, are capable of the most horrible atrocities. To Goldschmidt, there is no inherent nobility in the great masses of people, and most of them will do whatever is in accord with popular fashion. They are not in control of their own destiny or even, it would seem, of their own behavior. If their leaders say it is permissible to murder, they do not protest. If the leaders wish to intern an individual away from his family, so be it. Goldschmidt had little respect for the masses, but he had enormous respect for those noble individuals who could rise above popular prejudices.

Such an ideology is reflected in Goldschmidt's cell. The world of nature is harmonious, and Goldschmidt gloried throughout his autobiography on the bountiful beauty of nature. Likewise, the cell is a marvel of harmonious function. To Goldschmidt, the nucleus was the repository of the hereditary traits. As an intellectual aristocracy passed cultural traditions to a society, so the genes expressed their inherited potentials to the cell. The cytoplasm merely allowed these traits to be expressed. All the important elements in the cell were nuclear; the cytoplasm was merely a substrate. According to Goldschmidt, development was based on the harmonious reaction system comprised of all the genes. These genes were also enzymes, and they accomplished their cytoplasmic catalyses according to the law of mass action. In the case of the embryo, the nucleus acted as the enlightened monarch should, creating a well balanced cell that performed its proper function. In mutants, the monarch was less enlightened, and the instructions to the cytoplasm could cause the death of the cell or even the death or malformation of the entire organism. But sometimes, a mutation could arise that might change the fate of a cell to something even better. This was Goldschmidt's view of evolution and it was based on nuclear homoeotic mutations. While Just viewed evolution as the product of the cell cortex, Goldschmidt viewed evolution as being controlled totally by nuclear changes.

Epilogue: The Cell As Text

We know cells only through interpretation. Nobody has knowledge of the cellular structure and function except through technical (stains, microscopy) or literary devices (textbooks, articles). Just and Goldschmidt each interpreted cells in different ways and for different reasons. The creative interpretation of the cell can be seen in what is left out as well as what is included in their respective discussions. Just, for instance, never used the term "segregation" in his book even though it was widely used by his colleagues and was, in fact, Lillie's paradigm for early embryonic development. Just also did not discuss Hämmerling's *Acetabularia* experiments, although they were well publicized. These experiments would not fit his cytoplasmic model. Similarly, we see Goldschmidt's varied interpretation of the cell in his description of Morgan's theory of the gene as being "dead as the dodo" and his interpretation of Hämmerling's experiments in terms of nuclear dominance.

Landau has shown how scientists often use narrative structure in their texts to make political stories out of their data.[86] This is certainly true of Just and Goldschmidt. Just casts his book within two narratives. The first is the "Ugly Duckling" story wherein the despised character is belatedly recognized as the most treasured. The hero of this tale is alternately the cell surface and Just himself, the unrecognized genius. The second narrative is the "Emperor's New Clothes" wherein Just is seen as having the keen eyes, unclouded judgment, and personal integrity to shout that the genetic theory of mutations and development is a sham. Goldschmidt's narrative is more subtle, yet is just as pervasive. The reader of *Physiological Genetics* cannot help but see Goldschmidt's pioneering work described on nearly every page. He refers to "the gene of tomorrow" and to "tomorrow's theory of the germ plasm." He even discusses the chemistry of chromosome replication, but then says, "I shall not develop further this idea, to which I think the future belongs."[87] Goldschmidt portrayed himself as leader and prophet. Like Moses, he could lead people to the promised land, but he could not enter it himself. The American geneticists were given an opportunity to travel with him.

The Americans rejected it. They rejected both Just's and Goldschmidt's work as irrelevant. At the same time that Just and Goldschmidt published their syntheses, the British biologist Waddington published his own synthetic scheme, *Organisers and Genes*. This book reflected the dialectical Whiteheadian views of its author, who saw the nucleus and cytoplasm (and genetics and embryology) as mutually interacting partners. Moreover, neither genetics nor embryology was seen as sufficient to explain development. "A coherent theory of development cannot be founded on the known properties of genes. . . . No stimulus, nor single cause is itself an adequate explanation of anything."[88]

Waddington attempted to show that the geneticist's *genes* were the same as the embryologist's *organizers*. Although the attempt proved problematic, it led to a rapprochement between genetics and embryology. Each discipline explained part of development and neither could subsume the other. Thus, each could pursue its own program of research. The truce fit the American setting where the newly organized science of genetics was finding research funds and backing from sources quite separate from those of embryology (see Paul and

Kimmelman, this volume). While American embryology continued to have its roots in a European context, genetics had become a markedly different science and had separated itself from its European parentage.

Acknowledgments

I would like to thank several people for reviewing the manuscript and providing both perceptive comments and new leads to follow; N. J. Berrill, Marion Faber, Bentley Glass, Donna Haraway, John Harwood, Evelyn Fox Keller, Kenneth Manning, Jane Oppenheimer, Marsha Richmond, Jan Sapp, Peter Taylor, and the participants of the Friday Harbor conference all provided invaluable assistance.

Notes

1. Johannes Holtfreter, "Address in Honor of Viktor Hamburger," in Michael Locke, ed., *The Emergence of Order in Developing Systems. The Twenty-Seventh Symposium of the Society for Developmental Biology* (N.Y.: Academic Press,1968), p. xi.
2. Scott F. Gilbert, "The Embryological Origins of the Gene Theory," *Journal of the History of Biology,* 1978, *11:* 307–351.
3. T. H. Morgan, "Chromosomes and Heredity," *American Naturalist,* 1910, *44:* 449–496.
4. Jan Sapp, "The Struggle for Authority in the Field of Heredity, 1900–1932: New Perspectives on the Rise of Genetics," *J. Hist. Biol.,* 1983, *16:* 311–342, and Garland Allen, "T. H. Morgan and the Split between Embryology and Genetics, 1910–1935," in T. J. Horder, J. A. Witkowski, and C. C. Wylie, eds., *A History of Embryology* (Cambridge: Cambridge University Press, 1985), pp. 113–146.
5. T. H. Morgan, *The Theory of the Gene* (New Haven: Yale University Press, 1926), p. 27.
6. Boris Ephrussi, "The Cytoplasm and Somatic Cell Variation," *Journal of Cellular Comparative Physiology,* 1958, *52 (suppl.):* 35–54.
7. Frank R. Lillie, "The Gene and the Ontogenetic Process," *Science,* 1927, *66:* 361–368.
8. Ibid., pp. 362 and 365.
9. T. H. Morgan, *Embryology and Genetics* (New York: Columbia University Press, 1934), pp. vi and 9.
10. Ibid., pp. 9–10.
11. Hans Driesch, *Analytische Theorie Organischen Entwicklung* (Leipzig: Wilhelm Engelmann, 1894), quoted in Jane Oppenheimer, *Essays in the History of Embryology and Biology* (Cambridge, Mass.: MIT Press, 1967), p. 76.
12. Morgan, *Embryology and Genetics,* pp. 9 and 234.
13. Sapp, "The Struggle for Authority."
14. C. H. Waddington, *An Introduction to Modern Genetics,* (London: Allen and Unwin, 1939), p. 135. Waddington began the section on genetics and development: "Now that the mechanism of inheritance is known, in its main outlines at least, it is

possible to tackle the next question, of how the genes affect the developmental processes which connect the fertilized egg into the adult organism."

15. G. W. Beadle and Boris Ephrussi, "The Differentiation of Eye Pigments in *Drosophila* as Studied by Transplantation," *Genetics,* 1936, *21:* 225–247. The separation of physiological genetics from what we now call "developmental genetics" had not yet occurred. To wit, the title of Goldschmidt's work, *Physiological Genetics,* concerns itself with what we now would call developmental genetics. The term "physiological" connoted "modern," and was itself a put-down of what Goldschmidt considered the boring, static, descriptive genetics of the Morgan school. The Ephrussi papers did not, contrary to the interpretation found in contemporary genetics texts, attempt to integrate genetics with biochemistry. Rather, Ephrussi sought "to lay a bridge between causal embryology and genetics." His techniques were those of the embryologists (transplantation) rather than those of the geneticist (breeding), and he explicitly looked at these pigment phenomena in terms of development and differentiation, using the term "inducer" to refer to the substance made by the wild type *cn* gene. A study of Ephrussi's role in integrating genetics and developmental physiology is being prepared by Richard Burian, Doris Zallen, and J. Gayon.

16. E. E. Just, "A Single Theory for the Physiology of Development and Genetics," *Am. Nat.,* 1936, *70:* 267–312.

17. Ross G. Harrison, "Embryology and Its Relations," *Sci.,* 1937, *85:* 369–374.

18. Ibid., p. 372.

19. N. J. Berrill, personal communication, April 1985.

20. N. J. Berrill, "Foreword," *Growth,* 1927, *1 (suppl.):* i. In 1927, Lillie had reflected that "genetics has become quite a unitary science, and the physiology of development is at most a field of work." It was Berrill, however, who pronounced the wittiest put-down of the geneticists: "So much has been said of the role of genes that one is tempted to define them as statistically significant little devils collectively equivalent to one entelechy." This rejoinder was published in Berrill, "Spatial and Temporal Growth Patterns in Colonial Organisms," *Growth Symposium,* 1941, *3:* 110. For more information on this period, see Jane Oppenheimer, "The Growth and Development of Developmental Biology," in M. Locke, ed., *Major Problems in Developmental Biology. The Twenty-Fifth Symposium for the Society of Developmental Biology* (New York: Academic Press, 1966), pp. 1–28.

21. Berrill, "Foreword," p. i. For Schotte's version of how embryologists and geneticists alternatively represented the cell see Klaus Sander, "Genome Function in Sea-Urchin Embryos: Fundamental Insights of T. Boveri's Reflected in Recent Molecular Discoveries," in Horder, *A History of Embryology.*

22. J. Harwood, "The Reception of Morgan's Chromosome Theory in Germany: Debate over Cytoplasmic Inheritance," *Medical History,* 1984, *19:* 3–32.

23. Ibid., p. 24.

24. Philip J. Pauly, "American Biologists in Wilhelmian Germany: Another Look at Innocents Abroad" (unpublished manuscript).

25. T. H. Morgan, "Genetics and the Physiology of Development," *Am. Nat.,* 1926, *60:* 459–515.

26. Hans Spemann, *Embryonic Induction and Development* (New Haven: Yale University Press, 1938), pp. 210–211.

27. Paul Weiss, *Principles of Development* (New York: Holt, 1939).

28. Richard B. Goldschmidt, "Some Aspects of Evolution," *Sci.*, 1933, *78:* 539–547 and idem, *Physiological Genetics* (New York: McGraw-Hill, 1938).

29. Richard B. Goldschmidt, *The Material Basis of Evolution* (New Haven: Yale University Press, 1940), p. 212.

30. Goldschmidt, *Physiological Genetics*, p. 310.

31. Stephen Jay Gould, "Introduction," in Richard B. Goldschmidt, *Material Basis of Evolution*, 1940 (reprint; New Haven: Yale University Press, 1980), pp. xiv–xv.

32. Stephen Jay Gould, "The Return of Hopeful Monsters," *Natural History*, 1986, June/July: 22–30.

33. Richard B. Goldschmidt, *In and Out of the Ivory Tower* (Seattle: University of Washington, 1960), p. 76.

34. Paul Weindling, "Theories of the Cell State of Imperial Germany," in C. Webster, ed., *Biology, Medicine, and Society, 1840–1940* (Cambridge: Cambridge University Press, 1981), pp. 99–155.

35. Kenneth R. Manning, *Black Apollo of Science: The Life of Ernest Everett Just* (Oxford: Oxford University Press, 1983).

36. E. E. Just, "Die Rolle des kortikalen Cytoplasmas bei vitalen Erscheinungen," *Natur Wissenschaften*, 1931, *19:* 953–962, 980–984, 998–1000, and idem, *The Biology of the Cell Surface* (Philadelphia: P. Blakiston's Son & Co., 1939), pp. 337–339.

37. Just fared better among the Europeans than with the American researchers. Of the three Americans who emphasized the cell surface, R. Chambers, A. Tyler, and P. Weiss, only Tyler quoted Just, and only in reviews. At the 1940 meeting of the Cold Spring Harbor Symposium for Quantitative Biology, which focused on the cell periphery, Just's work was mentioned only once. However, Johannes Holtfreter's groundbreaking paper, "The Properties and Functions of the Surface Coat of Amphibian Embryos," *Journal of Experimental Zoology*, 1943, *93:* 251–323, cited Just's book in five separate places and also quoted two of his research papers approvingly. Moreover, Holtfreter agreed with Just's suggestion that sublethal cytolysis could activate the egg to undergo fertilization reactions. This may have been a source of Holtfreter's later view in 1947 that the induction of the neural tube is occasioned by sublethal cytolysis, for although he did not quote Just directly in the later paper, Holtfreter claimed that the concept of sublethal cytolysis "would link up the mechanism of neural induction with that of parthenogenesis where the response is likewise determined by specific intrinsic factors rather than by the nature of the external stimulus." This is the last evidence I have been able to find of Just's work stimulating new research. After that, it remained ignored until very recently. See also Holtfreter, "Neural Induction in Explants which Have Passed through a Sublethal Cytolysis," *J. Exp. Zool.*, 1947, *106:* 197–222.

38. Goldschmidt, *Ivory Tower*, pp. 7–8.

39. Manning, *Black Apollo*, p. 194.

40. Joseph Needham, *Order and Life* (New Haven: Yale University Press, 1936), p. 99.

41. Kenneth R. Manning, personal communication. This is also a persistent theme in Manning's book, n. 35.

42. Goldschmidt, *Physiological Genetics*, pp. 3 and 23–51.

43. Goldschmidt and Just present excellent case studies of outsiders in science. Their creativity probably did not derive from their respective Jewish or black back-

grounds. Rather, it may have been caused by their motives for entering biology, and development in particular. When they entered upon this science, neither of them could have hoped to become successful. Moreover, both could have gone to medical school where their chances of success were much greater. This suggests that they entered science for highly personal aesthetic or political reasons. Those reasons may not have been shared by their more mainstream colleagues.

44. "Nature has neither kernel nor shell, but is everywhere in everything." This quotation, from Just, *The Biology of the Cell Surface,* pp. 7–8, uses the same language to describe the nucleus and cortex of the cell.

45. Ibid., pp. 7–8.

46. Ibid., pp. 321 and 325.

47. Ibid., p. 321.

48. Ibid., pp. 57, 182, and 267.

49. Ibid., pp. 28–29, and 46.

50. Ibid., p. 21. This attitude is precisely the reverse of Curt Stern's verdict on Goldschmidt who, he said, "served Nature, but also tried to dominate her." Curt Stern, "Richard Benedict Goldschmidt," *National Academy of Sciences Biographical Memoirs,* 1967, *39:* 141–192.

51. Just, *The Biology of the Cell Surface,* pp. 18, and 30–31.

52. Goldschmidt, *The Material Basis of Evolution,* p. 2.

53. E. E. Just, "The Relationship of the First Cleavage-plane to the Entrance-point of the Sperm," *Biological Bulletin,* 1912, *22:* 239–252.

54. E. E. Just, "Initiation of Development in *Nereis,*" *Biol. Bull.,* 1915, *28:* 1–17. See also Scott F. Gilbert and J. P. Greenberg, "Intellectual Traditions in the Life Sciences. II. Stereocomplementarity," *Perspectives in Biology and Medicine* 1984, *28:* 18–34.

55. Although their differences concerning fertilization are enormous and they developed animosity toward each other, the views of Just and Loeb concerning the relationship of embryology and genetics appear rather similar. According to Philip Pauly, *Controlling Life: Jacques Loeb and the Engineering Ideal in Biology* (NY: Oxford University Press, 1987), pp. 148–149, Loeb was never quite sure where Morgan's genetics fit into dynamic biology, but he generally believed that the egg cytoplasm controlled most of development whereas the nuclear genes added the finishing touches.

56. Just, *The Biology of the Cell Surface,* pp. 6–7.

57. Frank R. Lillie, "Studies of Fertilization, VI. The mechanism in *Arbacia,*" *J. Exp. Zool.,* 1914, *16:* 523–590.

58. E. E. Just, "The Production of Filaments by Echinoderm Ova as a Response to Insemination, with Especial Reference to the Phenomenon as Exhibited by the Ova of the Genus *Asterias,*" *Biol. Bull.,* 1929, *57:* 311–325.

59. Just, *The Biology of the Cell Surface,* p. 146.

60. E. E. Just, "Experimental Production of Polyploidy in the Eggs of *Nereis limbata* by Means of Ultraviolet Radiation," *Anatomical Record,* 1926, *34:* 108.

61. E. E. Just, "On the Origin of Mutations," *Am. Nat.,* 1932, *66:* 61–74.

62. Just, *The Biology of the Cell Surface,* p. 70.

63. Ibid.

64. Ibid., pp. 326, 329. Just was not outside the biological mainstream when he

made this hypothesis. In 1938 Curt Stern, in his paper "During Which Stage in the Nuclear Cycle do the Genes Produce their Effects in the Cytoplasm?" *Am. Nat.*, 1938, *72:* 350–357, said the possibility that the nucleus specifically extracts materials from the cytoplasm during development cannot be ruled out. This again is a German, not American, approach to nucleocytoplasmic relationships. Just's debt to the German geneticists is also seen in his 1936 article, "A Single Theory for the Physiology of Development and Genetics," *Am. Nat.*, wherein he stated that his aim was to place genes "within the domain of physiology." This, of course, was also Goldschmidt's aim.

65. Ibid., p. 326.

66. Ibid., pp. 365–367. The concept of the organism as a society of cells was central to German biology. The transformation of this Zellenstaat from liberal republic to authoritarian dictatorship is documented in Weindling, "Theories of the Cell State in Imperial Germany."

67. W. C. Allee, *Animal Life and Social Growth* (Baltimore: William & Wilkins, 1932), pp. 157–159, 200. Allee claimed that primitive societies, like primitive organisms, are authoritarianly ruled from the brain. As both organisms and societies evolved, more interaction occurs between components and "social control tends to become less autocratic and more democratic. . . . The highest organizations even tend to dispense with leadership."

68. W. E. B. DuBois, *Darkwater: Voices from Within the Veil* (New York: Harcourt, Brace, 1921), p. 151.

69. W. E. B. DuBois, *The Souls of Black Folk* (1903), quoted in A. Chapman, *Black Voices* (New York: Mentor), p. 496.

70. This self-identification and love of a scientist with his or her object of study is not uncommon. For an uncommonly candid account of this phenomenon, see June Goodfield, *An Imagined World* (New York: Harper and Row, 1981), p. 229. This same theme is expressed in Evelyn Fox Keller, *A Feeling for the Organism* (San Francisco: W. H. Freeman, 1983), pp. 204–206.

71. Goldschmidt, *Physiological Genetics,* pp. v and 1.

72. Ibid., p. 200.

73. Ibid., p. 251.

74. Ibid., pp. 208–209.

75. Ibid., pp. 3 and 51.

76. J. Hämmerling, "Über formbildende Substanzen bei Acetabularia mediterranea, ihre ranmliche und zeitliche Verteilung und ihre Herkunft," *Archiv für Entwicklungmechanik,* 1934, *131:* 1–81.

77. Goldschmidt, *Physiological Genetics,* pp. 193 and 179.

78. Ibid., pp. 263, 266, and 280.

79. Richard Hertwig, "Die Protozoen und die Zelltheorie," *Archiv für Protistenkunde,* 1907, *1:* 1–40.

80. Goldschmidt, *Ivory Tower,* pp. 69 and 246.

81. Ibid., pp. 29, 33, 38, and 69.

82. Ibid., p. 3.

83. Ibid., p. 5.

84. Richard B. Goldschmidt, *Theoretical Genetics* (Berkeley: University of California Press, 1955), p. 247.

85. Ibid., p. 169.

86. Misia Landau, "Human Evolution as Narrative," *American Scientist,* 1984, *72:* 262–268.

87. Goldschmidt, *Physiological Genetics,* pp. 301 and 314.

88. C. H. Waddington, *Organisers and Genes* (Cambridge: Cambridge University Press, 1940), p. 3.

Bibliography

Abir-Am, Pnina. "The Discourse of Physical Power and Biological Knowledge in the 1930s: A Reappraisal of the Rockefeller Foundation's 'Policy' in Molecular Biology." *Social Studies of Science,* 1982, *12:* 341–382.

Aldrich, Michele L. "Women in Paleontology in the United States." *Earth Science History,* 1982, *1:* 14–22.

Allard, Dean Conrad, Jr. *Spencer Fullerton Baird and the U.S. Fish Commission.* New York: Arno Press, 1978.

Allen, Garland E. "The Eugenics Record Office at Cold Spring Harbor, 1910–1940." *Osiris,* n.s., 1986, *2:* 225–264.

———. "The Introduction of *Drosophila* into the Study of Heredity and Evolution, 1900–1910." *Isis,* 1975, *66:* 322–333.

———. *Life Science in the Twentieth Century.* Cambridge: Cambridge University Press, revised ed., 1978.

———. "The Misuse of Biological Hierarchies: The American Eugenics Movement, 1900–1940." *History and Philosophy of the Life Sciences,* 1983, *5:* 105–128.

———. "Naturalists and Experimentalists: The Genotype and the Phenotype." *Studies in History of Biology,* 1979, *3:* 179–209.

———. "T. H. Morgan and the Emergence of a New American Biology." *Quarterly Review of Biology,* 1969, *44:* 168–188.

———. "T. H. Morgan and the Split between Embryology and Genetics, 1910–1935." In T. J. Horder, J. A. Witkowski, and C. C. Wylie, eds., *A History of Embryology.* Cambridge: Cambridge University Press, 1985.

———. *Thomas Hunt Morgan: The Man and His Science.* Princeton: Princeton University Press, 1978.

———. "Thomas Hunt Morgan: Materialism and Experimentalism in the Development of Modern Genetics." *Social Research,* 1984, *51:* 709–738.

———. "The Transformation of a Science: T. H. Morgan and the Emergence of a New American Biology." In Alexandra Oleson and John Voss, eds., *The Organization of Knowledge in Modern America, 1860–1920.* Baltimore: Johns Hopkins University Press, 1979.

Appel, Toby A. "The American Physiological Society Archives." *Physiologist,* 1984, *27:* 131–132.

———. "Biological and Medical Societies and the Founding of the American Physiological Society." In Gerald L. Geison, ed., *Physiology in the American Context 1850–1940.* Bethesda, Md.: American Physiological Society, 1987.

———. "Science, Popular Culture, and Profit: Peale's Philadelphia Museum." *Journal of the Society for a Bibliography of Natural History,* 1980, *9:* 619–634.

Arnold, Lois Barber. *Four Lives in Science: Women's Education in Science.* New York: Schocken Books, 1984.

Atkinson, J. W. "E. G. Conklin on Evolution: The Popular Writings of an Evolutionist." *Journal of the History of Biology,* 1985, *18:* 31–50.

Banks, Edwin M. "Warder Clyde Allee and the Chicago School of Animal Behavior." *Journal of the History of the Behavioral Sciences,* 1985, *21:* 345–353.

Bates, Ralph S. *Scientific Societies in the U.S.* Cambridge, Mass.: MIT Press, 1965.

Baxter, Alice Levine. "E B. Wilson's 'Destruction' of the Germ-layer Theory." *Isis,* 1977, *68:* 363–374.

———. "Edmund B. Wilson as a Preformationist: Some Reasons for His Acceptance of the Chromosome Theory." *Journal of the History of Biology,* 1976, *9:* 29–57.

———. "Edmund Beecher Wilson and the Problem of Development." Ph.D. dissertation, Yale University, 1974.

Beaver, Donald De B. *The American Scientific Community, 1800–1860: A Statistical-Historical Study.* New York: Arno, 1980.

Bender, Thomas. *Towards an Urban Vision: Ideas and Institutions in Nineteenth-Century America.* Baltimore: Johns Hopkins University Press, 1975.

———. *Community and Social Change in America.* New Brunswick, N.J.: Rutgers University Press, 1978.

Benson, Keith R. "American Morphology in the Late Nineteenth Century: The Biology Department at Johns Hopkins University." *Journal of the History of Biology,* 1985, *18:* 163–205.

———. "The First Hundred Years: A Century of Natural History at the Burke Museum." *Landmarks,* 1985, *4:* 28–31.

———. "H. Newell Martin, W. K. Brooks, and the Reformation of American Biology." *American Zoologist,* 1987, *27:* 759–771.

———. "Laboratories on the New England Shore: The 'Somewhat Different Direction' of American Marine Biology." *New England Quarterly,* 1988 (in press).

———. "The Naples Stazione Zoologica and Its Impact on the Emergence of American Marine Biology: Entwicklungsmechanik and Cell-lineage Studies." *Journal of the History of Biology* (forthcoming).

———. "Problems of Individual Development: Descriptive Embryological Morphology in America at the Turn of the Century." *Journal of the History of Biology,* 1981, *14:* 115–128.

———. "William Keith Brooks (1848–1908): A Case Study in Morphology and the Development of American Biology." Ph.D. dissertation, Oregon State University, 1979.

———. "The Young Naturalists' Society: From Chess to Natural History Collections." *Pacific Northwest Quarterly,* 1986, *77:* 82–93.

———. "The Young Naturalists' Society and Natural History in the Northwest." *American Zoologist*, 1986, *26:* 351–361.

Berlan, Jean-Pierre, and Richard Lewontin. "Breeders' Rights and Patenting Life Forms." *Nature*, 1986, *322:* 785–788.

———. "The Political Economy of Hybrid Corn." *Monthly Review*, 1986, *38:* 35–47.

Blake, Lincoln C. "The Concept and Development of Science at the University of Chicago, 1890–1905." Ph.D. Dissertation, University of Chicago, 1966.

Bledstein, Burton J. *The Culture of Professionalism: The Middle Class and the Development of Higher Education in America*. New York: Norton, 1976.

Bluestein, Bonnie Ellen. "The Philadelphia Biological Society, 1857–1861: A Failed Experiment?" *Journal of the History of Medicine*, 1980, *35:* 188–202.

Boakes, Robert. *From Darwin to Behaviourism: Psychology and the Minds of Animals*. Cambridge: Cambridge University Press, 1984.

Bonner, Thomas Neville. *American Doctors and German Universities*. Lincoln, Neb.: University of Nebraska Press, 1963.

Boring, E. G. *A History of Experimental Psychology*, 2nd ed. New York: Appleton-Century-Crofts, 1950.

Bowler, Peter J. *The Eclipse of Darwinism: Anti-Darwinian Evolution Theories in the Decades Around 1900*. Baltimore: Johns Hopkins University Press, 1983.

———. "Edward Drinker Cope and the Changing Structure of Evolution Theory." *Isis*, 1977, *68:* 249–265.

———. *Evolution: The History of an Idea*. Berkeley: University of California Press, 1984.

———. *Theories of Human Evolution: A Century of Debate, 1844–1944*. Baltimore: Johns Hopkins University Press, 1986.

Brobeck, John R., Orr E. Reynolds, and Toby A. Appel, eds. *History of the American Physiological Society: The First Century, 1887–1987*. Bethesda, Md.: American Physiological Society, 1987.

Bruce, Robert V. *The Launching of American Science, 1846–1876*. New York: Knopf, 1987.

Brush, Stephen. "Nettie M. Stevens and the Discovery of Sex Determination by Chromosomes." *Isis*, 1978, *69:* 163–172.

Burke, Colin B. *American Collegiate Populations: A Test of the Traditional View*. New York: New York University Press, 1982.

Burkhardt, Richard W., Jr. "Darwin on Animal Behavior and Evolution." In David Kohn, ed., *The Darwinian Heritage*. Princeton: Princeton University Press, 1985.

———. "The Development of an Evolutionary Ethology." In D. S. Bendall, ed., *Evolution from Molecules to Men*. Cambridge: Cambridge University Press, 1983.

———. "*The Journal of Animal Behavior* and the Early History of Animal Behavior Studies in America." *Journal of Comparative Psychology*, 1987, *101:* 223–230.

———. "Lamarckism in Britain and the United States." In Ernst Mayr and William B. Provine, eds. *The Evolutionary Synthesis: Perspectives on the Unification of Biology*. Cambridge, Mass.: Harvard University Press, 1980.

———. "On the Emergence of Ethology as a Scientific Discipline." *Conspectus of History*, 1981, *1:* 62–81.

Burnham, John C. "On the Origins of Behaviorism." *Journal of the History of the Behavioral Sciences,* 1968, *4:* 143–151.

———. "Psychiatry, Psychology, and the Progressive Movement." *American Quarterly,* 1960, *12:* 457–465.

———, ed. *Science in America.* New York: Rinehart and Winston, 1971.

Carlson, Elof A. "The *Drosophila* Group: The Transition from Mendelian Unit to Individual Gene." *Journal of the History of Biology,* 1974, *7:* 31–48.

———. *The Gene: A Critical History.* Philadelphia: Saunders, 1966.

———. *Genes, Radiation, and Society: The Life and Work of H. J. Muller.* Ithaca: Cornell University Press, 1981.

———. "H. J. Muller: The Role of the Scientist in Creating and Applying Knowledge." *Social Research,* 1984, *51:* 763–782.

Cassedy, James H. *American Medicine and Statistical Thinking, 1800–1860.* Cambridge, Mass.: Harvard University Press, 1984.

———. "The Microscope in American Medical Science." *Isis,* 1976, *67:* 76–97.

Caullery, Maurice. *Universities and Scientific Life in the United States.* Cambridge, Mass.: Harvard University Press, 1922.

Chittenden, Russell H. *History of the Sheffield Scientific School of Yale University, 1848–1922,* 2 vols. New Haven: Yale University Press, 1929.

Churchill, Frederick. "Chabry, Roux, and the Experimental Method in Nineteenth-century Embryology." In R. N. Giere and R. S. Westfall, eds., *Foundations of Scientific Method: The Nineteenth Century.* Bloomington, Ind.: Indiana University Press, 1973.

———. "Hertwig, Weismann, and the Meaning of Reduction Division circa 1890." *Isis,* 1970, *61:* 429–457.

———. "In Search of the New Biology: An Epilogue." *Journal of the History of Biology,* 1981, *14:* 177–191.

———. "William Johannsen [1857–1927] and the Genotype Concept." *Journal of the History of Biology,* 1974, *7:* 5–30.

Cittadino, Eugene. "Ecology and the Professionalization of Botany in America, 1890–1905." *Studies in History of Biology,* 1980, *4:* 171–198.

Clarke, Adele. "Research Materials and Reproductive Physiology in the United States, 1910–1940." In Gerald L. Geison, ed., *Physiology in the American Context, 1850–1940.* Bethesda, Md.: American Physiological Society, 1987.

Cock, A. G. "William Bateson's Rejection and Eventual Acceptance of Chromosome Theory." *Annals of Science,* 1983, *40:* 19–59.

Cohen, Barnett. *Chronicles of the Society of American Bacteriologists, 1899–1950.* Baltimore: Society of American Bacteriologists, 1950.

Cohen, Michael P. *The Pathless Way: John Muir and American Wilderness.* Madison: University of Wisconsin Press, 1984.

Cohen, Seymour S. "Some Struggles of Jacques Loeb, Albert Mathews, and Ernest Just at the Marine Biological Laboratory." *Biological Bulletin,* 1985, *168* (supplement): 127–136.

Colbert, Edwin H. *A Fossil-Hunter's Notebook: My Life with Dinosaurs and Other Friends.* New York: E. P. Dutton, 1980.

Coleman, William. *Biology in the Nineteenth Century.* New York: John Wiley, 1971.

Colwell, Robert K. "The Evolution of Ecology." *American Zoologist,* 1985, *25:* 771–777.

Cox, Thomas R. "Americans and Their Forests: Romanticism, Progress, and Science in the 19th Century." *Journal of Forest History,* 1985, *29:* 156–168.

Cravens, Hamilton. "The Role of Universities in the Rise of Experimental Biology." *Science Teacher,* 1977, *44:* 33–37.

———. *The Triumph of Evolution: American Scientists and the Heredity-Environment Controversy, 1900–1941.* Philadelphia: University of Pennsylvania Press, 1978.

Danbom, David. *The Resisted Revolution: Urban America and the Industrialization of Agriculture, 1900–1930.* Ames: Iowa State University Press, 1979.

Daniels, George H. "The Process of Professionalization in American Science, 1820–1860." *Isis,* 1967, *58:* 151–166.

———, ed. *Darwinism Comes to America.* Waltham, Mass.: Blaisdell, 1968.

———. *Nineteenth Century American Science: A Reappraisal.* Evanston: Northwestern University Press, 1972.

Deiss, William A. "Spencer F. Baird and His Collectors." *Journal of the Society for the Bibliography of Natural History,* 1980, *9:* 635–645.

Dexter, Ralph W. "Agassiz on Zoological Classification and Nomenclature." *BIOS,* 1979, *50:* 218–222.

———. "The Annisquam Laboratory of Alpheus Hyatt." *Scientific Monthly,* 1952, *74:* 112–116.

———. "The Annisquam Sea-Side Laboratory of Alpheus Hyatt, Predecessor of the Marine Biological Laboratory at Woods Hole, 1880–1886." In Mary Sears and Daniel Merriam, eds. *Oceanography: The Past.* New York: Springer-Verlag, 1980.

———. "C[harles] O[tis] Whitman [1842–1910] and the American Society of Zoologists." *American Zoologist,* 1979, *19:* 1251–1253.

———. "From Penikese to the Marine Biological Laboratory at Woods Hole—The Role of Agassiz's Students," *Essex Institute Historical Collection,* 1974, *110:* 151–161.

———. "Natural History at the Essex Institute, 1848–1898." *Essex Institute Historical Collection,* 1980, *116:* 21–33.

———. "The Salem Secession of Agassiz Zoologists." *Essex Institute Historical Collection,* 1965, *101:* 27–39.

Dornfeld, E. J. "The Allis Lake Laboratory, 1886–1893." *Marquette Medical Review,* 1956, *21:* 115–144.

Dupree, A. Hunter. *Asa Gray, 1810–1888.* New York: Athenaeum, 1968.

———. *Science and the Emergence of Modern America, 1865–1916.* Chicago: Rand McNally, 1963.

———. *Science in the Federal Government,* 2nd ed. Baltimore: Johns Hopkins University Press, 1986.

Durant, John R. "Innate Character in Animals and Man: A Perspective on the Origins of Ethology." In Charles Webster, ed., *Biology, Medicine and Society 1840–1940.* Cambridge: Cambridge University Press, 1981.

Ebert, James D. "Laying the Ghost: Embryonic Development, in Plain Words." *Biological Bulletin,* 1985, *168* (supplement): 62–79.

Eckel, Edwin B. *The Geological Society of America: Life History of a Learned Society.* Memoir 155. Boulder, Colo.: Geological Society of America, 1982.

Egerton, Frank N. "The History of Ecology: Achievements and Opportunities, Part One." *Journal of the History of Biology,* 1983, *16:* 259–311; "Part Two." *Journal of the History of Biology,* 1985, *18:* 103–144.

———, ed. *History of American Ecology.* New York: Arno Press, 1977.

Engel, J. Ronald. *Sacred Sands: The Struggle for Community in the Indiana Dunes.* Middletown, Conn.: Wesleyan University Press, 1983.

———. "Social Democracy, the Roots of Ecology, and the Preservation of the Indiana Dunes." *Journal of Forest Research,* 1984, *28:* 4–13.

Evans, Mary Alice and Howard Ensign Evans. *William Morton Wheeler, Biologist.* Cambridge, Mass.: Harvard University Press, 1970.

Ewan, Joseph, ed. *A Short History of Botany in the United States.* New York: Hafner Publishing Company, 1969.

Farber, Paul L. "Discussion Paper: The Transformation of Natural History in the Nineteenth Century." *Journal of the History of Biology,* 1982, *15:* 145–152.

Farley, John. *Gametes and Spores: Ideas about Sexual Reproduction, 1750–1914.* Baltimore: Johns Hopkins University Press, 1982.

Figlio, Karl. "The Historiography of Scientific Medicine: An Invitation to the Human Sciences." *Comparative Studies in Society and History,* 1977, *19:* 262–286.

Fine, William F. *Progressive Evolutionism and American Sociology, 1890–1920.* Ann Arbor: UMI Research Press, 1979.

Fitzgerald, Deborah K. "The Business of Breeding: Public and Private Development of Hybrid Corn in Illinois, 1890–1940." Ph.D. dissertation, University of Pennsylvania, 1985.

Frederickson, George Marsh. *The Inner Civil War: Northern Intellectuals and the Crisis of the Union.* New York: Harper and Row, 1965.

Fruton, Joseph S. "The Interplay of Chemistry and Biology at the Turn of the Century." In Carl Gustaf Bernhard, Elisabeth Crawford, and Per Sörbom, eds., *Science, Technology, and Society in the Time of Alfred Nobel.* Oxford: Pergamon Press, 1982.

Fye, W. Bruce. *The Development of American Physiology: Scientific Medicine in the Nineteenth Century.* Baltimore: Johns Hopkins University Press, 1987.

———. "H. Newell Martin: A Remarkable Career Destroyed by Neurasthenia and Alcoholism." *Journal of the History of Medicine and Allied Sciences,* 1985, *40:* 133–166.

———. "Why a Physiologist? The Case of Henry P. Bowditch." *Bulletin of the History of Medicine,* 1982, *56:* 19–29.

Galtsoff, Paul C. *The Story of the Bureau of Commercial Fisheries Biological Laboratory, Woods Hole, Massachusetts.* U.S. Department of the Interior, circular no. 145, 1962.

Geiger, Roger L. *To Advance Knowledge: The Growth of American Research Universities, 1900–1940.* New York: Oxford University Press, 1986.

Geison, Gerald L. *Michael Foster and the Cambridge School of Physiology: The Scientific Enterprise in Late Victorian Society.* Princeton: Princeton University Press, 1978.

———, ed. *Physiology in the American Context, 1850–1940.* Bethesda, Md.: American Physiological Society, 1987.

Gerson, Elihu. "Scientific Work and Social Worlds." *Knowledge: Creation, Diffusion, Utilization,* 1983, *4:* 357–377.

Gilbert, Scott F. "The Embryological Origins of the Gene Theory." *Journal of the History of Biology,* 1978, *11:* 307–351.

———. "Intellectual Traditions in the Life Sciences: Molecular Biology and Biochemistry." *Perspectives in Biology and Medicine,* 1982, *26:* 151–162.

Gilbert, Scott F., and J. P. Greenberg. "Intellectual Traditions in the Life Sciences. II. Stereocomplementarity." *Perspectives in Biology and Medicine,* 1984, *28:* 18–34.

Gillespie, Neal C. "Preparing for Darwin: Conchology and Natural Theology in Anglo-American Natural History." *Studies in History of Biology,* 1984, *7:* 93–145.

Goetzmann, William H. *Exploration and Empire: The Explorer and the Scientist in the Winning of the American West.* New York: Norton, 1978.

Goodfield, June. *An Imagined World.* New York: Harper and Row, 1981.

Gould, Stephen Jay. "Dollo on Dollo's Law: Irreversibility and the Status of Evolutionary Laws." *Journal of the History of Biology,* 1970, *3:* 189–212.

———. "G. G. Simpson, Paleontology, and the Modern Synthesis." In Ernst Mayr and William B. Provine, eds., *The Evolutionary Synthesis: Perspectives on the Unification of Biology.* Cambridge, Mass.: Harvard University Press, 1980.

———. *Ontogeny and Phylogeny.* Cambridge, Mass.: Belknap Press of Harvard University Press, 1977.

Greene, John C. *American Science in the Age of Jefferson.* Ames: Iowa State University Press, 1984.

Gregory, Joseph T. "North American Vertebrate Paleontology, 1776–1976." In Cecil J. Schneer, ed., *Two Hundred Years of Geology in America; The Proceedings of the New Hampshire Bicentennial Conference on the History of Geology.* Hanover, N.H.: University Press of New England, 1979.

Guralnick, Stanley M. *Science and the Ante-Bellum American College.* Philadelphia: American Philosophical Society, 1975.

———. "Sources of Misconception on the Roles of Science in the Nineteenth-Century American College." *Isis,* 1975, *65:* 352–366.

Gussin, A. E. S. "Jacques Loeb: The Man and His Tropism Theory of Animal Conduct." *Journal of the History of Medicine and Allied Sciences,* 1963, *18:* 333.

Hagen, Joel B. "The Development of Experimental Methods in Plant Taxonomy." *Taxon,* 1983, *32:* 406–416.

———. "Ecologists and Taxonomists: Divergent Traditions in Twentieth-Century Plant Geography." *Journal of the History of Biology,* 1986, *19:* 197–214.

———. "Experimentalists and Naturalists in Twentieth-Century Botany: Experimental Taxonomy, 1920–1950." *Journal of the History of Biology,* 1984, *17:* 249–270.

Hall, Peter. *The Organization of American Culture, 1700–1900. Private Institutions, Elites, and the Origins of American Nationality.* New York: New York University Press, 1982.

Haller, Mark. *Eugenics: Hereditarian Attitudes in American Thought.* New Brunswick, N.J.: Rutgers University Press, 1963.

Hamburger, Viktor. *The Heritage of Experimental Embryology.* New York: Oxford University Press, 1987.

Hanley, Wayne. *Natural History in America, from Mark Catesby to Rachel Carson.* New York: Quadrangle, 1977.

Hannaway, Owen. "The German Model of Chemical Education in America: Ira Remsen at Johns Hopkins (1876–1913)." *Ambix,* 1976, *23:* 145–163.

Haraway, Donna J. *Crystals, Fabrics, and Fields.* New Haven: Yale University Press, 1976.

———. "The Marine Biological Laboratory of Woods-Hole: An Ideology of Biological Expansion." Unpublished paper, 1975.

———. "Reinterpretation or Rehabilitation: An Exercise in Contemporary Marxist History of Science (A Review)." *Studies in the History of Biology,* 1978, *2:* 193–209.

———. "Teddy Bear Patriarchy: Taxidermy in the Garden of Eden, New York City, 1908–1936." *Social Text,* 1984, *5:* 20–64.

Hawkins, Hugh. *Pioneer: A History of the Johns Hopkins University, 1874–1889.* Ithaca: Cornell University Press, 1960.

———. "Transatlantic Discipleship: Two American Biologists and Their German Mentor." *Isis,* 1980, *71:* 197–210.

———, ed. *The Emerging University and Industrial America.* Lexington, Mass.: D. C. Heath, 1970.

Hearnshaw, Leslie. *Cyril Burt, Psychologist.* Ithaca: Cornell University Press, 1979.

Herbst, Jurgen. "Diversification in American Higher Education." In Konrad Jarusch, ed., *The Transformation of Higher Learning, 1860–1930: Expansion, Diversification, Social Opening, and Professionalism in England, Germany, Russia, and the United States.* Chicago: University of Chicago Press, 1983.

Hinsley, Curtis J., Jr. *Savages and Scientists: The Smithsonian Institution and the Development of American Anthropology, 1846–1910.* Washington, D.C.: Smithsonian Institution Press, 1980.

Hoeveler, J. David, Jr. *James McCosh and the Scottish Intellectual Tradition.* Princeton: Princeton University Press, 1981.

Hofstadter, Richard. *Social Darwinism in American Thought,* 2nd ed. New York: George Braziller, 1955.

Horder, T. J., J. A. Witkowski, and C. C. Wylie, eds. *A History of Embryology.* Cambridge: Cambridge University Press, 1985.

Horowitz, Helen Lefkowitz. *Alma Mater: Design and Experience in the Women's Colleges from their Nineteenth Century Beginnings to the 1930s.* New York: Knopf, 1984.

———. "Animal and Man in the New York Zoological Park." *New York History,* 1975, *55:* 426–453.

———. *Culture and the City: Cultural Philanthropy in Chicago from the 1880s to 1917.* Lexington: University of Kentucky Press, 1976.

Hovencamp, Herbert. *Science and Religion in America, 1800–1860.* Philadelphia, University of Pennsylvania Press, 1978.

Hutchinson, George Evelyn. *The Kindly Fruits of the Earth: The Development of an Embryo Ecologist.* New Haven: Yale University Press, 1979.

Jarusch, Konrad, ed. *The Transformation of Higher Learning, 1860–1930: Expansion, Diversification, Social Opening, and Professionalism in England, Germany, Russia, and the United States*. Chicago: University of Chicago Press, 1983.

Jones, Howard Mumford. *The Age of Energy: Varieties of the American Experience 1865–1915*. New York: Viking Press, 1971.

Judson, Horace Freeland. *The Eighth Day of Creation: Makers of the Revolution in Biology*. New York: Simon & Schuster, 1979.

Kalikow, Theodora J. "Konrad Lorenz's Ethological Theory: Explanation and Ideology, 1938–1943." *Journal of the History of Biology,* 1983, *16:* 39–73.

Kay, Lily E. "Conceptual Models and Analytical Tools: The Biology of Physicist Max Delbruck," *Journal of the History of Biology,* 1985, *18:* 207–246.

———. "W. M. Stanley's Crystallization of the Tobacco Mosaic Virus, 1930–1940." *Isis,* 1986, *77:* 450–472.

Keenan, Katherine. "Lilian Vaughan Morgan (1870–1952): Her Life and Work." *American Zoologist,* 1983, *23:* 867–876.

Keller, Evelyn Fox. *A Feeling for the Organism*. San Francisco: W. H. Freeman, 1983.

Kennedy, John Michael. "Philanthropy and Science in New York City: The American Museum of Natural History, 1868–1968." Ph.D. dissertation, Yale University, 1968.

Kevles, Daniel. "Genetics in the United States and Great Britain, 1890–1930: A Review with Speculations." In Charles Webster, ed., *Biology, Medicine and Society 1840–1940*. Cambridge: Cambridge University Press, 1981.

———. *In the Name of Eugenics: Genetics and the Uses of Human Heredity*. New York: Knopf, 1985.

Kevles, Daniel, Jeffrey L. Sturchio and Thomas P. Carroll. "The Sciences in America, circa 1880." *Science,* 1980, *209:* 27–32.

Kimmelman, Barbara A. "The American Breeders Association: Genetics and Eugenics in an Agricultural Context." *Social Studies of Science,* 1983, *13:* 163–204.

———. "A Progressive Era Discipline: Genetics and American Agricultural Colleges and Experiment Stations, 1890–1920." Ph.D. dissertation, University of Pennsylvania, 1987.

Kingsland, Sharon. *Modeling Nature: Episodes in the History of Population Ecology*. Chicago: University of Chicago Press, 1985.

———. "Raymond Pearl: On the Frontier in the 1920s." *Human Biology,* 1984, *56:* 1–18.

———. "The Refractory Model: The Logistic Curve and the History of Population Ecology." *Quarterly Review of Biology,* 1982, *52:* 29–52.

Kloppenburg, Jack. *First the Seed: The Political Economy of Plant Biotechnology*. Cambridge: Cambridge University Press, 1988.

Kohler, Robert. *From Medical Chemistry to Biochemistry. The Making of a Biomedical Discipline*. Cambridge: Cambridge University Press, 1982.

———. "A Policy for the Advancement of Science: The Rockefeller Foundation, 1924–29." *Minerva,* 1978, *16:* 480–515.

———. "Science and Philanthropy: Wickliffe Rose and the International Education Board." *Minerva,* 1985, *23:* 75–95.

———. "Warren Weaver and the Rockefeller Foundation Program in Molecular Biology: A Case Study in the Management of Science." In Nathan Reingold, ed., *The Sciences in the American Context: New Perspectives.* Washington, D.C.: Smithsonian Institution Press, 1979.

Kohlstedt, Sally Gregory. "Collectors, Cabinets, and Summer Camp: Natural History in the Public Life of Nineteenth-century Worcester." *Museum Studies Journal,* 1985, *2:* 10–23.

———. *The Formation of the American Scientific Community.* Urbana: University of Illinois Press, 1976.

———. "From Learned Society to Public Museum: The Boston Society of Natural History." In Alexandra Oleson and John Voss, eds., *The Organization of Knowledge in Modern America, 1860–1920.* Baltimore: Johns Hopkins University Press, 1979.

———. "Henry A. Ward: The Merchant Naturalist and American Museum Development." *Journal of the Society for the Bibliography of Natural History,* 1980, *9:* 647–661.

———. "Historical Resources in Natural History Museums." *History of Science in America: News and Views,* 1985, *3:* 5–6.

———. "In from the Periphery: American Women in Science, 1830–1880." *Signs,* 1978, *4:* 81–96.

———. "Institutional History." *Osiris,* n.s., 1985, *1:* 17–36.

———. "International Exchange and National Style: A View of Natural History Museums in the United States, 1860–1900." In Nathan Reingold and Marc Rothenberg, eds., *Scientific Colonialism: A Cross-National Comparison.* Washington, D.C.: Smithsonian Institution Press, 1986.

———. "The Nineteenth-Century Amateur Tradition: The Case of the Boston Society of Natural History." In Gerald Holton and W. A. Blanpied, eds., *Science and Its Public.* Dordrecht: D. Reidel Company, 1976.

———. "*Science:* The Struggle for Survival, 1880 to 1894." *Science,* 1980, *209:* 33–42.

Kohlstedt, Sally Gregory, and Margaret Rossiter, eds. *Historical Writing on American Science: Perspectives and Prospects.* Baltimore: Johns Hopkins Press, 1986.

Kopper, Philip. *The National Museum of Natural History.* New York: Abrams/Smithsonian, 1982.

Kuritz, Hyman. "The Popularization of Science in Nineteenth-Century America." *History of Education Quarterly,* 1981, *21:* 259–274.

Lanham, Url. *The Bone Hunters.* New York: Columbia University Press, 1973.

Laporte, Leo F. "Simpson's *Tempo and Mode in Evolution* Revisited." *Proceedings of the American Philosophical Society,* 1983, *127:* 365–416.

Lederer, Susan E. "The Controversy Over Animal Experimentation in America, 1880–1914." In Nicolaas A. Rupke, ed., *Vivisection in Historical Perspective.* London: Croom Helm, 1987.

Levine, David O. *The American College and the Culture of Aspiration, 1915–1940.* Ithaca: Cornell University Press, 1986.

Lillie, Frank R. *The Woods Hole Marine Biological Laboratory.* Chicago: University of Chicago Press, 1944.

Long, Diana E. "Physiological Identity of American Sex Researchers between the two

World Wars." In Gerald L. Geison, ed., *Physiology in the American Context, 1850–1940*. Bethesda, Md.: American Physiological Society, 1987.

Ludmerer, Kenneth M. *Genetics and American Society*. Baltimore: Johns Hopkins University Press, 1974.

———. *Learning to Heal: The Development of American Medical Education*. New York: Basic Books, 1985.

———. "Reform at Harvard Medical School, 1869–1909." *Bulletin of the History of Medicine*, 1981, *55:* 343–370.

Lurie, Edward. *The Founding of the Museum of Comparative Zoology*. Cambridge, Mass.: Museum of Comparative Zoology, 1960.

———. *Louis Agassiz: A Life in Science*. Chicago: University of Chicago Press, 1960.

———. *Nature and the American Mind: Louis Agassiz and the Culture of Science*. New York: Science History, 1974.

Maienschein, Jane. "Agassiz, Hyatt, Whitman and the Birth of the Marine Biological Laboratory." *Biological Bulletin*, 1985, *168* (supplement): 26–34.

———. "Cell Lineage, Ancestral Reminiscence, and the Biogenetic Law." *Journal of the History of Biology*, 1978, *11:* 129–153.

———. "Early Struggles at the Marine Biological Laboratory over Mission and Money." *Biological Bulletin*, 1985, *168* (supplement): 192–196.

———. "Experimental Biology in Transition: Harrison's Embryology." *Studies in History of Biology*, 1983, *6:* 107–127.

———. "History of Biology." *Osiris*, n.s., 1985, *1:* 147–162.

———. "Physiology, Biology, and the Advent of Physiological Morphology." In Gerald L. Geison, ed., *Physiology in the American Context, 1850–1940*. Bethesda, Md.: American Physiological Society, 1987.

———. "Preformation or New Formation—Or Neither or Both?" In T. J. Horder, J. A. Witkowsky, and C. C. Wylie, eds., *A History of Embryology*. Cambridge: Cambridge University Press, 1986.

———. "Shifting Assumptions in American Biology: Embryology, 1890–1910." *Journal of the History of Biology*, 1981, *14:* 89–113.

———. "What Determines Sex? A Study of Converging Approaches, 1880–1916." *Isis*, 1984, *75:* 457–480.

———, ed. *Defining Biology: Lectures from the 1890s*. Cambridge, Mass.: Harvard University Press, 1986.

Maienschein, Jane, James P. Collins, and John Beatty, eds. "Reflections on Ecology and Evolution." *Journal of the History of Biology*, 1986, *19:* 167–312.

Maienschein, Jane, Ronald Rainger, and Keith R. Benson. "Introduction: Were American Morphologists in Revolt?" *Journal of the History of Biology*, 1981, *14:* 83–87.

Manier, Edward. "The Experimental Method in Biology: T. H. Morgan and the Theory of the Gene." *Synthese*, 1969, *20:* 185–205.

Manning, Kenneth R. *Black Apollo of Science: The Life of Ernest Everett Just*. New York: Oxford University Press, 1983.

Manning, Thomas G. *Government in Science: The U.S. Geological Survey, 1867–1894*. Lexington: University of Kentucky Press, 1967.

Marcell, David. *Progress and Pragmatism*. Westport, Conn.: Greenwood Press, 1974.

Marcus, Alan I. *Agricultural Science and the Quest for Legitimacy: Farmers, Agricul-*

tural Colleges, and Experiment Stations, 1870–1890. Ames: Iowa State University Press, 1985.

May, Robert M., and Jon Seger. "Ideas in Ecology." *American Scientist,* 1986, *74:* 256–267.

Mayr, Ernst. *The Growth of Biological Thought.* Cambridge, Mass.: Belknap Press of Harvard University Press, 1982.

Mayr, Ernst, and William B. Provine, eds. *The Evolutionary Synthesis: Perspectives on the Unification of Biology.* Cambridge, Mass.: Harvard University Press, 1980.

McCaughey, Robert A. "The Transformation of American Academic Life: Harvard University 1821–1892." *Perspectives in American History,* 1974, *8:* 229–334.

McClung, L. S. "The American Society for Microbiology/Society of American Bacteriologists: A Brief History." *A S M News,* 1978, *44:* 446–451.

McCullough, Dennis. "W. K. Brooks' Role in the History of American Biology." *Journal of the History of Biology,* 1968, *2:* 411–438.

McIntosh, Robert P. *The Background of Ecology: Concept and Theory.* New York: Cambridge University Press, 1985.

Meisel, Max. *A Bibliography of American Natural History: The Pioneer Century: 1769–1865,* 3 vols. New York: Hafner Publishing Company, 1967.

Merchant, Carolyn. "Women of the Progressive Conservation Movement." *Environmental Review,* 1984, *8:* 57–85.

Miller, Howard. *Dollars for Research: Science and Its Patrons in Nineteenth-Century America.* Seattle: University of Washington Press, 1970.

Moore, James R. *The Post-Darwinian Controversies: A Study of the Protestant Struggle to Come to Terms with Darwin in Great Britain and America, 1870–1900.* Cambridge: Cambridge University Press, 1979.

Moore, John A. "Zoology of the Pacific Railroad Surveys." *American Zoologist,* 1986, *26:* 311–341.

Morantz-Sanchez, Regina Markell. *Sympathy and Science: Women Physicians in American Medicine.* New York: Oxford University Press, 1985.

Morrison, Theodore. *Chautauqua: A Center for Education, Religion, and the Arts in America.* Chicago: University of Chicago Press, 1974.

Nash, Roderick. *Wilderness and the American Mind,* 3rd ed. New Haven: Yale University Press, 1982.

Nelkin, Dorothy. *The Creation Controversy: Science or Scripture in the Schools.* New York: Norton, 1982.

Nelson, Clifford M., and Ellis L. Yochelson. "Organizing Federal Paleontology in the United States, 1858–1907." *Journal of the Society for the Bibliography of Natural History,* 1980, *9:* 607–618.

Newan, Louise Michele, ed. *Men's Ideas/Women's Realities: Popular Science, 1870–1915.* New York: Pergamon Press, 1985.

Noble, David W. *The Paradox of Progressive Thought.* Minneapolis: University of Minnesota Press, 1958.

Numbers, Ronald L., ed. *The Education of American Physicians.* Berkeley: University of California Press, 1980.

Nyhart, Lynn Keller. "The *American Naturalist,* 1867–1886: A Case Study of the Relationship Between Amateur and Professional Naturalists in Nineteenth-Century America." Senior thesis, Princeton University, 1979.

O'Donnell, John M. *The Origins of Behaviorism: American Psychology, 1870–1920.* New York: New York University Press, 1985.

Ogilvie, Marilyn, and Clifford Choquette. "Nettie Marie Stevens (1861–1912): Her Life and Contributions to Cytogenetics." *Proceedings of the American Philosophical Society,* 1981, *125:* 292–311.

Oleson, Alexandra, and Sanford C. Brown, eds. *The Pursuit of Knowledge in the Early American Republic: American Scientific and Learned Societies from Colonial Times to the Civil War.* Baltimore: Johns Hopkins University Press, 1976.

Oleson, Alexandra, and John Voss, eds. *The Organization of Knowledge in Modern America, 1860–1920.* Baltimore: Johns Hopkins University Press, 1979.

Oppenheimer, Jane. "Basic Embryology and Clinical Medicine: A Case History in Serendipity." *Bulletin of the History of Medicine,* 1984, *58:* 236–240.

———. *Essays in the History of Embryology and Biology.* Cambridge, Mass.: MIT Press, 1967.

Osborn, Henry Fairfield. *Cope: Master Naturalist.* Princeton: Princeton University Press, 1931.

Overfield, Richard. "Charles E. Bessey: The Impact of the 'New Botany' in American Agriculture, 1880–1910." *Technology and Culture,* 1975, *16:* 162–181.

Owen, Ray David. "Genetics in the 20th Century." *Journal of Heredity,* 1983, *74:* 314–319.

Owens, Larry. "Pure and Sound Government: Laboratories, Gymnasia, and Playing Fields in Nineteenth-Century America." *Isis,* 1985, *76:* 182–194.

Paul, Diane B. "Eugenics and the Left." *Journal of the History of Ideas,* 1984, *45:* 567–590.

———. "Textbook Treatments of the Genetics of Intelligence." *Quarterly Review of Biology,* 1985, *60:* 317–326.

Pauly, Philip. "American Biologists in Wilhelmian Germany: Another Look at Innocents Abroad." Unpublished manuscript.

———. "The Appearance of Academic Biology in Late Nineteenth-Century America." *Journal of the History of Biology,* 1984, *17:* 369–397.

———. *Controlling Life. Jacques Loeb and the Engineering Ideal in Biology.* New York: Oxford University Press, 1987.

———. "G. Stanley Hall and His Successors: A History of the First Half-Century of Psychology at Johns Hopkins University." In Stewart H. Hulse and Bert F. Green, Jr., eds., *One Hundred Years of Experimental Psychology: G. Stanley Hall and the Hopkins Tradition.* Baltimore: Johns Hopkins University Press, 1986.

———. "General Physiology and the 'Discipline' of Physiology, 1890–1930." In Gerald L. Geison, ed., *Physiology in the American Context, 1850–1940.* Bethesda, Md.: American Physiological Society, 1987.

———. "The Loeb-Jennings Debate and the Science of Animal Behavior." *Journal of the History of the Behavioral Sciences,* 1981, *17:* 504–515.

———. "Psychology at Hopkins: Its Rise and Fall and Rise and Fall and . . . ." *Johns Hopkins Magazine,* 1979, *30:* 36–41.

———. "The World and All That is in It: The National Geographical Society, 1888–1918." *American Quarterly,* 1979, *31:* 517–532.

Pfeifer, Edward J. "The Genesis of American Neo-Lamarckism." *Isis,* 1965, *56:* 156–167.

Phillips, D. C. "Organicism in the Late Nineteenth and Early Twentieth Centuries." *Journal of the History of Ideas*, 1970, *31:* 413–432.

Porter, Charlotte M. "The Concussion of Revolution: Publications and Reform at the Early Academy of Natural Sciences, Philadelphia, 1812–1842." *Journal of the History of Biology*, 1979, *12:* 273–292.

———. *The Eagle's Nest. Natural History and American Ideas, 1812–1842*. University, Ala.: University of Alabama Press, 1986.

———. "The Rise of Parnassus: Henry Fairfield Osborn and the Hall of the Age of Man." *Museum Studies Journal*, 1983, *1:* 26–34.

Provine, William B. "Francis B. Sumner and the Evolutionary Synthesis." *Studies in History of Biology*, 1979, *3:* 211–240.

———. *The Origins of Theoretical Population Genetics*. Chicago: University of Chicago Press, 1971.

———. *Sewall Wright: Geneticist and Evolutionist*. Chicago: University of Chicago Press, 1987.

Pyne, Stephen J. *Fire in America: A Cultural History of Wildland and Rural Fire*. Princeton: Princeton University Press, 1982.

Quinn, Brother C. Edward. "Ancestry and Beginnings: The Early History of the American Society of Zoologists." *American Zoologist*, 1982, *22:* 735–748.

———. "The Beginnings of the American Society of Zoologists." *American Zoologist*, 1979, *19:* 1247–1249.

Rainger, Ronald. "The Continuation of the Morphological Tradition: American Paleontology, 1880–1910." *Journal of the History of Biology*, 1981, *14:* 129–158.

———. "Just Before Simpson: William Diller Matthew's Understanding of Evolution." *Proceedings of the American Philosophical Society*, 1986, *130:* 453–474.

———. "Paleontology and Philosophy: A Critique." *Journal of the History of Biology*, 1985, *18:* 267–287.

———. "What's the Use: William King Gregory and the Functional Morphology of Fossil Vertebrates." *Journal of the History of Biology* (forthcoming).

Raitt, Helen, and Beatrice Moulton. *Scripps Institution of Oceanography*. La Jolla, Calif.: The Ward Ritchie Press, 1967.

Reingold, Nathan. "Definitions and Speculations: The Professionalization of Science in America in the Nineteenth Century." In Alexandra Oleson and Sanborn C. Brown, eds., *The Pursuit of Knowledge in the Early American Republic: American Scientific and Learned Societies from Colonial Times to the Civil War*. Baltimore: Johns Hopkins University Press, 1976.

———. "Jacques Loeb, the Scientist: His Papers and His Era." *Library of Congress Quarterly Journal of Current Acquisitions*, 1962, *19:* 119–130.

———, ed. *Science in America since 1820*. New York: Science History Publications, 1976.

———, ed. *Science in Nineteenth-Century America: A Documentary History*. New York: Hill & Wang, 1964.

———, ed. *The Sciences in the American Context: New Perspectives*. Washington, D.C.: Smithsonian Institution Press, 1979.

Reingold, Nathan, and Ida H. Reingold, eds. *Science in America: A Documentary History, 1900–1939*. Chicago: University of Chicago Press, 1981.

Reingold, Nathan, and Marc Rothenberg, eds. *Scientific Colonialism: A Cross-National Comparison.* Washington, D.C.: Smithsonian Institution Press, 1986.

Richardson, Jonathan L. "The Organismic Community: Resilience of an Embattled Concept." *Bioscience,* 1980, *30:* 465–471.

Rosen, George. "Carl Ludwig and His American Students." *Bulletin of the History of Medicine,* 1936, *4:* 609–650.

Rosenberg, Charles E. "The Adams Act: Politics and the Cause of Scientific Research." In Charles E. Rosenberg, *No Other Gods: On Science and American Social Thought.* Baltimore: Johns Hopkins University Press, 1976.

———. "Factors in the Development of Genetics in the United States: Some Suggestions." *Journal of the History of Medicine and Allied Sciences,* 1967, *22:* 27–46.

———. *No Other Gods: On Science and American Social Thought.* Baltimore: Johns Hopkins University Press, 1976.

———. "Science, Technology, and Economic Growth: The Case of the Agricultural Experiment Station Scientist, 1875–1914." In Charles E. Rosenberg, ed., *No Other Gods: On Science and American Social Thought.* Baltimore: Johns Hopkins University Press, 1976.

Rosenberg, Rosalind. *Beyond Separate Spheres: The Intellectual Roots of Modern Feminism.* New Haven: Yale University Press, 1982.

Ross, Dorothy. *G. Stanley Hall: The Psychologist as Prophet.* Chicago: University of Chicago Press, 1972.

Rossiter, Margaret W. "Benjamin Silliman and the Lowell Institute: The Popularization of Science in Nineteenth-Century America." *New England Quarterly,* 1971, *44:* 602–626.

———. *The Emergence of Agricultural Science: Justus Liebig and the Americans, 1840–1880.* New Haven: Yale University Press, 1975.

———. "The Organization of the Agricultural Sciences." In Alexandra Oleson and John Voss, eds. *The Organization of Knowledge in Modern America, 1860–1920.* Baltimore: Johns Hopkins University Press, 1979.

———. *Women Scientists in America: Struggles and Strategies to 1940.* Baltimore: Johns Hopkins University Press, 1982.

———. " 'Women's Work' in Science, 1880–1910." *Isis,* 1980, *71:* 381–398.

Rudolph, Emanuel D. "The Introduction of the Natural System of Classification of Plants to 19th Century American Students." *Archives of Natural History,* 1982, *10:* 461–468.

———. "Women in 19th Century American Botany: A Generally Unrecognized Constituency." *American Journal of Botany,* 1982, *69:* 1346–1355.

Russell-Hunter, W. D. "An Evolutionary Century at Woods Hole: Instruction in Invertebrate Zoology." *Biological Bulletin,* 1985, *168* (supplement): 88–98.

Russett, Cynthia Eagle. *The Concept of Equilibrium in American Social Thought.* New Haven: Yale University Press, 1966.

———. *Darwin in America.* San Francisco: W. H. Freeman, 1976.

Samelson, Franz. "The Struggle for Scientific Authority: The Reception of Watson's Behaviorism." *Journal of the History of the Behavioral Sciences,* 1981, *17:* 399–425.

Sandler, Iris, and Laurence Sandler. "A Conceptual Ambiguity that Contributed to the Neglect of Mendel's Paper." *History and Philosophy of the Life Sciences,* 1985, *7:* 3–70.

Sapp, Jan. *Beyond the Gene: Cytoplasmic Inheritance and the Struggle for Authority in Genetics.* New York: Oxford University Press, 1987.

———. "The Struggle for Authority in the Field of Heredity, 1900–1932: New Perspectives on the Rise of Genetics." *Journal of the History of Biology,* 1983, *16:* 311–342.

Schlee, Susan. *The Edge of an Unfamiliar World: A History of Oceanography.* New York: Dutton, 1973.

Schmitt, Peter J. *Back to Nature: The Arcadian Myth in Urban America.* New York: Oxford University Press, 1969.

Schuchert, Charles, and Clara M. LeVene. *O. C. Marsh: Pioneer in Paleontology.* New Haven: Yale University Press, 1940.

Sears, Mary, and Daniel Merriam, eds. *Oceanography: The Past.* New York: Springer, 1980.

Sellers, Charles Coleman. *Mr. Peale's Museum: Charles Willson Peale and the First Popular Museum of Natural Science and Art.* New York: Norton, 1980.

Servos, John C. "Physical Chemistry in America, 1890–1933: Origins, Growth, and Definitions." Ph.D. dissertation, Johns Hopkins University, 1979.

Sheets-Pyenson, Susan. "Henry Augustus Ward and Museum Development in the Hinterland." *University of Rochester Library Bulletin,* 1985, *38:* 38–59.

Shils, Edward. "The Order of Learning in the United States from 1865–1920: The Ascendancy of the Universities." In Alexandra Oleson and John Voss, eds., *The Organization of Knowledge in Modern America, 1860–1920.* Baltimore: Johns Hopkins University Press, 1979.

Shor, Elizabeth Noble. *The Fossil Feud Between E. D. Cope and O. C. Marsh.* Hicksville, N.Y.: Exposition Press, 1974.

———. *Fossils and Flies: The Life of a Compleat Scientist Samuel Wendell Williston (1851–1918).* Norman: University of Oklahoma Press, 1971.

Simpson, George Gaylord. *Concession to the Improbable: An Unconventional Autobiography.* New Haven: Yale University Press, 1978.

Skinner, B. F. *The Shaping of a Behaviorist: Part Two of an Autobiography.* New York: Knopf, 1979.

Sloan, Douglas. "Science in New York City, 1867–1907." *Isis,* 1980, *71:* 35–76.

Sokal, Michael. "Companions in Zealous Research, 1886–1986." *American Scientist,* 1986, *74:* 486–509.

———, ed. *An Education in Psychology: James McKeen Cattell's Journal and Letters from Germany and England, 1880–1888.* Cambridge, Mass.: MIT Press, 1981.

Spencer, Frank, ed. *A History of American Physical Anthropology, 1930–1980.* New York: Academic Press, 1982.

Stearns, Stephen C. "The Emergence of Evolutionary and Community Ecology as Experimental Sciences." *Perspectives in Biology and Medicine,* 1982, *25:* 621–648.

Steere, William Campbell. *Biological Abstracts/BIOSIS. The First Fifty Years: The Evolution of a Major Science Information Service.* New York: Plenum Press, 1976.

———, ed. *Fifty Years of Botany: Golden Jubilee Volume of the Botanical Society of America.* New York: McGraw-Hill, 1958.

Stephens, Lester D. "Joseph LeConte and the Development of the Physiology and Psychology of Vision in the United States." *Annals of Science,* 1980, *37:* 303–321.

———. *Joseph LeConte, Gentle Prophet of Evolution.* Baton Rouge: Louisiana State University Press, 1982.

Sterling, Keir Brooks. *The Last of the Naturalists: The Career of C. Hart Merriam.* New York: Arno Press, 1977.

Stocking, George W., Jr. "Franz Boas and the Founding of the American Anthropological Association." *American Anthropologist,* 1960, *62:* 1–17.

———. *Race, Culture, and Evolution: Essays in the History of Anthropology.* New York: Free Press, 1968.

Storr, Richard J. *Harper's University: The Beginnings.* Chicago: University of Chicago Press, 1966.

Stott, R. Jeffrey. "The Historical Origins of the Zoological Park in American Thought." *Environmental Review,* 1981, *5:* 52–65.

Thorpe, W. H. *The Origins and Rise of Ethology.* London: Heinemann, 1979.

Tippo, Oswald. "The Early History of the Botanical Society of America." In William Campbell Steere, ed., *Fifty Years of Botany: Golden Jubilee volume of the Botanical Society of America.* New York: McGraw-Hill, 1958.

Tobey, Ronald C. *Saving the Prairies: The Life Cycle of the Founding School of American Plant Ecology, 1895–1955.* Berkeley: University of California Press, 1981.

Veysey, Laurence R. *The Emergence of the American University.* Chicago: University of Chicago Press, 1965.

Viola, Herman J., and Carolyn Margolis, eds. *Magnificent Voyagers: The U.S. Exploring Expedition, 1838–1842.* Washington, D.C.: Smithsonian Institution Press, 1985.

Warner, Deborah. "Science Education for Women in Antebellum America." *Isis,* 1978, *69:* 58–67.

Warner, John Harley. " 'Exploring the Inner Labyrinths of Creation': Popular Microscopy in 19th Century America." *Journal of the History of Medicine,* 1982, *37:* 7–33.

———. "Physiology." In Ronald L. Numbers, ed., *The Education of American Physicians.* Berkeley: University of California Press, 1980.

Watterson, Ray Leighton. "The Striking Influence of the Leadership, Research, and Teaching of Frank R. Lillie (1870–1947) in Zoology, Embryology, and other Biological Sciences." *American Zoologist,* 1979, *19:* 1275–1287.

Weindling, Paul. "Theories of the Cell State in Imperial Germany." In Charles Webster, ed., *Biology, Medicine, and Society, 1840–1940.* Cambridge: Cambridge University Press, 1981.

Weinstein, James. *The Corporate Ideal in the Liberal State, 1890–1918.* Boston: Beacon Press, 1968.

Werdinger, Jeffrey. "Embryology at Woods Hole: The Emergence of a New American Biology." Ph.D. dissertation, Indiana University, 1980.

Whalen, Matthew D., and Mary F. Tobin. "Periodicals and the Popularization of Science in America, 1860–1910." *Journal of American Culture,* 1980, *3:* 195–203.

Wiebe, Robert H. *The Search for Order 1877–1920.* New York: Hill and Wang, 1967.

Willier, B. H., and J. M. Oppenheimer, eds. *Foundations of Experimental Embryology.* Englewood Cliffs, N.J.: Prentice-Hall, 1964.

Wilson, Leonard. "The Emergence of Geology as Science in the United States." *Journal of World History,* 1967, *10:* 416–437.

Winsor, Mary P. "Louis Agassiz and the Species Question." *Studies in History of Biology,* 1979, *3:* 89–117.

Worster, Donald. *Nature's Economy: The Roots of Ecology.* San Francisco: Sierra Club Books, 1977.

Index

Academy of Natural Sciences, 29, 54, 56

Adams, Charles C.: and animal behavior, 186, 211; and Craig, 197, 200–201; and midwestern ecology, 259; and Yerkes, 211

Agassiz, Alexander: and Anderson Scientific School, 64, 128; and Brooks, 64; and Fish Commission laboratory, 129; and laboratory research, 64, 65, 73; and Museum of Comparative Zoology, 44n.67, 170; Newport Laboratory of, 64–65, 128; and Whitman, 128, 130, 170

Agassiz, Louis, 52, 90; and Anderson Scientific School, 64, 73, 128; *Essay on Classification,* 21; and Harvard, 16–17, 27, 46n.87, 151; and Lowell Institute, 55; and museum collections, 16–17; and Museum of Comparative Zoology, 17, 22, 27, 52, 55, 59, 60; and natural history education, 17, 42n.53, 57, 59, 60, 62; and natural theology, 57, 60; and *Principles of Zoology,* 57, 76; students of, 42n.52, 53, 55, 60, 62–63

Agriculture: and biology, 7, 138–40, 282–85, 290, 293, 300–302; and ecology, 259; and genetics, 139–40, 281–82, 290–93, 295–302; plant breeding and, 287–90, 293, 296–97, 300–302; at state colleges, 282, 290, 293; at state experiment stations, 290–93; USDA and, 282, 286–93; and wheat overproduction, 287–88

Alexander, Annie Montague, 32–33; and Grinnell, 32–33, 47n.99; and Museum of Vertebrate Zoology, 29, 32–33, 47n.99

Allee, Warder Clyde: and animal behavior studies, 211, 218n.66; and Chicago, 174; concept of physiological democracy, 333, 345n.67; and Whitman, 218n.66

Allis, Edward Phelps: and *Journal of Morphology,* 101, 170; and Whitman, 130, 132, 146n.32, 170

Allis Lake Laboratory: and Whitman, 70, 132, 158, 162, 170, 188

American Association for the Advancement of Science: amateurs and, 54; and American Society of Naturalists, 98, 102–6, 111; American Society of Zoologists and, 104–6, 111; and biological disciplinary societies, 89–90, 102–6, 108–11; and Botanical Society of America, 97, 102–3, 111; and convocation week, 102; and Federation of American Societies for Experimental Biology, 108, 111; and science teaching, 50; and Society of Naturalists of Eastern United States, 91; and Society for Plant Morphology and Physiology, 103

American Association of Anatomists, 88–89; and American Society of Naturalists, 95, 99, 106, 108, 111; and American Society of Zoologists, 108, 111; centennial of, 88; and Congress of American Physicians and Surgeons,

American Association of Anatomists (*continued*)
93–95; and Federation of American Societies for Experimental Biology, 111; founding of, 94–95; journals of, 101–2, 107; membership of, 94–95; and Wistar Institute, 101–2, 107
American Breeders Association: and eugenics, 281–82; and genetics, 281–83, 285; membership of, 281–85; and plant breeding, 281–85
American Chemical Society: and disciplinary unity, 87–88; and Mathews's Plan, 107; successes of, 107; and unity of biology, 87–88, 110, 112
American Morphological Society, 89; and American Society of Naturalists, 96, 99; and American Society of Zoologists, 89, 103, 118 n.53; and founding of, 96; and *Journal of Morphology,* 100; membership of, 96, 103; and morphology, 103; and systematic biology, 104; Whitman and, 96, 100, 188. *See also* American Society of Zoologists
American Museum of Natural History, 26, 29; and Columbia, 229, 237, 245; cooperative research at, 7, 230–45; Department of Comparative Anatomy, 237; Department of Vertebrate Paleontology, 7, 221, 229–45, 251 n.42, 255–56, n.84; and Gregory, 221, 237–45; and Low, 229; and Matthew, 221, 230–37, 238, 243–45; and Osborn, 7, 221, 228–33, 236–38, 243–45, 247 n.13, 250 n.39, 251 n.42; resources of, 229; and Simpson, 245, 255–56 n.84
American Naturalist, 68, 98, 100, 138, 224
American Physiological Society, 87–89; and American Association for Advancement of Science, 108; and American Association of Anatomists, 93–94, 111; and *American Journal of Physiology,* 101; and American Society of Naturalists, 94, 99, 106, 108, 111; centennial of, 88; and Congress of American Physicians and Surgeons, 93–94; and Federation of American Societies for Experimental Biology, 108, 111; founding of, 93–94; and Mathews's Plan, 107–10; membership of, 94, 107–8
American Psychological Association, 96, 99, 100–101, 210
American Society of Naturalists, 31, 87; and American Association for the Advancement of Science, 89–91, 102, 104–6, 108, 111; and American Association of Anatomists, 93–95, 106, 111; and American Morphological Society, 96; and *American Naturalist,* 98, 100; and American Physiological Society, 93–94, 106, 108; and American Psychological Association, 96; and American Society of Zoologists, 49, 104–6, 108, 111–12; and biological disciplinary societies, 89–90, 93–99, 102, 104–6, 108, 111–12, 127, 138; and Botanical Society of America, 97, 111; and core of biology, 112, 127; and Federation of American Societies for Experimental Biology, 108, 111; meetings of, 92, 98, 127; membership of, 91–92, 127, 138, 139; and Society of Naturalists of Eastern United States, 90–92, 127; and Society for Plant Morphology and Physiology, 97; and unity of biology, 98–99, 104–6, 108–12, 127, 138
American Society of Zoologists, xi, xii, 89; and American Association for Advancement of Science, 104–6, 108, 111; and American Morphological Society, 49, 89, 103, 118 n.53; and American Society of Naturalists, 49, 104–6, 108, 111–12; centennial of, xi, 49, 50, 88; and core of biology, 112; and experimental biology, 89; and morphology, 89, 103, 118 n.53; objectives of, 49, 50; origins of, 49–50, 77; and professionalization of biology, 49, 50, 77; and Society of Naturalists of Eastern United States, 50, 77; and systematic biology, 104; and unity of biology, 104–6, 111–12
Amherst College, 17, 18, 38 n.10
Anatomy: and biology, 4, 9, 125, 127; discipline of, 94–95, 101, 127; and embryology, 101; journals of, 101–2, 107; and morphology, 101; societies, 93–95, 106, 108, 111; and Wistar Institute, 101–2, 107
Anderson School of Natural History, 64, 73, 77, 128, 168
Andrews, Ethan Allen, 68, 125

Animal behavior: Adams and, 186, 200–201, 211; Allee and, 211, 218 n.66; biology and, 4, 8, 77, 186–87, 203–6, 209–11, 218 n.66; at California (Berkeley), 218 n.66; Carnegie Institution and, 173, 195, 202, 207–8; at Chicago, 8, 168, 172–73, 195, 204, 208, 218 n.66; at Columbia, 218 n.66; Craig and, 8, 173, 185, 187, 196–206, 209, 213 n.32, 215 n.33, 216 n.41; Holmes and, 185, 209–10, 218 n.66; at Hopkins, 218 n.66; Jennings and, 204, 207, 209–10, 218 n.66; *Journal of Animal Behavior* and, 186, 205–6, 217 n.60; Lashley and, 208, 211; Lillie and, 189, 218 n.66; Loeb and, 185–86, 187, 213 n.13; Lorenz and, 8, 185, 196–97, 211, 213–14 n.13, 215 n.32; Mast and, 210, 218 n.66; at MBL, 137, 172, 194; Morgan and, 218 n.66; Nice and, 196, 211; Noble and, 211; and problems of disciplinary identification, 8, 203–6; and problems of institutional support, 8, 200–209; psychology and, 8, 186, 210; researches in, 186, 209; Riddle and, 173, 185, 195; Shelford and, 186, 218 n.66; Watson and, 186, 204, 207–8, 210; Wheeler and, 185, 186, 197, 209; Whitman and, 8, 185–96, 198, 203–5, 208–9, 213 n.7, 213–14 n.13, 218 n.66; Yerkes and, 186, 204–5, 207, 210–11

Bailey, Liberty Hyde, 97; and Bateson's *Mendel's Principles of Heredity,* 283; and genetics, 283–84, 286, 295; and plant breeding, 295; and USDA, 286

Baird, Spencer Fullerton, 92; and A. Agassiz, 128–29; Brooks and, 73; and Fish Commission laboratory, 64, 73, 128–29; fish propagation studies of, 64; Henry and, 53–54; and Smithsonian Institution, 53–54; and Whitman, 130. *See also* U.S. Fish Commission laboratory; Marine Biological Laboratory

Barnes, Charles, 271

Bateson, William: and Conference on Hybridization and Crossbreeding, 294–95; and Conference on Plant Breeding and Hybridization, 283–84; and evolution, 281–82, 293–94; and genetics, 281–84, 286, 293–95; and *Mendel's Principles of Heredity,* 283; and plant breeding, 281–84, 294–95; success of, 281, 283, 294–95

Baur, George, 157, 164

Berrill, N. J.: and embryology, 318; and geneticists, 317, 342 n.20

Bessey, Charles, 264, 274

Biochemical Bulletin: and Mathews's Plan for American Biology, 109–10

Biochemistry, 139; and biology, 4, 9; and embryology, 327–28

Biological disciplinary societies, 3–4, 7–8, 88–89; and American Association for Advancement of Science, 89–91, 102–6, 108, 111; and American Society of Naturalists, 89–99, 102–13; and biomedical societies, 89, 113; and chemistry, 87, 110; and government scientists, 91, 113; and Mathews's Plan, 105–10; meetings of, 99–102; and museum scientists, 91, 97, 113; and natural history, 89, 97; and practical biology, 89; publications of, 99–102; and systematic biology, 89, 97, 104, 113, 118 n.54; and unity of biology, 87–89, 98–99, 104–13, 120 n.87, 122

Biology: and agriculture, 7, 139–40, 282–85, 290, 293, 300–302; and American Breeders Association, 281–84; and American Chemical Society, 87–88, 110, 122; at American Museum, 7, 228–45; American Society of Naturalists and, 98–99, 112, 122, 127; and anatomy, 4, 9, 127; animal behavior and, 4, 8, 186, 187, 203–6, 209–11, 218 n.66; and biochemistry, 4, 9; and botany, 9, 120 n.87, 126–27; boundaries of, 5–8, 9; at Bryn Mawr, 121, 151; and chemistry, 87–88, 110, 112; at Chicago, 4, 5, 7, 8, 121, 122, 125, 151–77, 179–80 n.43, 182 n.90; at Clark, 151; collaborative research in, 7, 10, 171–75, 231–45; at Columbia, 121, 125, 151, 228, 237, 245, 250 n.35; core of, 4–7, 112, 123, 228; and Craig, 8; definition of, 5–9, 49, 71–73, 142; disciplinary societies, 3–4, 7, 8, 87–120; discipline of, 4–6, 9, 69, 77, 87–89, 110, 112–13, 122, 123, 137–42; ecology and, 258–59, 260–62, 274–76; formaliza-

Biology (*continued*)
tion of, 141–42; and genetics, 282–85, 293–95, 301–2, 304n.18, 312, 315–20; Goldschmidt and, 8, 311, 312, 320–22, 325, 340; at Harvard, 72, 151; at Hopkins, 4, 5, 63–72, 76, 122, 125, 151; ideal of, 121–22; instrumentation and, 4–9, 70–71, 75; intellectual ambiguity of, 121, 137; and invertebrate paleontology, 220; journals of, 68, 100–102, 224; Just and, 8, 311–12, 320, 322–25, 340; and laboratory research, 4, 7, 10, 63–77; and marine zoology, 5, 64–65, 127–28; and masculinity crisis, 126, 140, 144n.18; at MBL, 4, 5, 7, 8, 74–76, 121, 128–42; and medicine, 5, 7, 69, 123, 136; and museums, 4, 5, 15–16, 31–32, 70, 77; and nationalism, 9, 10, 113n.5; and natural history, 4–7, 15–16, 31–32, 46–47n.96, 50, 63–64, 69–73, 75–77; at Pennsylvania, 69, 125; and physiology, 4, 9, 66–68, 70–71, 261–62, 274; and plant breeding, 282–85, 290–95, 301–2; professionalization of, 5, 6, 9, 50, 69, 77, 88; and relationship of embryology and genetics, 315–20, 339–41; and seedsmen, 281–82, 283, 285, 295–96; at state agriculture colleges, 282–87, 290, 293, 301–2; at state experiment stations, 290–93; unity of, 87–89, 98–99, 104–13, 120n.87, 122, 127, 136, 137, 138–39; in urban centers, 125–26; and USDA, 282, 284, 290, 293, 301–2; and vertebrate paleontology, 164, 219–21, 224–45; Whitman and, 8; Whitman on, 130–32, 158–60, 186–87; and women, 126; World War I and, 140; at Yale, 151
Blackman, F. F., 271
Bonnier, Gaston, 261, 265
Boston Society of Natural History: amateurs and, 54; and Brooks, 52; and Massachusetts Institute of Technology, 29; museum of, 29, 52, 55; and natural history, 54–56, 79n.37
Botanical Society of America, 89; and American Association for Advancement of Science, 97, 102–3; and American Society of Naturalists, 97, 103, 111; Farlow and, 97; founding of, 97; journals, 101; membership of, 116n.36; presidents of, 97, 117n.37; and Society for Plant Morphology and Physiology, 97, 102–3
Botany: agriculture and, 139; Bessey and, 263; and biology, 9, 120n.87, 126–27, 259–63, 270–76; Carnegie Institution and, 139; Clements and, 139, 260, 262–63, 270–76; and Clements's *Plant Succession,* 272–74; and Clements's *Research Methods,* 270–72; Cowles and, 260; discipline of, 96–97, 127; and ecology, 258, 260–63, 270, 274–76; Gray and, 57; journals, 101; and masculinity crisis, 126, 144n.8; at MBL, 74, 132; and natural history, 259–63, 274–76; and physiology, 270–72, 274–76; religion and, 57; societies, 97, 103; and systematics, 97; women and, 126
Boveri, Theodore, 322, 331
Brooks, William Keith: and A. Agassiz, 64; and Baird, 73; and biology at Hopkins, 4, 64–65, 67–74, 76; and Boston Society of Natural History, 52; and Chesapeake Zoological Laboratory, 64–65, 67, 72–74, 126; and Fish Commission laboratory, 73; and Gilman, 64; and Hyatt, 52; and laboratory research, 64, 67–73, 76; and Martin, 67–72; and masculinity crisis, 126; and MBL, 73–74; and Newport Laboratory, 64–65; students of, 64–65, 67–69, 71–74. *See also* Gilman, Daniel Coit; Johns Hopkins University
Brown University, 18, 19, 26
Bryn Mawr College, 121, 151

California, University of, at Berkeley: Alexander and, 29, 32–33, 47n.99; and animal behavior, 218n.66; Grinnell and, 32–33, 47n.99; and laboratory research, 76; museums of, 15, 29, 35–36
Camp, Charles L., 239–40, 245
Carnegie Institution of Washington, 20; Advisory Committee on Zoology, 207; and animal behavior, 173, 195, 202, 207–9; Castle and, 206–7,

208–9; Clements and, 209, 259, 262; Craig and, 202, 209; Davenport and, 195, 207; Experimental Station at Cold Springs Harbor, 195, 207; Mayer and, 207; and MBL, 132, 139, 162; MBL at Bird Key, 207–8; T. H. Morgan and, 209; Riddle and, 195; Watson and, 207–8; Whitman and, 132, 162, 195–96; Woodward and, 195, 210
Castle, William E., 206–7, 208–9, 284
Cattell, J. McKeen, 75–76, 100–101
Cell-lineage studies. *See* Embryology
Chesapeake Zoological Laboratory. *See* Brooks, William Keith
Chicago, University of: and academic reform, 65, 66, 76; and animal behavior, 8, 172–73, 195, 218 n.66; biology at, 5, 7, 8, 69, 122, 125, 151–60; and collaborative research, 7, 168, 173–75; Culver and, 161–62, 179–80 n.43; and cytology, 168; Davenport and, 164; and discipline of biology, 4, 5, 175; and ecology, 174, 257, 259–60; and embryology, 168, 172–74, 175; and evolution, 168, 172–74, 175; and Fish Commission laboratory, 129; Harper and, 153–57, 160–65; Herrick and, 154, 157; inbreeding at, 182 n.90; Lillie and, 163, 164, 165, 174–75; Mall and, 156–57, 159, 165, 178 n.17; and MBL, 132–33, 139, 162, 166, 173–74; and museums, 28, 29; Newman and, 167, 172; and organicism, 168, 172–73; origins of, 151–53; and paleontology, 164, 219; Rockefeller and, 153; Twitty and, 182 n.89; Watase and, 163–64; Wheeler and, 163–64; Whitman and, 152–53, 154–57, 159, 160–65, 167–68, 172–75, 188; women and, 166; zoology at, 160–75, 182 n.90; zoology Ph.D.s at, 175–77
Clark University: and academic reform, 65, 66; biology at, 69, 121, 151; and J. G. Clark, 155; and Fish Commission laboratory, 129; and G. S. Hall, 65, 155–57; Harper and, 155, 157; Mall and, 156–57; problems at, 155–57; and Whitman, 155–57, 159, 167, 188
Clarke, Samuel F., 68, 90–91
Clements, Frederic, 26; Barnes on, 271; and Bessey, 263, 274; Blackman on, 271; and botany, 139, 260–63, 270–76; and E. S. Clements, 267; and Comte, 263, 278–79 n.31; and Cowles, 258, 260–62, 264, 270–71; and Drude, 263; and ecology, 7, 8, 139, 258–76; Elton and, 276; Gleason on, 273–74; and natural history, 259–63, 272–76; and organismal concept, 8, 258–59, 262–76, 278 n.31; and physiology, 258–59, 261–76; and *Phytogeography of Nebraska,* 263; and plant geography, 261; and *Plant Succession,* 272–74; and Pound, 263, 264; and *Research Methods,* 262–75; and Shelford, 276; and Spencer, 263, 278–79 n.31
College museum collections: administrative interest in, 19, 22–23, 25–28, 44 n.68; administrative responsibility for, 26–28, 30; and biology, 4, 15, 31, 32, 46–47 n.96; at California (Berkeley), 15, 29, 32–33, 35–36, 47 n.99; changing role of, 15, 16, 21, 22–33, 46–47 n.96; at Chicago, 28, 29; at Cornell, 19–21, 26, 40 n.27; and evolution, 16, 31; faculty and, 18–19, 21–28, 31; growth of, 18–19, 22, 27, 28; at Harvard, 17, 27, 44 n.67, 44 n.68, 46 n.87; and higher education, 15–17, 21–28, 31–33, 72; at Hopkins, 29, 37 n.3; at Kansas, 26–27; maintenance of, 23, 27–28; at Mount Union, 22–23; and natural history dealers, 18, 19–23, 24; at Northwestern, 25, 27–28; at Pennsylvania, 29; and public education, 26; at Rochester, 19, 22; standardization of, 23–28; and state scientific surveys, 15, 30; strategies for building, 18–24; student societies and, 16, 24–25; students and, 16, 18, 21, 24–28, 31–32, 39 n.14; at Syracuse, 25, 30–31; and systematics, 15–16, 31–33, 46–47 n.96; and urban museums, 26, 29; at Vassar, 19–20, 22; at Virginia, 20, 22; Ward and, 19–20, 22, 24, 26; at Yale, 22, 38 n.9. *See also* Natural history; Natural history museums
College natural history societies, 16, 24–25, 33–34

College science buildings, 18, 34–35
Columbia University, 302; and American Museum, 229; and animal behavior, 218 n.66; biology at, 69, 121, 122, 125, 151, 154, 228–29, 250 n.35; and Fish Commission laboratory, 129; Gregory and, 237, 253 n.62; Low and, 228–29; and MBL, 129, 131, 138, 250 n.35; and T. H. Morgan, 218 n.66, 313; and Osborn, 228–29, 237, 245, 250 n.35, 253 n.62; and science education, 60; and vertebrate paleontology, 229, 237, 245, 255–56 n.84
Comte, August: and organismal concept, 147 n.34, 263, 278–79 n.31
Congress of American Physicians and Surgeons, 93–94
Conklin, Edwin Grant, 69; on biological societies, 109; and biology in World War I, 140; and embryology, 166, 171; and eugenics, 134, 149 n.60; and Fish Commission laboratory, 147 n.38, 171; and MBL, 134, 147 n.38, 166, 171, 180 n.60. *See also* Whitman, Charles Otis; Wilson, E. B.
Cope, Edward Drinker: and *American Naturalist,* 224; and evolution, 219, 220, 226–27, 232, 234, 244; and Marsh, 222, 247 n.13; and Osborn, 221, 226, 247 n.13; and Scott, 226–28; and vertebrate paleontology, 219–21
Cornell University: and academic reform, 65, 66, 76; and faculty collections, 19–20, 40 n.27; and laboratory research, 25, 63; museum, 20–21, 26; natural history society, 25; and White, 20, 26; and Wilder, 20, 25
Correns, Carl F., 312, 318
Cowles, Henry Chandler: and Clements, 258, 260–62, 264; on Clements's *Research Methods,* 270–71; and ecology, 258, 260–62, 276; on evolution, 261; and experimental method, 261–62; and plant geography, 261
Craig, Wallace: and Adams, 197, 200–201, and animal behavior, 173, 185, 187, 196–206, 209, 215 n.32, 216 n.41; career, 197–203; and Carnegie Institution, 202, 209; at Chicago, 173, 198, 204; and Davenport, 201–2; at Harvard, 203–4; and Lorenz, 185, 196–97, 203, 215 n.32; at Maine, 199–202, 204, 206; and Pearl, 202, 217 n.53; problems of disciplinary identification, 173, 203–4, 206; problems of institutional support, 205, 206, 209; professional affiliations, 217 n.58; publications, 198–203, 216–17 n.50; and Whitman, 173, 197, 198, 204; and Yerkes, 202
Crane, Charles R.: and MBL, 122–23, 138, 140–41, 149 n.59
Culver, Helen: and biology at Chicago, 161–62, 178–79 n.43
Cytology, 7, 31, 71, 168

Dana, James Dwight, 21, 22, 38 n.9, 92
Darwin, Charles: evolutionary theory of, 293–94; on hybrid vigor, 298; and systematics, 37 n.4; on variation, 191, 293, 295, 303 n.5; Whitman and, 191–93. *See also* Evolution
Davenport, Charles B., 103, 298; and American Breeders Association, 281–82, 284; and biology at Chicago, 164, 174; and Carnegie Institution, 201–2, 207; and Craig, 202; and genetics, 284; on Whitman, 195
Development. *See* Embryology
de Vries, Hugo, 261, 284; mutation theory of, 281, 293, 297; and plant breeding, 282, 294
Driesch, Hans, 314, 315, 331, 333, 336
Drude, Oscar, 261, 263
DuBois, W. E. B.: and Just's cellular federalism, 333

East, Edward M.: and commercial agriculture, 300–302; and genetics, 284, 299–301; and hybrid corn, 296–302
Ecology: and agriculture, 259; Bessey and, 274; biology and, 4, 8, 77, 258–61, 262–71, 274–76; and botany, 139, 257, 258–64, 270, 272–76; Clements and, 7–8, 139, 258–76; and Clements's *Plant Succession,* 272–74; and Clements's *Research Methods,* 262–76; Cowles and, 258, 260–62, 276; discipline of, 258–60, 262, 270, 274–76; and evolution,

259–60, 261; and forestry, 259, 273–74; and geographical distribution, 259–60, 261, 263; and Midwest, 257, 259–60, 276; and natural history, 258, 260–63, 270, 274–75; and organismal concept, 257–76; and physiology, 258–59, 261–76; and succession, 257–61, 264, 268–76, 277n.3
Eliot, Charles W.: and higher education reform, 65, 66
Elton, Charles, 276
Embryology: and anatomy, 101; and animal behavior, 168, 172–73, 188; and Berrill, 317–18; and Brooks, 67, 70–71, 171; and cell-lineage studies, 136, 166, 170–71, 180n.60; at Chicago, 166, 168, 172–75; club, 96, 116n.32, 224, 248n.21; and Conklin, 166, 171, 180n.60; and core of biology, 4, 5, 7; discipline of, 77, 316–20; and Driesch, 314, 315, 331, 333, 336; Ephrussi and, 313, 316, 342n.15; and evolution, 5, 168, 171–73, 188; experimental, 313, 323; and fertilization, 323–24, 328–31; and genetics, 5, 172–73, 311, 312–20, 326–27, 331–36, 338–41; in Germany, 318–19; Goldschmidt and, 334–37, 339; at Growth Symposium, 318; Harrison and, 316–17, 323, 326; and Holtfreter, 311, 343n.37; and Just, 316–17, 319, 320, 323–34, 340, 344–45n.64; Lillie and, 313–14, 323–24, 326, 328–29; and Loeb, 323–24, 329; at MBL, 70–71, 75, 135–37, 166, 170–72, 180n.60; T. H. Morgan and, 312–14, 315, 319, 336; at Museum of Comparative Zoology, 70; and organicism, 147n.34, 168, 171–72; researches in, 316, 318, 323; Waddington and, 318, 340; and Whitman, 70–71, 166, 168–73, 180n.60, 188; and Wilson, 166, 170–71, 180n.60
Ephrussi, Boris: Morgan's *Embryology and Genetics,* 313; and relationship of embryology and genetics, 316, 342n.15
Ethology. *See* Animal behavior
Evolution: and agriculture, 293–95; Bateson and, 293–94; at Chicago, 168, 172–74, 175; Cope and, 219, 220, 226–27; and core of biology, 5, 7, 112; de Vries and, 293; and genetics, 243–45, 282, 293–95, 320–21; and geographical distribution, 231–34; Goldschmidt and, 320–21, 339; Gray and, 57; Gregory and, 240–45; Hurst and, 294; Hyatt and, 219, 220; Just and, 339; Matthew and, 234–37, 241, 244–45; at MBL, 171; and museum collections, 16, 31, 70; neo-Lamarckian theories of, 70, 226–27; orthogenetic theories of, 191–92, 226–27, 230, 235, 242–45; Osborn and, 221, 225–27, 230–31, 241–45; Scott and, 225–27, 230, 235, 253n.56; Simpson and, 245, 255–56n.84; and unity of biology, 121; Whitman and, 171–73, 189–94

Farlow, William G., 98; on biological instruction, 158; and Botanical Society of America, 97, 117n.37; and masculinity crisis, 126; and Society for Plant Morphology and Physiology, 97
Fay, Joseph Story: and Woods Hole, 145n.24
Federation of American Societies for Experimental Biology: and unity of biology, 108, 111
Fertilization. *See* Embryology
Field Museum of Natural History, 26, 29, 52, 219, 220
Functional morphology: Camp and, 239–40; Gregory and, 237–41, 244–45; Kovalevsky and, 237; Matthew and, 238; Miner and, 239, 240; Osborn and, 237–38, 244; Romer and, 240

Ganong, William Francis, 260, 270
Genetics: and agriculture, 139–40, 282–85, 290–93, 295–97, 300–302; and American Breeders Association, 281–82, 283, 285; Bailey on, 295; Bateson and, 281–84, 286, 293–96, Berrill on, 317; and biology, 4, 5, 7, 77, 282–85, 293–95, 301–2, 304n.18; Castle and, 206–7, 208–9, 284; cytoplasmic, 312, 318–19; Davenport

Genetics (*continued*)
and, 284, 298; de Vries and, 284; discipline of, 312–13, 315–20, 340–41; East and, 284, 299–302; and embryology, 5, 311, 312–20, 326–27, 331–38, 340–41; and evolution, 5, 236–37, 243–45, 282, 293–95, 320–21; in *Experiment Station Record,* 286; German tradition in, 312, 318–19, 343n.64; Goldschmidt and, 311–12, 320–21, 334–37, 339–40; Gregory and, 243–45; Harrison and, 317, 326; Hays and, 281, 284; of hybrid corn, 296–302; Johannsen and, 297, 312–13; Jones and, 299–302; Just and, 311–12, 319, 326–27, 331–32; Lillie and, 314, 326; Matthew and, 236–37, 243–45; at MBL, 136; Morgan and, 136, 236, 243, 311, 312–15, 319; Morgan school of, 311, 318, 319, 320, 334, 340; naturalists and, 284, 304n.18; and plant breeding, 281–85, 290–95, 300–302; and seedsmen, 281–82, 283, 285, 295–96, 300–302; Shull and, 284, 296–97; Simpson and, 245, 255–56n.84; Spemann and, 319; and state agricultural colleges, 282–85, 290, 293, 301–2; and state experiment stations, 282–85, 290–93, 301–2; Stevens and, 136, 284; success of, 281–82, 283, 293–95, 301–2, 315–16, 340–41; and USDA, 282, 283, 285–86, 290–93, 301–2; and von Tschermak, 286; Waddington and, 340; Weiss and, 320; Wilson and, 136, 284

Genetics Society of America, 316

Genteel resorts, 124–26; Woods Hole, 122–23, 127–36

Geographical distribution: Clements and, 261, 263; Cowles and, 261; ecology and, 260–61, 263; Matthew and, 231–34, 237, 244–45, 252n.47; Osborn and, 221, 231–33, 244, 252n.47

Geological Society of America, 95, 96, 99

Geology, 56, 57, 219–20, 245–46n.2

Gilman, Daniel Coit: Brooks and, 64; Eliot and, 65, 66; and higher education reform, 65–66, 67, 69; and Hopkins, 29, 65–69; Martin and, 67, 71; and museum collection, 29

Gleason, Henry Allan, 273–74

Goldschmidt, Richard B.: anti-Semitism and, 321–23; as aristocrat, 337–340; background of, 311, 320–21, 338; Boveri on, 322; conception of cytoplasm in development, 336–37, 339; conception of gene, 311, 321, 334, 336–37; and discipline of biology, 8, 311–12, 320–22, 325; and evolution, 320–21, 339; and genetics, 311–12, 320–21, 334–37, 339–40, 342n.15; and German tradition in genetics, 311–12, 319–20, 325, 340; and Hammerling, 336, 340; and Hertwig, 321, 337; and Just, 311–12, 322–25, 339–40; and Kaiser-Wilhelm Institute for Biology, 311, 322, 337; and Loeb, 320; and Morgan's genetics, 311, 320, 334, 340, 342n.15; and Nazi Party, 311, 320, 338, 339; neglect of his work, 320–21, 340; as outsider, 8, 311, 320–22, 343–44n.43; and *Physiological Genetics,* 320, 334–40, 342n.15; on proletariat, 338–39; on relationship of embryology and genetics, 312, 319, 320, 321, 334–37, 339, 340; on relationship of nucleus to cytoplasm, 335–39; on role of nucleus, 335–37, 339; Stern on, 344n.50

Goode, George Brown, 22, 30, 46n.89

Gray, Asa, 55, 57, 152

Gregory, William King: and American Museum, 237, 238, 240, 244–45; and Camp, 239–40, 245; and Columbia, 237, 245, 253n.62; on evolution, 240–45; and functional morphology, 237–41, 244–45; and genetics, 243–45; and Matthew, 238, 240, 241, 244–45; and Miner, 240; and Noble, 240; and Osborn, 221, 230, 240, 241–45, 253n.62; and Romer, 240, 245; and Simpson, 245; students of, 245

Grinnell, Joseph: and Alexander, 32–33, 47n.99; and Museum of Vertebrate Zoology, 32–33; Sumner on, 47n.100

Growth Symposium, 317–18

Hall, G. Stanley, 92; and American Psychological Association, 96; and Clark University, 65, 155–57; and Harper, 157, 178n.20; on Whitman, 167
Hall, James, 30, 92
Hammerling, J., 336, 340
Harnack, Adolf, 322
Harper, William Rainey: and Chicago, 65, 153–57, 160–64; and Clark University, 155, 157; and C. L. Herrick, 154, 157, 178–79n.24; and Mall, 156–57, 160, 178n.17; and Watase, 163–64; and Wheeler, 163–64; and Whitman, 154–57, 160–65
Harrison, Ross G.: and Chicago, 182n.89; and embryology, 316, 317, 323; and Federation of American Societies for Experimental Biology, 108, 111; on genetics, 317, 326; and Hopkins, 69; and *Journal of Experimental Zoology,* 101; and unity of biology, 108, 111
Harvard University: and academic reform, 65–66, 76; biology at, 69, 70, 121, 151; Eliot and, 65–66; and laboratory research, 63; Lawrence Scientific School of, 16–17; and museum collections, 27, 44n.68, 46n.87; Museum of Comparative Zoology, 17, 22, 27, 44n.67, 70; and natural history, 72; and paleontology, 219; Peabody Museum, 29. *See also* Agassiz, Louis; Museum of Comparative Zoology
Hatch Act, 287, 290, 291
Hays, Willet: and American Breeders Association, 281, 285; and genetics, 281, 284; and plant breeding, 294
Heinroth, Oscar, 185, 203
Henry, Joseph, 53–54
Heredity. *See* Genetics
Herrick, Clarence Luther: and biology at Chicago, 154, 157; career, 154; and Harper, 154, 157, 178–79n.24; and students, 24, 42n.52
Hertwig, Richard, 321, 337
Higher education: L. Agassiz and, 16–17, 60, 62, 69; and agricultural reform, 286–87, 293; and biology, 3–4, 31–32, 69, 77, 88, 125; and California (Berkeley), 15, 32–33, 35–36, 76; at Chicago, 65–66, 76, 151–57, 159–68, 173–75; and Clark, 65–66, 155; and college science, 16–17, 21–26, 56–72; and Columbia, 60, 121, 151, 228–29, 250n.35; and Cornell, 65–66, 76; criticisms of, 58; and Dartmouth, 60; defense of, 59; Eliot and, 65–66; and faculty collections, 16, 21–28; Farlow and, 158; and German science, 61–63, 65–66; Gilman and, 65–66; Hall and, 65, 155; Harper and, 65, 153–57; and Harvard, 16–17, 65–66, 76; at Hopkins, 61, 63, 65–72, 76; impact on student societies, 24–25; LeConte and, 17; Low and, 228–29; and MBL, 129–130; and medicine, 69, 76; Morrill Act and, 15, 22, 61, 286–87; and museums, 15–16, 17, 21–33, 72, 77; and natural history, 16–17, 21–33, 46–47n.96, 56–63; and natural theology, 56–57, 59; problems of, 59; redefinition of, 21–28, 40n.30; and reform, 59–72, 76–77; and research, 58–77; and science for women, 17, 38n.12; standardization of, 21–28; and student societies, 16, 24–25; White and, 59, 65–66; Whitman and, 158–60; Whitney and, 15; Willard and, 17
Hitchcock, Edward, 38n.12, 56–57
Holism. *See* Organicism
Holmes, Samuel J., 185, 209, 210, 218n.66
Holtfreter, Johannes, 311, 343n.37
Howard University, 322, 324
Hurst, C. C., 284, 294
Huxley, Thomas Henry, 129, 222, 226
Hyatt, Alpheus, 52, 90–91, 100; and evolution, 219, 220, 227; and MBL, 74; and school at Annisquam, 74, 128; and scientific reform, 63
Hybrid corn: commercial importance of, 296–97, 300–302; interpretations of, 297–302; success of, 296, 299–300, 301, 308–9n.80, 310n.103. *See also* East, Edward M.; Jones, Donald F.; Shull, George Harrison

Inheritance. *See* Genetics
International Conference on Plant Breeding and Hybridization, 282–84
Invertebrate paleontology, 220

Jennings, Herbert Spencer: and American Association for Advancement of Science, 104–6; and American Society of Naturalists, 106; and American Society of Zoologists, 104–6; and animal behavior, 185, 209–10, 218 n.66; on biology, 137, 142; and genetics, 209–10; and Marine Biological Laboratory at Bird Key, 207; and problems of disciplinary identification, 204; and systematic biology, 104, 118 n.54
Johannsen, Wilhelm: and genotype/phenotype distinction, 312–13; and pure lines, 297. *See also* East, Edward M.; Morgan, Thomas Hunt; Shull, George Harrison
Johns Hopkins University: and academic reform, 61, 65–71, 77; and animal behavior, 218 n.66; biology at, 4, 5, 66–72, 121, 125, 151; Brooks and, 4, 64, 65, 67–72; and discipline of biology, 4, 66–69; doctoral degrees awarded, 68–69; faculty of, 66; Gilman and, 29, 65–66, 68, 69; Martin and, 4, 66, 67–69, 70, 71–72; and Maryland Academy of Science, 29; museums and, 29, 37 n.3; and natural history, 63, 66, 67, 72
Jones, Donald F.: and double-cross method of breeding, 299, 300, 310 n.99; and East, 299–302; and hybrid corn, 299–302. *See also* Agriculture; Genetics; Plant breeding
Jordan, David Starr: and genetics, 305 n.26; and museum collections, 26, 43 n.62; and Stanford, 177 n.7; and Ward, 26
Jordan, Edwin O., 115 n.31, 157
Just, Ernest Everett, 152; and Allee, 333; and American scientific community, 322–23, 324; background of, 312, 322; and biochemistry, 327–28; and biology, 8, 322–25; and *Biology of Cell Surface,* 320, 322, 325–28, 330–34, 340; and cell cortex, 324–34, 339–40; and cellular federalism, 332–34; and cellular holism, 325–26; segregation, 340; and DuBois, 333; on fertilization, 322–24, 327–31, 334; and genetics, 311–12, 319, 326–27, 331–32; and German tradition in genetics, 311–12, 319–20, 324, 325, 344–45 n.64, 345 n.66; and Goldschmidt, 311–12, 322–25, 339–40; and Hammerling, 340; and Harlem Renaissance, 332; and Howard University, 322, 324; and Lillie, 322–24, 328–29; and Loeb, 323–24, 329; as naturalist, 328; neglect of his work, 322–23, 340, 343 n.37; as outsider, 8, 320, 322–24, 343–44 n.43; racism and, 322–23; and reductionism, 325–28; on relationship of embryology and genetics, 311, 312, 316, 317, 319–20, 324–26, 327, 331–34, 344 n.55; on relationship of nucleus to cytoplasm, 324–27, 331–34

Kansas, University of, 24, 26–27
Kennicott, Robert, 27–28
Knower, H. McE., 105
Kuhn, Alfred, 312, 318

Laboratories: and biology at Hopkins, 63–77; and museums, 31–33, 46–47 n.96; and natural history, 63–64, 77; and specialization, 72–77
Lashley, Karl, 208, 211
LeConte, Joseph, 17, 57
Leidy, Joseph, 92, 95; natural history collection of, 44 n.72; and vertebrate paleontology, 219, 220, 225, 244
Lillie, Frank R.: on biology, 142; and biology at Chicago, 152, 157, 163, 164, 165, 174–75; on fertilization, 323–24, 328–29; on genetics, 314, 326, 342 n.20; Just and, 322–24, 328–29; and Loeb, 329; and MBL, 135, 136, 140–41, 149 n.59, 162, 166; on relationship of embryology and genetics, 313–14, 342 n.20; and Whitman, 166, 189, 218 n.66
Loeb, Jacques: and animal behavior, 185, 186, 187, 213 n.13; and fertilization, 323–24, 329; and Goldschmidt, 320; and Just, 323–24, 329; and MBL, 70, 136; and relationship of embryology and genetics, 344 n.55; Whitman on, 187, 213 n.13
Lorenz, Konrad: and animal behavior, 8, 185, 196–97, 203, 213–14 n.13, 215 n.32; and Craig, 185, 196–97, 203, 215 n.32; and Heinroth, 185,

203; and Whitman, 185, 213–14n.13
Low, Seth, 229. *See also* Osborn, Henry Fairfield
Lowell Institute, 54–55

Mall, Franklin Paine: and *American Journal of Anatomy,* 101; and American Morphological Society, 95, 96; and Association of American Anatomists, 95; on biology, 137; and biology at Chicago, 156–57, 159, 178n.17; and Harper, 156–57, 165; and Hopkins, 94, 178n.17; and Society of American Bacteriologists, 115n.31; and Whitman, 156–57, 159, 165
Marine Biological Laboratory: and applied biology, 136; as biological community, 122, 130–32, 137–38, 323; biological work at, 8, 74–75, 135, 136; Brooks and, 73, 74; and Carnegie Institution, 132, 139, 162; and Chicago, 132, 138, 139, 162, 166, 173–74; Cornelia Clapp on, 166; and Clark, 131; and collaborative research, 7, 166, 171, 180n.60; Conklin and, 134, 147n.38, 166, 171, 180n.60; and Columbia, 131, 138, 250n.35; Crane and, 122–23, 138, 140, 141, 149n.59; and cytology, 7, 70–71, 75; and discipline of biology, 4, 5, 74–75, 123, 133, 136–42; and embryology, 70, 75, 135–36, 166, 170–71; facilities of, 131, 134–35; and Fish Commission laboratory, 128–31, 131, 136, 147n.38; growth of, 131–33, 139–41; investigators at, 133, 137–38; Just and, 323; and Lillie, 135, 136, 138–41, 142, 149n.59, 162, 166; Loeb and, 70, 136; and marine organisms, 135; and microscopy, 71; and Morgan, 136; and Naples Zoological Station, 123; origins of, 74–75, 128–29, 145n.27; Osborn and, 250n.35; and physiology, 70, 75; as resort, 122–23, 134–36; and Stevens, 136; and teaching, 74–75, 132; and unity of biology, 122, 136–42; and Van Vleck, 74; Whitman and, 74–75, 129–32, 135, 139, 146n.32, 147n.38, 149n.59, 159, 160, 162, 165–66, 170, 171, 172, 187–88, 194; Wilson and, 136, 147n.38, 166, 170–71, 180n.60; and Woman's Education Association, 74, 75, 128–29, 132; women and, 74, 128–29, 134
Marine zoology, 64–65, 127–28
Mark, Edward L.: and American Society of Zoologists, 103; and biology journal, 224; and embryological club, 96, 224, 248n.21; and masculinity crisis, 126; on microscopical techniques, 71
Marsh, O. C.: and Cope, 222, 247n.13; and Osborn, 221, 225, 244, 247n.13; and Peabody Museum, 22, 41n.35, 219; and vertebrate paleontology, 219, 220, 221, 225, 244
Martin, H. Newell: and American Physiological Society, 93–94; and American Society of Naturalists, 98; and biology, 4; and biology at Hopkins, 67–72; and physiology, 67, 70–71
Mast, S. O., 210, 218n.66
Mathews, Albert P.: Plan for biology society, 105–9; on Whitman, 195
Matthew, William Diller: background of, 231–32; on evolution, 234–37, 241, 245; and functional morphology, 238; and genetics, 236–37, 244–45; on geographical distribution, 231–34, 237, 244–45, 252n.47; and Gregory, 238, 240, 241, 244–45; and T. H. Morgan, 236; and Osborn, 221, 230–36, 244–45, 252n.47; and Simpson, 245, 255–56n.84; and Sumner, 236
Mayer, Alfred G., 207
McCosh, James, 222, 226
McMurrich, J. Playfair, 96, 104–5
Medicine: and biological disciplinary societies, 89, 113; and biology, 5, 7, 69, 125, 136, 140; and laboratory research, 69, 140
Mendel, Gregor, 281. *See also* Genetics; Plant breeding
Mendelian genetics. *See* Genetics
Mendelism. *See* Genetics
Metcalf, Maynard M., 69, 103
Microscopy, 7, 70–71, 169–70
Miner, Roy W., 239, 240
Minot, Charles Sedgwick: and *American Journal of Anatomy,* 101; and American Society of Naturalists, 98, 99, 102; and MBL, 145n.27; and Society of Naturalists of Eastern United States, 91, 92

Mitchell, S. Weir, 93, 94
Morgan, J. Pierpont, 220, 229, 250n.39
Morgan, Thomas Hunt, 69, 105, 243; and animal behavior, 218n.66; and Driesch, 314, 315, 336; and embryology, 312–14, 315, 319; and *Embryology and Genetics,* 313–14, 315, 316; and genetics, 236, 284, 311–15, 319; Goldschmidt and, 311, 320, 334, 336, 340; and MBL, 136; Matthew and, 236
Morrill Act (1890), 22, 41n.37, 287
Morrill Land Grant Act, 15, 16, 22, 286
Mount Union College, 22, 23
Museum of Comparative Zoology, 22, 59; A. Agassiz and, 44n.67, 64; L. Agassiz and, 17, 27, 52, 55, 60; and Harvard, 27, 44n.68, 46n.87; and Massachusetts legislature, 17, 27; and natural history societies, 55; research at, 70; students at, 52
Museum of Vertebrate Zoology, 32–33, 35–36, 47n.99

Nachsteim, H., 319
Naples Zoological Station, 7, 123, 128, 322
National Research Council, 111, 140, 141, 209
Natural history: amateurs and, 54; and biological disciplinary societies, 89, 97; and biology, 4–7, 15–16, 31–32, 50, 63–64, 69–70, 73, 75, 76–77, 275; and biology at Hopkins, 63, 66, 67, 72; and Chautauqua, 51; criticisms of, 58–60, 62; defense of, 58–59; definition of, 49–50; and ecology, 258, 260–63, 274–75; and European science, 57–60, 62, 63; and government expeditions, 53–54; and higher education, 16–17, 21–33, 56–63, 77; journals, 52–53; and laboratory research, 63–64, 72; and lyceums, 51; methods, 51, 56–64, 72, 77; at MBL, 74; moral and practical benefits of, 51, 53, 56–58; dealers, 18, 19–24, 52–53; and museums, 4, 7, 15–33, 51–52, and natural theology, 51, 56–58; and physiology, 261, 274; and popular culture, 50–53, 54–55, 57; and public education, 51; societies, 54–55; and women, 126, 129–30
Natural history museums: and biology, 15, 31–32, 46–47n.96, 70, 72, 77; and embryology, 70; and fieldwork, 7; and higher education, 15, 16, 56–72; and job opportunities, 52; and paleontology, 70, 219–20, 228–45; and popular culture, 51–52; and systematics, 15–16, 31–33, 37n.4, 70
Natural history societies, 51–52, 55–56. *See also* College natural history societies
Needham, Joseph, 318, 323, 331–32
Neo-Lamarckism. *See* Evolution
Newman, H. H., 167, 172
Nice, Margaret Morse, 196, 211
Noble, G. K., 211, 240
Northwestern University, 25, 27–28
Nunn, L. L., 194, 214n.26

Odum, Eugene P., 276
Organicism, 142, 150n.69; at Chicago, 168, 172–73; Clements and, 257–76, 278–79n.31; Comte and, 147n.34, 263, 278–79n.31; Leuckart and, 147n.34; Spencer and, 147n.34, 263, 278–79n.31; Whitman and, 130, 131, 142, 147n.34, 168–72, 189–92
Orthogenesis. *See* Evolution
Osborn, Henry Fairfield, 91, 96, 98; and American Museum, 7, 228–30; background of, 221–22; and Bickmore, 229, 247n.13; and biological journal, 224; and biology, 8, 221–29, 248n.22; and collaborative research, 7; and Columbia, 228–29, 230, 237, 245, 250n.35, 253n.62; and Cope, 221, 222, 226–27, 244, 247n.13; and W. E. Dodge, 229, 250–51n.39; and embryology, 221–22; 248n.17; and embryology club, 224, 248n.21; on evolution, 225–27, 230–31, 234–36, 241–45; and fieldwork, 225, 230, 249n.24; and functional morphology, 237–38, 244; and geographical distribution, 221, 231–33, 244, 252n.47; and Gregory, 221, 230, 237, 238, 240–45, 253n.62; and Huxley, 222, 226; and Leidy, 225, 244; and Low, 228–29; and Mark, 224; and Marsh,

221, 222, 225, 244, 247 n.13; and Matthew, 221, 230–33, 243–45, 252 n.47; and MBL, 139, 250 n.35; and McCosh, 222, 226; and J. P. Morgan, 229, 250 n.39; and W. H. Osborn, 223, 247–48 n.16, 248 n.17; at Princeton, 221–28, 248 n.22; and Scott, 221–22, 224–28, 248 n.22; and Simpson, 245, 255–56 n.84; and vertebrate paleontology, 7, 8, 221, 222, 224–45, 248 n.22, 251 n.42; and Whitman, 250 n.35; and Wilson, 224
Osborn, William Henry. *See* Osborn, Henry Fairfield

Packard, Alpheus Spring, 63, 91, 100
Paleontology. *See* Invertebrate paleontology; Vertebrate paleontology
Parker, George Howard, 135, 149 n.60, 187
Peabody Museum: at Harvard, 29; at Salem, 52; at Yale, 22, 29, 38 n.9
Pearl, Raymond: and biology in WWI, 140; and Craig, 202, 217 n.53; and Institute of Biological Research at Hopkins, 142; on unity of biology, 110
Pennsylvania, University of, 25, 29, 69, 125
Physiology: and biology, 4, 9, 70, 71, 77, 125, 127, 139; and biology at Hopkins, 66–72, 125; Clements and, 258–59, 261–76; discipline of, 94; and ecology, 258–59, 261–76; journals, 101; and Loeb, 70; Martin and, 67–71; MBL, 70, 75, 132; and natural history, 261, 274; societies, 93–94
Plant breeding: and agriculture, 281, 282, 287–90, 293–95, 300–302; and American Breeders Association, 281–85; Bailey and, 295; Bateson and, 281–84, 293–95; and biology, 282–85, 293–95, 300–302; conferences, 282–84, 294; and crop diversification, 288–89; de Vries and, 282, 294; East and, 299–302; and evolution, 293–95; and genetics, 281–85, 293–96, 300–302; Hays and, 281, 284, 294; Hurst and, 294; and seedsmen, 283, 285, 295–96, 300–302, 303 n.14; and Shull, 297–98, 301–2; and state agricultural colleges, 282–87, 290, 293, 301–2; at state experiment stations, 282–84, 287, 290–93, 301–2; A. C. True and, 291–92; and USDA, 282, 284, 287–93, 301–2; Webber and, 294; and wheat overproduction, 287–88
Pound, Roscoe, 263
Princeton University, 24, 168; E. M. Museum, 28, 44 n.73, 222; H. F. Osborn and, 221–28, 248 n.22; W. H. Osborn and, 223, 247–48 n.16; vertebrate paleontology at, 224–28
Psychology: and animal behavior, 8, 186, 204–5, 206, 210; journals, 100–101, 205–6, 210; Watson and, 204; Yerkes and, 204–5

Riddle, Oscar, 173, 185, 195, 207
Ritter, William Emerson, 55, 76, 142
Rochester, University of, 19, 22, 26, 52
Rockefeller, John D. Jr., 122–23, 140–41, 153
Rockefeller Foundation, 141
Rockefeller Institute for Medical Research, 140
Romer, Alfred Sherwood, 240, 245
Roosevelt, Theodore, 126
Royal Horticultural Society Conference on Hybridization and Crossbreeding, 294

Schimper, Andreas, 260, 261, 271
Science education. *See* Higher education
Scott, William Berryman, 91, 98, 223; background, 221–22; and biology, 223, 226–28; and biology journal, 224; on evolution, 225–28, 235, 253 n.36; and geographical distribution, 232, 252 n.56; and Huxley, 222, 226; and McCosh, 222, 226; and H. F. Osborn, 221–28
Sedgwick, W. T., 68, 98, 145 n.27
Seedsmen, 283, 285, 295–96, 300–302, 303 n.14
Shelford, Victor E.: and animal behavior, 186, 218 n.66; and Clements, 276; and midwestern ecology, 259
Shull, George Harrison, 284; and hybrid corn, 297–98, 301–2
Silliman, Benjamin, 56–57
Simpson, George Gaylord, 245, 255–56 n.84

Smithsonian Institution: Baird and, 53–54; collections, 53–54; and college museum collections, 23–24, 38–39n.12, 42n.48; Henry and, 53
Snow, Francis, 26–27
Society of American Bacteriologists, 98, 106, 115–16n.31
Society of Naturalists of Eastern United States: and American Society of Zoologists, 50, 77; founding of, 90–91; justification for, 91; membership of, 91–92, 114n.17; meetings, 92; and microscopy, 62. *See also* American Society of Naturalists
Society for Plant Morphology and Physiology, 89; and Botanical Society of America, 97, 102–3
Spemann, Hans, 316, 319, 323
Spencer, Herbert. *See* Organicism
Spillman, W. J., 284, 286, 295
Stanford University, 141, 142, 154; D. S. Jordan and, 177n.7; and Whitman, 155, 178n.14
State agricultural colleges. *See* Genetics; Higher education; Plant breeding
State experiment stations. *See* Genetics; Plant breeding; USDA
Stern, Curt, 318, 320; and relationship of nucleus and cytoplasm, 345n.64
Stevens, Nettie, 136, 284
Student scientific societies. *See* College natural history societies
Sumner, Francis B., 47n.100, 236
Sutton, Walter, 284, 312
Syracuse University: museum, 25, 26, 30–31
Systematics: and biological disciplinary societies, 89, 97, 104, 113, 118n.54; and botany, 97; and evolution, 16, 37n.4; and experimental biology, 31; and museum collections, 15–16, 31, 32, 33, 46–47n.96, 70; women and, 126

Tansley, Alfred G., 260, 271
True, Alfred C., 291–92

Union of American Biological Societies, 89, 112
U.S. Department of Agriculture, 282, 284, 285; and biology, 290, 293, 301–2; and crop diversification, 288–89; divisions of, 288–90, 306n.39; *Experimental Station Record,* 286, 289–90, 295; and genetics, 285–86, 290–94, 301–2; Graduate School of Agriculture, 286; and Hatch Act, 287–90, 291; and Morrill Acts, 286–87; Office of Experiment Stations, 290–93; and plant breeding, 282, 284, 287–93, 301–2; and state agricultural colleges, 282, 287, 290, 301–2; and state experiment stations, 282, 287, 290–93, 301–2; and wheat overproduction, 282, 287–88
U.S. Fish Commission laboratory: A. Agassiz and, 128–29; Baird and, 64, 73, 128–29; biology departments and, 129–30; Brooks and, 72–73, 129; Conklin and, 147n.38, 171; facilities of, 134–35; investigation of, 129, 130, 148n.55; origins of, 72–73, 128; relationship to MBL, 128–30, 131, 147n.38, 148n.55; Whitman and, 129, 130
United States National Museum, 219. *See also* Smithsonian Institution

Van Vleck, B. H., 74
Vassar College, museum, 19–20, 22
Vertebrate paleontology: at American Museum, 7, 221, 228–45, 251n.42; and biology, 7, 8, 219, 220, 221, 230–45; and biology at Columbia, 229, 237, 244–45; Cope and, 219, 220, 221, 226–27, 232, 244, 247n.13; and geology, 219, 245–46n.2; Gregory and, 237–45; Leidy and, 219, 220, 225, 244; Marsh and, 219, 220, 221, 224, 244, 247n.13; Matthew and, 234–37, 244–45; at museums, 219–20; Osborn and, 221, 222, 224–45, 248n.22, 251n.42; Scott and, 221, 224–28; Simpson and, 244–45, 255–56n.84
Virginia, University of, 20, 22
von Tschermak, Erich, 286
von Wettstein, Fritz, 312, 318–19

Waddington, C. H., 318, 340
Ward, Henry A., 92; and college museum collections, 19, 20, 23–24, 26;

and Field Museum, 52; D. S. Jordan and, 26, 52; natural history business of, 52
Ward, Lester Frank, 263
Ward's Natural History Establishment. *See* Ward, Henry A.
Wärming, Eugenius, 257, 260, 271
Washington, University of, 72, 106
Watase, Shosaburo, 157, 163–64
Watson, John B.: and animal behavior, 186, 204, 210; and Carnegie Institution, 207–8; problems of disciplinary identification, 204; and psychology, 205, 210
Webber, Herbert J., 284, 286, 290, 294
Weiss, Paul, 320, 323
Wheeler, William Morton, 142; and animal behavior, 186, 197, 209; and biology at Chicago, 152, 157, 163–64
White, Andrew Dickson: and Cornell museum, 20, 26; and higher education reform, 65–66; on natural history instruction, 59
Whitman, Charles Otis: as administrator, 160–65, 174; and A. Agassiz, 130; Allee and, 218n.66; and Allis, 130, 132, 146n.32, 170; and Allis Lake Laboratory, 158, 162; and American Morphological Society, 96, 100, 188; and animal behavior, 8, 168, 172–73, 185–94, 208, 209, 213n.7, 213–14n.13, 214n.17, 214n.26; and Baird, 130; and Baur, 164; and biological farm, 160, 163, 175, 186, 187, 189, 194, 207; on biology, 4, 8, 71, 74–75, 116n.33, 130–32, 142, 155, 158–60, 170, 181n.76, 186–89; career, 129–30, 146n.32, 157–58, 168, 187–88; Carnegie Institution and, 132, 162, 195–96; and Chicago, 130, 152–53, 154–68, 172–75, 181n.63, 188, 195, 208; Clapp and, 166; and Clark, 130, 155–57, 159, 167, 188; and Comte, 147n.34; and Conklin, 147n.38, 166, 171, 180n.60; and Craig, 173, 185, 187, 197–98, 216n.41; and cytology, 70–71, 168; and Darwin, 191, 193; and Davenport, 164, 195; and embryological club, 96, 97; and embryology, 70–71, 135–36, 166–72, 173, 180n.66, 188; and evolution, 129, 168, 171–73, 189, 190–92, 213n.13; and Fish Commission laboratory, 129–30, 147n.38; and German model for biology, 130, 158–60, 165; and Harper, 154–57, 160–65, 167; and inheritance, 171–73, 190–92; and inland lake laboratory, 160–63, 175; and *Journal of Morphology,* 100, 101, 170, 188; and Leuckart, 130, 147n.34, 158, 165, 168, 169, 188; and Lillie, 163, 164, 166, 189, 194, 218n.66; and Loeb, 186, 187, 213n.13; Lorenz and, 185, 213–14n.13; Mall and, 156–57, 159; Mathews on, 195; and MBL, 74–75, 129–32, 135, 139, 142, 146n.32, 149n.59, 159, 160–62, 165–66, 170–72, 187–88, 194; and microscopy, 70–71, 168–70; Newman and, 167, 172; and Nunn, 194, 214n.26; and organicism, 130, 131, 142, 147n.34, 168, 171–72, 188, 190, 192; Osborn and, 250n.35; and problems of disciplinary identification, 8, 203; and problems of institutional support, 8, 195, 205, 206, 214n.26; on psychology, 193; publications of, 193–94, 214n.22; and Riddle, 173, 185, 195–96; and Spencer, 147n.34; and Stanford, 155, 178n.14; students of, 185; as teacher, 165–68, 172, 174, 180n.60, 181n.63; and Watase, 163–64; and Wheeler, 163–64, 185, 186; E. Whitman and, 194, 195, 214n.26; and Wilson, 166, 170–71, 180n.60; and Woman's Education Association, 74, 128–29, 132; Woodward on, 195
Whitney, Josiah Dwight, 15
Wilder, Burt G., 19, 20, 25
Willard, Emma, 17
Wilson, E. B., 68, 96, 103; and biology at MBL, 136, 139, 166, 170–71; and Conklin, 147n.38, 171; and genetics, 136, 284; and Osborn's biology journal, 224; and Whitman, 166, 170–71, 180n.60
Wistar Institute, 101–2, 107
Woman's Education Association: and Boston Society of Natural History, 79n.39; and Hyatt's school at Annisquam, 128; and Fish Commission

Woman's Education Association (*continued*)
laboratory, 128–29; and MBL, 74–75, 128–29, 132
Women: and biology, 126; and biology at Chicago, 166; and botany, 126; and genteel resorts, 122, 124; and MBL, 128–29, 134; and science education, 17, 38n.12
Woods Hole: as genteel resort, 122, 123, 128, 145n.25. *See also* Marine Biological Laboratory
Woodward, Robert S., 195, 208–9
Wyman, Jeffries, 90; anatomy and physiology collection of, 17, 39n.19; and European research methods, 62; and natural history, 55
Yale University, 15, 151; natural history societies at, 55; Peabody Museum, 22, 29, 38n.9, 219; and science education, 60; and Sheffield Scientific School, 17, 59
Yerkes, Robert M.: and animal behavior, 186, 204–5, 210; and Craig, 202; *Journal of Animal Behavior* and, 205; primate laboratory of, 211; problems of disciplinary identification, 204–5; and problems of institutional support, 204–5; and psychology, 205; students of, 210

Zoology: and biology, 127; at Chicago, 165–77. *See also* Biology